*Structured Digital Design
Including MSI/LSI Components
and Microprocessors*

PRENTICE-HALL COMPUTER APPLICATIONS IN ELECTRICAL ENGINEERING SERIES

FRANKLIN F. KUO, editor

ABRAMSON and KUO, Computer-Communication Networks
CADZOW, Discrete Time Systems: An Introduction with Interdisciplinary Applications
CADZOW and MARTENS, Discrete-Time and Computer Control Systems
HUELSMAN, Basic Circuit Theory
KLINE, Digital Computer Design
KLINE, Structured Digital Design Including MSI/LSI Components and Microprocessors
KOCHENBURGER, Computer Simulation of Dynamic Systems
KUO (ed.), Protocols and Techniques for Data Communication Networks
LIN, An Introduction to Error-Correcting Codes
LIN and COSTELLO, Error Control Coding: Fundamentals and Applications
NAGLE, CARROLL, and IRWIN, An Introduction to Computer Logic
RHYNE, Fundamentals of Digital Systems Design
SIFFERLEN and VARTANIAN, Digital Electronics with Engineering Applications
VAN VALKENBURG and KINARIWALA, Linear Circuits

Structured Digital Design Including MSI/LSI Components and Microprocessors

RAYMOND M. KLINE

Washington University
St. Louis, Missouri

Department of Electronic and
Electrical Engineering
University of Strathclyde
Royal College Building
204 George Street
Glasgow G1 1XW

PRENTICE-HALL, INC.
Englewood Cliffs, New Jersey 07632

Library of Congress Cataloging in Publication Data

KLINE, RAYMOND M. (date)
  Structured digital design including MSI/LSI components and microprocessors.

  (Prentice-Hall computer applications in electrical engineering series)
  Rev. ed. of: Digital computer design. c1977.
  Includes bibliographies and index.
  1. Electronic digital computers—Design and construction. I. Title. II. Series.
  TK7888.3.K56 1983   621.3819′582   82-15077
  ISBN 0-13-854554-5

Editorial/production supervision: *Ros Herion,*
  Oscar Ocampo, Natalie Krivanek, and Chris Stengel
Interior design: *Oscar Ocampo*
Cover design: *Photo Plus Art, Celine Brandes*
Manufacturing buyer: *Anthony Caruso*

To *Marian, Dave, Kathy, Sue,* and *John*

© 1983, 1977 by **PRENTICE-HALL, INC.**, Englewood Cliffs, N.J. 07632

Previous edition published under the title
  Digital Computer Design.

All rights reserved. No part of this book
may be reproduced in any form or
by any means without permission in writing
from the publisher.

Printed in the United States of America

10 9 8 7 6 5 4 3 2 1

ISBN 0-13-854554-5

PRENTICE-HALL INTERNATIONAL, INC., *London*
PRENTICE-HALL OF AUSTRALIA PTY. LIMITED, *Sydney*
EDITORA PRENTICE-HALL DO BRASIL, LTDA, *Rio de Janeiro*
PRENTICE-HALL CANADA LTD., *Toronto*
PRENTICE-HALL OF INDIA PRIVATE LIMITED, *New Delhi*
PRENTICE-HALL OF JAPAN, INC., *Tokyo*
PRENTICE-HALL OF SOUTHEAST ASIA PTE. LTD., *Singapore*
WHITEHALL BOOKS LIMITED, *Wellington, New Zealand*

Contents

Preface *ix*

PART I
PRINCIPLES OF DIGITAL SYSTEMS *1*

1. ***A Preview*** *3*
   1.1 The Evolution of Digital Computers *7*
   1.2 The Growth of Computer Technology *11*
   1.3 Classification of Computers *14*
   1.4 Important Digital Computer Characteristics *17*
   1.5 Prologue to the Remainder of the Text *19*
   1.6 Bibliography *20*
   1.7 Topics for Further Study and Student Reports *21*

2. ***Number Systems*** *22*
   2.1 Arabic Notation *23*
   2.2 Conversion of Integers *24*
   2.3 Conversion of Fractions *26*
   2.4 Addition and Subtraction *27*
   2.5 Multiplication and Division *30*
   2.6 Codes *32*
   2.7 Bibliography *41*
   2.8 Exercises *42*

3. **Software** 46
   3.1 Basic Machine Organization *48*
   3.2 Machine Language Programming *50*
   3.3 Programming Languages *57*
   3.4 Microprocessor Programming *60*
   3.5 Bibliography *74*
   3.6 Exercises *74*

4. **Logical Design** 81
   4.1 Basic Logical Operations *81*
   4.2 Theorems *84*
   4.3 The Canonical Expansion of Functions *89*
   4.4 The Karnaugh Map *92*
   4.5 Logic Diagrams *99*
   4.6 NAND-NOR Logic *100*
   4.7 EXCLUSIVE-OR Logic *106*
   4.8 Implementation of Computer Circuits *109*
   4.9 Bibliography *114*
   4.10 Exercises *114*

5. **Sequential Circuits** 117
   5.1 Some Basic Flip-Flops *118*
   5.2 Analysis *121*
   5.3 Description of Sequential Circuits *122*
   5.4 Circuit Synthesis *129*
   5.5 Synthesis of Counters *133*
   5.6 Heuristic Design Methods *141*
   5.7 Nonideal Components *149*
   5.8 Flip-Flop Realization Methods *153*
   5.9 The Ripple Method *161*
   5.10 Dependency Notation *164*
   5.11 Implementation of Digital Logic *169*
   5.12 Bibliography *175*
   5.13 Exercises *176*

6. **Computer Architecture: Register Transfer Logic, Organization, and Subsystems** 183
   6.1 Register Transfer Logic *184*
   6.2 Applications of RTL *190*
   6.3 Computer Organization *201*
   6.4 Further Development of RTL *206*
   6.5 The Bus Subsystem *208*
   6.6 The Arithmetic and Logic Subsystem *217*
   6.7 The Memory Subsystem *222*
   6.8 Semiconductor Memories *223*
   6.9 Bibliography *231*
   6.10 Exercises *232*

PART II
MODERN STRUCTURED LOGIC DESIGN 237

7. **Logic Design with MSI/LSI Components** 239
   7.1 Fixed MSI/LSI Components: Higher-Level Primitives *240*
   7.2 Structured Design with Fixed Components *246*
   7.3 Programmable MSI/LSI Components *255*
   7.4 Applications to Structured Design *265*
   7.5 Control Memories *268*
   7.6 Bibliography *276*
   7.7 Exercises *277*

8. **Logic Design with Microprocessor Systems** 280
   8.1 An Introduction to Microprocessors *282*
   8.2 Microprocessor Architecture *283*
   8.3 Memory Addressing *288*
   8.4 Interfacing Principles *293*
   8.5 Structured Microsystem Design *303*
   8.6 Microprocessor versus Random Logic Design *307*
   8.7 Bibliography *313*
   8.8 Exercises *314*

9. **Microprogrammable Microprocessors** 319
   9.1 Microprocessor Classes *319*
   9.2 Microinstructions and Microprogramming *322*
   9.3 An Elementary Microprocessor System *327*
   9.4 A Macroinstruction Application *330*
   9.5 Structured Design Revisited *336*
   9.6 Bibliography *338*
   9.7 Exercises *339*

PART III
ADVANCED TOPICS 341

10. **Circuits for Computer Arithmetic** 343
    10.1 Fast Adders *343*
    10.2 Multiplication *348*
    10.3 Fast Multiplication *356*
    10.4 Division *368*
    10.5 Overflow *376*
    10.6 Bibliography *378*
    10.7 Exercises *378*

11. Computer Interface Design   *380*
    11.1 Data Transfer Techniques   *381*
    11.2 Programmed Transfers   *381*
    11.3 Interrupt Transfers   *385*
    11.4 Direct Memory Access   *388*
    11.5 Input/Output System Variations   *391*
    11.6 Bibliography   *397*
    11.7 Exercises   *398*

12. Digital Electronics   *400*
    12.1 Diode Gates   *400*
    12.2 Transistor Gates   *411*
    12.3 MOSFET Circuits   *429*
    12.4 Bibliography   *436*
    12.5 Exercises   *436*

Appendixes
    A. Summary of Pseudo-LINC I/O Instructions   *440*
    B. Answers to Selected Exercises   *441*

Index   *447*

Preface

As digital technology continues to develop at a rapid pace, it is mandatory that an introductory computer engineering text at the university level be as modern as possible both by including the most significant recent developments and by emphasizing fundamentals so that the reader will be adequately prepared to participate in the next generation of advancements.

This text offers, as does an earlier version of the material, a broad though substantive introduction to computer hardware, yet it provides a good balance between theory and engineering design techniques. The fundamentals of digital systems, microprocessor principles, computer architecture, modern logic design, introductory machine language programming, and advanced digital circuits are the major areas considered. It is an appropriate text both for introductory computer hardware courses and for self-study. Curriculum requirements listed by articles in journals of the Association of Computing Machinery, the Institute of Electrical and Electronic Engineers, and elsewhere are adequately met. At Washington University, we have successfully employed the first version of the text for more than ten semesters in teaching an introductory course in computer engineering for both electrical engineering and computer science majors. Several features make the book attractive for self-study: The answers to many of the exercises are listed in Appendix B, a bibliography is available at the end of each chapter to suggest supplementary reading, and the book's flexibility allows selective chapters to be considered without reading all previous sections.

I follow the same general philosophy and make the same assumptions that were made in the earlier version of the text; i.e., it is expected that the typical reader will be a second- or third-year university student in either electrical engineering or computer science. The revisions, in both the material and the order in which topics are presented, are extensive, particularly from Chapter 5 on, but not so drastic that users of the earlier version will feel lost or uncomfortable. The new version, however, contains many advancements that reflect both long-term trends in computer design and developments in the teaching of computer engineering. A large number of new exercises have been added, and for continuity, many of the more interesting old ones have been retained. Answers are given in Appendix B for about 35% of the exercises.

MAJOR TOPICS

In addition to the usual classic subjects such as Boolean algebra, Karnaugh maps, and computer organization, the following are the most important topics retained from the first version:

Register Transfer Logic, which permits the action of higher-level digital circuits to be precisely described and analyzed.

Microprocessor basics, particularly the material on control memories and microprogrammable processors.

Heuristic logic design techniques as a supplement to the usual sequential logic synthesis methods.

New to the present work are several modern topics that give the book both depth and breadth. These encompass:

A thorough study of microprocessor hardware (and an introduction to the corresponding software), including a section in which design roles for microprocessors and random logic are compared and examples presented.

The introduction of Structured Hardware Design as a means for engineering reliable, maintainable, and cost-effective systems. (This process has original aspects but also parallels some of the techniques employed in structured software, such as top–down design.)

The integration of logic design, employing higher-level primitives, into a unified strategy. (The primitives considered are microprocessors, microprogrammable processors, and MSI/LSI components, such as programmable logic arrays.)

An extensive revision of Chapter 5, Sequential Circuits, to provide greater theoretical insights, to introduce Dependency Notation, and to supply additional practical topics. (Dependency Notation is gaining wide acceptance as a process for concisely representing the detailed functional characteristics of a wide range of digital modules, from flip-flops to microprocessors.)

Classification, study, and application of MSI/LSI fixed primitives (such as multiplexers) to digital logic.

A large addition to the section on the use of bus structures in computers.

ALTERNATE WAYS FOR EMPLOYING THE TEXT

My goals are the development of a broad selection of fundamental, substantive topics emphasizing structured digital system design but with coupling between sections loose enough to permit the instructor *wide flexibility in tailoring a course* to meet the exact needs of the students. To achieve these goals, as many chapter sections as possible were written in such a way as to make their inclusion optional or to permit alternatives to the printed numerical chapter sequence. For example, instead of the given numerical sequence, Sections 3.1 and 3.2 may be read after Chapter 5 and the remainder of Chapter 3 omitted entirely unless Chapter 8 is to be covered in detail. (In that latter case, Section 3.4 should be read before Chapter 8.) Moreover, one aid in employing the available flexibility is the division of the text into three multichapter parts. Part I mainly considers fundamentals; Part II presents structured design, including such modern components as microprocessors and programmable logic arrays; and Part III involves advanced topics (e.g., high speed arithmetic circuits and special interface systems). Another aid for employing the book's flexibility is the marking of optional sections with a double dagger (††). This method is first used at the beginning of Section 2.3, and it continues in all remaining chapters. (A further explanation of the method is found where it is first used in Chapter 2.)

The time spent on Chapters 2 and 3 (Number Systems and Software) will depend on the nature of any previous student experience. Most instructors will want to include in their course the fundamentals of Boolean algebra and elementary logic design from Chapter 4, but the degree of emphasis given to Boolean theorems, map techniques, and tabular methods is optional. In presenting modern digital system design to university students, the main values of the latter three topics are *not* in circuit optimization at the gate level but rather in providing: (1) a strong theoretical base for true understanding; (2) a systematic method for determining logic circuits (functions) required by various data processing operations (e.g., code conversion); (3) a design tool at the chip (MSI/LSI) level; and (4) an aid in promoting structured design. (A useful presentation of these topics can be achieved in as few as four 1-hour class periods.)

Chapter 5 covers sequential circuits; it parallels the combinational material in Chapter 4 and includes more applications-oriented topics, such as the nonideal characteristics of digital components, dependency notation, advantages of edge-triggered and master-slave flip-flops, design of sequential circuits, and integrated-circuit implementation of digital logic. The contents of Sections 5.4 and 5.5 are somewhat more theoretical than the other sections, yet the synthesis ideas presented represent a valuable new facet of computer science for most students. Moreover, these sections include considerable structure and mathematical substance, which is important for the proper development of engineers. Again, there is flexibility in selecting topics, yet Chapter 5 in its entirety can be presented rather quickly if desired. It is likely that Chapter 6 on Computer Architecture will be studied in detail because this is where fundamental material such as bus structures and register transfer logic are developed. Here too, however, there are ample opportunities to omit topics or to modify the writer's original emphasis.

Many teachers will want to place their emphasis on Part II, Modern Structured Logic Design, which is composed of Chapters 7, 8, and 9. Moreover, Part II has the potential of even greater flexibility than is possible in Part I. Some alternatives are:

1. Skip immediately from Chapter 6 to Chapter 8, and devote three or four class weeks to the design of microprocessor systems.

2. Present Section 7.4, the control memory; then go immediately to Chapter 9, where the remaining details on microprogramming and bit-slice architectures are discussed.
3. Thoroughly cover Chapters 7, 8, and 9 with the intention of illustrating the detailed advantages, disadvantages, and trade-offs between the MSI/LSI and the microprocessor approaches to structured design.

The chapters in Part III, Advanced Topics, are entirely independent of each other; thus, one can cover them in any desired order. Chapter 10, however, requires slightly more knowledge of mathematics and Chapter 12 slightly more knowledge of electrical circuits than the earlier chapters. Although several earlier topics are expanded in depth in Part III and could form good tutorial combinations, the most likely pairs are: Chapter 8 (Microprocessors) with Chapter 11 (Computer Interface Design), and Section 6.6 (The Arithmetic and Logic Subsystem) with Chapter 10 (Circuits for Computer Arithmetic).

FURTHER TEACHING SUGGESTIONS

The text contains the depth required by those readers majoring in computer engineering, yet it is flexible enough for those in other areas of engineering or science, requiring only a limited introduction to digital systems. It includes more than sufficient material for a 3-hour course of one semester or two quarters in length; and it may be employed for a two-semester course with only a minor amount of supplementing from the Bibliography or by including a few "built-in" laboratory sessions. This latter alternative is also strongly recommended for shorter courses even when it results in an abbreviation of some other important material.

Undoubtedly, many students will later take specialized courses in the various topics covered here. However, I believe that a broad introduction to digital computers, with depth in some areas, gives one the perspective and background required to assimilate more advanced courses. This text, therefore, is intended to foster the maturity and the experience that these students require to make an advanced course more meaningful. As a specific example, most of our computer engineering students at Washington University follow their introduction to computer hardware by a 3-hour course devoted to switching theory. Chapters 4 and 5 on logical design, although not intended to be exhaustive treatments, are complete enough that our switching course is able to progress at a more rapid pace and on a more advanced level.

I have observed that introductory computer courses often do not present the same challenge to college students that many of their other courses do; thus students frequently do not learn as much and are not as highly motivated. Certainly, there is no need to create artificial challenges, but it is one of my objectives to present the material in sufficient depth to be useful in practical design situations as well as to command the interest of the most promising students.

To aid those teaching from this book, an *Instructor's Manual* containing suggestions for presenting the material as well as solutions for many of the exercises is available from the publisher. Also, I will be happy to receive correspondence on any aspect of the text.

ACKNOWLEDGMENTS

Professor Mark Franklin made several valuable suggestions during the preparation of the manuscript. In addition, other colleagues and students, too numerous to list individually, have been my teachers during preparation of both versions of this text; their questions, suggestions, and ideas were indispensable. Also, encouragement during the manuscript preparation and editing by Professor Donald Snyder, Chairman of our Department of Electrical Engineering, is greatly appreciated. Finally, I want to acknowledge again the unique climate for computer research, development, and teaching that is present in the Department of Electrical Engineering, the Department of Computer Science, the Computer Systems Laboratory, and the Biomedical Computer Laboratory at Washington University for strongly influencing my thinking.

RAYMOND M. KLINE
Washington University

Part I

PRINCIPLES
OF DIGITAL SYSTEMS

Chapter 1

A Preview

A popular technique for designing computer systems is called the *top-down approach*. Here, the engineer first studies the overall requirements, then carefully plans the system implementation from a broad perspective before getting into the details of the actual circuit design. A similar approach to presenting computer information and related techniques is taken in this text. Thus, the current chapter is a brief but broad overview of the history, capabilities, and limitations of computer technology, while in all the following chapters, the emphasis is on detailed engineering analysis and design. In addition to the traditional computer topics, several innovative features will be found in later chapters. Typical members of this group are the dependency notation and heuristic design methods presented in Chapter 5, the register transfer logic developed in Chapter 6, the treatment of standard components versus the microprocessor for logic design applications in Chapters 7 and 8, the structural hardware design concepts of chapters 7 through 9, and the study of microprogrammable microprocessors in Chapter 9. Our initial preview of computer technology should make it easier for the reader to maintain a broad prospective concerning the complete system while delving into the details of various components and subsystems.

Rapid advances are currently being made in all phases of digital computer technology. A typical example is Bell Laboratories' MAC-4 one-chip computer with a photomicrograph of the actual circuit shown in Figure 1.1. Here, a thin slice of silicon (a so-called chip) much smaller than a paper clip contains a complete computer with many powerful processing features. The light and dark patterns are thousands of transistors and their connections, which have been deposited on the silicon slice by means of a several-step photolithographic process. The functional areas, such as the Arithmetic and Logic Unit ALU, shown on the right side, are labeled on the photo-

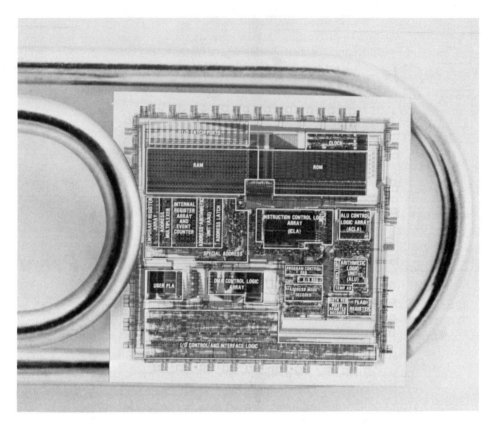

(a)

The MAC-4 one-chip computer, developed for a variety of telecommunications applications, is compared to a standard-sized paper clip. The chip's numerous functional areas are labeled.

(b)

Figure 1.1 Photomicrograph of the MAC-4 One-Chip Computer. (Courtesy of Bell Laboratories.)

graph. Most of these areas are the subjects of later chapters and the reader should return to review this figure after, say, Chapter 8 has been considered. The MAC-4 performs numerous telecommunication tasks in the Bell System, yet it consumes very little electrical power. This very small computer, a *microprocessor*, is a close relative of the pocket calculator electronics on which all of us have so quickly learned to depend.

Digital computers have assumed a prominent place in our society, and their value in business and industry is without question. However, many people still have a distorted view of them. Some picture computers exclusively as big adding machines, when in reality they currently perform extremely complex logic; general nonnumerical tasks are also becoming increasingly important.

Typical of this trend is the application of computers in medical research and treatment. An interesting case in point is the development of "artificial hearing" for tens of thousands of deaf people who cannot be helped by conventional hearing aids. In an experimental program, scientists from the Ear Research Institute in Los Angeles and the University of Utah in Salt Lake City reported that a deaf subject was able to sense "sounds" produced by a PDP-8/F minicomputer. Figure 1.2 shows a 62-year-old engineer named Joe, deaf since infancy, who is participating in experiments to develop an artificial ear. Touching a key or combination of keys on the keypad on his lap, Joe causes the PDP-8/F, left side of photo, to send signals at selected frequencies to electrodes implanted in his inner ear. The computer provides a flexible system for conducting a wide variety of tests, eliminating the need for building special equipment for each experiment. When the experiments are completed, a portable artificial ear will likely be constructed using a subminiature computer, a *microprocessor*, not much larger than a conventional hearing aid.

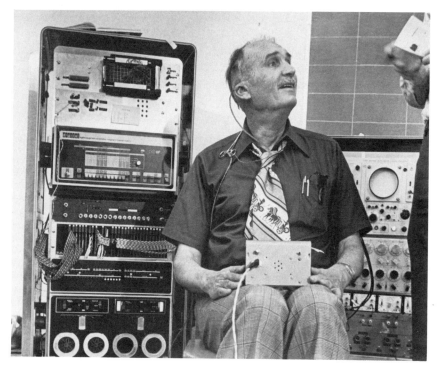

Figure 1.2 Patient and Computer-Controlled Artificial Hearing System. (Courtesy of Digital Equipment Corporation.)

Another example of an important nonnumerical application is the computerized knitting machine shown in Figure 1.3(a). One of the problems in the textile industry is the long time required for development of new fabric patterns on conventional knitting machines. By using a PDP-11/20 minicomputer [background at the right of Figure 1.3(a)] to control one of its knitting machines (left foreground), the Uxbridge Knitting Mills of Massachusetts have been able to reduce the fabric development

Figure 1.3 Computer-Aided Fabric Pattern Design System: (a) knitting machine controlled by a PDP-11/20; (b) operator's console and television display. (Courtesy of Digital Equipment Corporation.)

time to a matter of hours, or even minutes for less complex tasks. The computer-aided pattern design begins with sketches made, at the operator's console, on an electronic tablet coupled to a color television monitor by the computer [see Figure 1.3(b)]. As the work progresses, the computer transforms the information displayed on the monitor into stitch data recorded in the core memory of the PDP-11/20. When the designer wants to examine a piece of the new fabric, a brief set of instructions is given through the console, and the stitch data in the memory provides control of the knitting machine to produce a sample. Should corrections be required, the designer can recall the pattern to the television monitor and indicate the adjustments with the electronic tablet. When the designer is satisfied with the modified pattern, he allows the computer to resume production. A much more complicated procedure is required with older methods.

Other representative nonnumerical tasks are automated typesetting, computer-aided drafting, derivation of theorems, and weather prediction. The reader should consult Reference 1 for details concerning these applications.

Although the use of advanced pocket calculators and home computers has eliminated some of the myths about computers, popular magazine and newspaper articles still frequently imply that computers are some kind of magical device. In reality, they are based on well-known engineering principles, and the quality of the results is critically dependent on the quality of the computer's logic design, the input data, and the skill of the programmer. (There is a well-known motto in computer engineering which aptly expresses this: Garbage in ... garbage out!)

In the last 10 years computers, in various forms, have become commonplace in all areas of our society. Small general-purpose machines (*minicomputers*) with a display, 16,000 words of fast memory, a disk for permanent storage, and a BASIC compiler are currently available for less than $1000, and surprisingly the prices are coming down. Applications for these low-cost minicomputer versions range from electronic hobbies to a large variety of uses in small businesses. More expensive minicomputers with much greater capabilities are being applied to such tasks as automatic monitoring and control of industrial processes, rapid testing of complex systems, and controlling the sequence of programs at television stations. (In 1980, more than 300,000 minicomputers were operating in the United States and this does not include hobby computers.) Dedicated computers (as opposed to general-purpose machines) usually in the form of microprocessors are being employed in a seemingly unlimited number of applications ranging from function controllers for microwave ovens to computerized ignition systems in automobiles.

1.1 THE EVOLUTION OF DIGITAL COMPUTERS[2]

Although the digital computer itself is a quite recent development, it has had a complex evolution that goes far back into the history of technology. One of the earliest sources of digital concepts occurred thousands of years ago in Hindu-Arabic arithmetic notation. Somewhat later, yet before the birth of Christ, the odometer was invented, and this eventually led to the development of the mechanical adder and multiplier in the 17th century. Then, improvements in machine tools and other developments finally led to the first mass-produced desk calculator in 1911.

Meanwhile, in approximately 1800, Joseph Jacquard perfected a punched-card process* for automatically programming the location of threads in cloth-weaving looms. The punched-card method was so successful in automating the weaving process that Herman Hollerith decided to consider a variation of it to solve the massive data-processing problem (for that time) of tabulating statistics for the 1890 census. The result was a great improvement in tabulating speed over the previous manual methods. (In 1896, Hollerith formed his own company to manufacture punched-card equipment, and this later became part of IBM.)

Of all the contributions to the computer, by far the most remarkable was the contribution of Charles Babbage[4,8] (1791–1871), who was a professor of mathematics at Cambridge University. He conceived many of the basic ideas of the modern digital computer over 100 years before they were perfected. His *difference engine*, shown in Figure 1.4 was intended to prepare mathematical tables of various types by accumulating differences through mechanical means. The apparatus shown in the picture is capable of producing tables to 20-place accuracy. In the process of his work, however, Babbage conceived of a much more general machine, the *analytical engine*, having several of the properties of a modern computer, Although his theory was sound, the mechanical technology of his day was not capable of supporting his very ambitious ideas, and his analytical engine was never completed.

Figure 1.4 Babbage's Difference Engine. (Courtesy of International Business Machines Corporation.)

*Actually, it was a sequence of cards joined together in a long belt.

Although there were improvements in desk calculators, in punch card tabulating machines, and in other calculating devices between 1900 and 1937, no real breakthroughs occurred. Then, from 1937 to 1950, revolutionary changes occurred in such rapid succession, by so many different people, and in so many different places that it is impossible, in this short chapter, to give credit to all those concerned. Some of the major contributors were:

1. John Atanasoff and a colleague, Clifford Berry, at Iowa State College during the period 1937–1942 demonstrated an experimental special-purpose electronic digital computer for solving simultaneous linear equations.
2. During the early 1940s George Stibitz, at Bell Telephone Laboratories, built several special-purpose digital computers based on relay technology. He was one of the first to implement floating-point arithmetic.
3. The Mark I relay computer was completed in 1944 by Howard Aiken at Harvard. It is considered to be the first general-purpose machine.
4. The ENIAC, Electronic Numerical Integrator And Calculator, was completed in 1946 by J. Presper Eckert and John W. Mauchly of the Moore School of Electrical Engineering of the University of Pennsylvania. Although somewhat similar in structure to the Mark I, it employed vacuum tubes instead of relays; thus, it became the first operational electronic computer.
5. The EDSAC, Electronic Delay Storage Automatic Computer, the first stored-program computer, began operating in 1949 at the University of Manchester in England.

An abbreviated family tree for the computer is shown in Figure 1.5. It includes the history just described, some important events since 1950, and certain related facts not previously mentioned. Most of the individual entries are self-explanatory; a few, such as core memory and assembly language, for which the reader may only have a vague definition, will be explained in considerable detail later. More historical details may be found by consulting the bibliography.[3,4,6,8,10] (Because of the multiplicity of commercial computers, only a few of those which represent landmarks in the development of the field appear.)

From the density of dots in Figure 1.5, it may appear that the activity peak in the development of computers was reached around 1970. This is misleading because a vantage point in the 1980s does not allow one the proper historical perspective to evaluate the long-range importance of most of the techniques which were originated since the early 1970s; hence, only such obvious developments as Very Large Scale Integration, VLSI, have been added to the latter portion of the chart.

A number of items, such as the slide rule and Babbage's difference engine, are not directly connected into the mainstream. Although they had a role in the development of computer science, their contribution to the digital computer can not be identified through a specific single path.

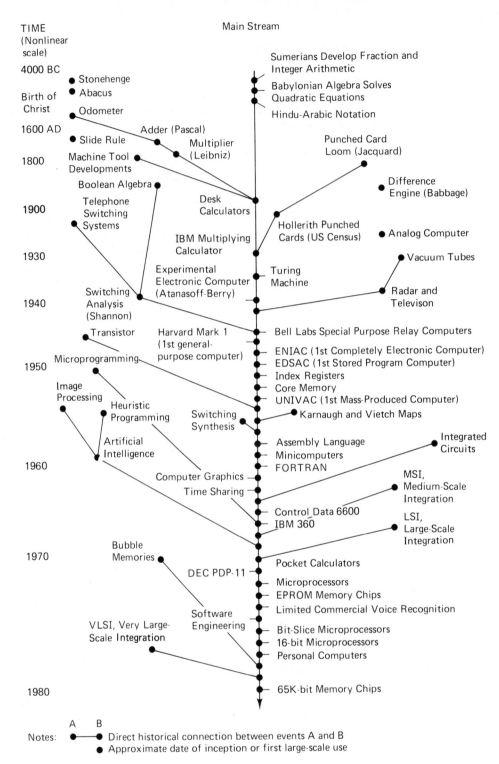

Figure 1.5 Brief Chronology for Digital Computers.

1.2 THE GROWTH OF COMPUTER TECHNOLOGY

Having just considered the general evolution of computers, we shall now emphasize recent developments. We do not intend to present a lot of potentially boring statistics, but at this point it does seem that quantitative information is one of the best ways to give the reader a perspective concerning the rapid changes that have recently occurred in the industry.

The comparison between the IBM 650 computer and the TI-59 calculator in Table 1.1 is one interesting way to visualize the 25 years of remarkable progress in computer development between 1955 and the 1980s. The most striking factors, cost and weight, show an improvement of nearly three orders of magnitude. Since the TI-59 design is optimized for calculator functions, the trade-off is low cost and weight for execution time, yet even here the multiply operation shows a factor-of-5 improvement, and the add operation shows almost a factor-of-10 improvement. The memory capacity–secondary values could be misleading since a large, heavy magnetic drum is employed in the IBM 650, compared to the small magnetic card employed in the TI-59. In reality, secondary memory capacity has improved several orders of magnitude since 1955.

TABLE 1.1 25 YEARS OF "MAINFRAME" PROGRESS

	IBM 650 Computer (1955)	TI-59 Calculator (1980)
Components	2000 tubes	166,500 transistor equivalents
Power (kVA)	17.7	0.0018
Volume (ft^2)	270	0.017
Weight (lb)	5650	0.67
Air conditioning	5 to 10 tons	None
Memory capacity		
Primary	3000 bits	7680 bits
Secondary	100,000 bits	40,000 bits
Execution time (milliseconds)		
Add	0.75	0.070
Multiply	20.0	4.0
Price	$200,000 (1955 dollars)	$299.95 (1980 dollars)

Courtesy of Texas Instruments, Inc.

Table 1.2 presents an historic overview of computer technology as measured by various economic and engineering parameters. The first parameter, computer generations, is a popular means for indicating the major periods of development. To a large extent, these intervals are determined by the available electronic components. (See the second line in the table.) There are other factors, however, which are unique to the various generations. One of these is programming languages. During the first generation, only elementary languages were available, but in later generations very sophis-

TABLE 1.2 GROWTH OF COMPUTER TECHNOLOGY*

Parameter	1950	1960	1965	1970	1975	1980
Computer generation	First	Second	Early third	Late third	Fourth	Fifth
Electronic components	Vacuum tubes	Transistors	Integrated circuits	Some medium-scale integrated circuits	Medium- and large-scale integrated circuits	Large-scale integrated circuits
Value of computers shipped (billions of dollars)	—	0.8	2.5	7.2	14.6	—
General-purpose digital computers in the world (thousands)	$\ll 1$	5	35	70	200	1000
Dollar cost per 10,000 ten-digit multiplications	300 (desk calculator)	1.40	0.12	0.01	0.008	0.0006
Cost of a minicomputer with a display and 4000-word memory (thousands of dollars)	Not in production	80	25	7	2.2	0.5
Semiconductor components per package (chips): "Moore's Law"	—	—	64	2000	64,000	2×10^6
Logic delay—delay of one stage of digital logic (nanoseconds)	1000	30	10	3	1.5	0.7
Cost of main memory (cents per binary digit)	—	100	10	2.5	0.3	0.04

*The tabular entries are approximate state-of-the-art values. The data are intended to give the reader a feeling for the magnitude of computer technology. Anyone desiring the most up-to-date information and data for which the parameters have been completely qualified should consult statistical abstracts or other specialized sources.

ticated programming systems have been employed. (Some of these will be considered in Chapter 3.) An extensive discussion of the various computer generations is found in the literature.[3]

Observe that during the decade from 1960 to 1970 there was approximately an order-of-magnitude increase in both the economic importance of computers (e.g., note the increased value of computers shipped) and their performance (e.g., note the decrease in logic delay). The table shows that this trend continued in the 1970–1980 decade. Moreover, these were years of generally rising manufacturing costs; yet the prices for multiplications, minicomputers, and memory all showed very sharp decreases. A major key to the great economic and technological progress of the computer industry is the number of semiconductor components per integrated circuit chip shown in the table. Early in the development of large integrated circuits Gordon E. Moore of the Intel Corporation recognized that the production data show an annual doubling in the components feasible per integrated circuit package. For this reason, the doubling relation has come to be known as Moore's Law. Eventually the relation must fail, but many knowledgeable people in the semiconductor industry feel that it will continue in force for at least the next several years.

Another aspect of the progress in computer hardware can be obtained from the photograph in Figure 1.6, which shows comparable circuitry in each of the first three computer generations. The original vacuum-tube circuitry, a transistor version, and an integrated-circuit version are all easily identified. Actually, the reduction in size is

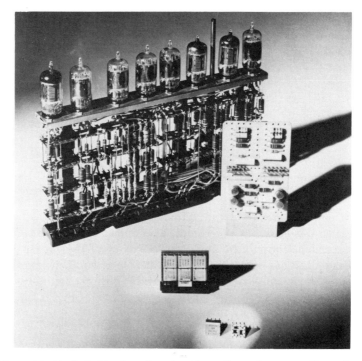

Figure 1.6 Typical Circuitry for the First Three Computer Generations. (Courtesy of International Business Machines Corporation.)

compounded in that there is a corresponding change in power supplies, cooling, and other support functions.

Integrated circuits like those in Figure 1.6 can be employed to implement systems ranging from microcomputers to very large scale digital processors such as the IBM 3081 system shown in Figure 1.7. This modern computer contains more than $\tfrac{3}{4}$ million logic circuits within the relatively small volume of 4 cubic feet (a cube with sides of about 1.6 ft). It can execute approximately 9 million instructions per second and sells for 3.7 million dollars. The reader should now compare Figure 1.7 with Figure 1.1. The size contrast observed between the 3081 and the MAC-4 will clearly establish the fact that practical computers exist in an extremely wide variety of sizes and shapes. (Of course, a proportional contrast in capabilities between the large and small systems is also expected.)

Figure 1.7 An Example of a Modern Large-Scale Computer: the IBM 3081 System. (Courtesy of International Business Machines Corporation.)

Continued growth of the computer industry far into the future seems assured both from the viewpoint of technical developments, which are continually emerging from research laboratories, and from estimates of the current business prospects. For example, it has been estimated that the more than 1 million computers currently in use represent less than 10% of the possible business applications.

1.3 CLASSIFICATION OF COMPUTERS

In order to understand computers, it is necessary to be aware of their two main classes—analog and digital. Typically, the magnitude of an electric voltage is used to represent each variable in an analog computer, but the key to the classification is

TABLE 1.3 COMPARISON OF TYPICAL DIGITAL AND ANALOG COMPUTATIONAL METHODS

Method	Form of Number	Accuracy	Minimum System Cost	Time to Obtain Result	Ease in Programming	Major Strengths
Digital	Two-level signal (e.g., 0 and +5 volts)	Very high (almost unlimited)	Practical limit for a commercial general-purpose system is about $1000; but special-purpose units, microprocessors, are now available for under $10	Finite—nanoseconds to hours	Excellent (special problem- and procedure-oriented languages are available)	*Versatility* in the solution of a wide variety of problems; *fast, accurate* (see Section 1.4)
Analog	Continuous signal (e.g., 0 to +5 volts)	Approximately 1%	Proportional to size of computation; may be only a few dollars	Output is continuously available, but transients may require 1 second or longer to die out	Fair (extensive manual, electrical, and/or mechanical connections are usually required)	*Simulation* of electrical, mechanical, and chemical systems; solutions to *differential equations*

that the quantity representing the variable must be continuous in both time and magnitude. This should be contrasted with a digital signal, where the boltage (or other quantity) representing the variable is discrete (quantized) in both time and magnitude.

A slide rule is one of the most common types of analog calculators. Here, numbers are represented as continuous distances along the scales, and multiplication (division) is achieved by the addition (subtraction) of logarithmic scale lengths.

Table 1.3 presents a summary of the basic properties of digital and analog computers. Note that digital methods have the advantage of accuracy, ease of programming, and versatility, while analog methods may be cheaper if only a simple calculation is required. The cost parameter is difficult to generalize, particularly since the birth of large-scale integrated circuits. It is still true, however, that there is a lower practical limit on the cost of a digital computer, even to perform a simple task—a type of overhead charge must be paid. Once the basic digital system is available, however, a variety of other tasks can be carried out with little or no additional cost. (For more information concerning analog computers, consult Reference 11 in the bibliography.)

The field of digital systems can be further divided into *general-purpose* and *special-purpose* machines. The general-purpose computers are those such as the IBM

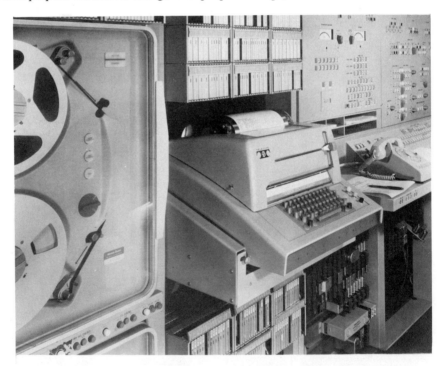

Figure 1.8 Master Control Center in an ESS Telephone Exchange. (Control and lamps on the panels at the far right are employed for various maintenance and administrative functions. The teletypewriter supplements the panel controls. The tape drive on the left is used for automatic message accounting recordings. (Courtesy of Bell Laboratories.)

3033 series, which is completely programmable and can be used to perform an almost unlimited variety of numerical calculations in the engineering, scientific, and business areas. A special-purpose machine is designed for a single type of application and frequently has most of its program prewired. The navigation computers aboard space vehicles are examples of this class of machine. Another very interesting example, shown in Figure 1.8, is the Electronic Switching System, ESS, being employed by the Bell Telephone Company to replace its older electromechanical systems in telephone exchanges. ESS allows a considerable improvement in the performance of the exchange and permits a number of new features to be sold to telephone customers. A few of these features are two-digit dialing for frequently called numbers, a simpler method for implementing conference calls, and a call-back option for busy numbers.

While not actual computers, there are a large number of special-purpose digital devices which use many of the same principles and require the same design techniques as an actual computer. Examples are digital voltmeters, counters, and various other types of automated digital instruments.

1.4 IMPORTANT DIGITAL COMPUTER CHARACTERISTICS

In addition to the fact that analog processing of data is sometimes more appropriate, the digital computer, like any other tool, has its disadvantages. Perhaps the chief disadvantage is the number of decimal places printed and the format employed sometimes give an apparent accuracy to results which is entirely erroneous. Thus, the printed output may show 10-place numbers when in reality, due to inaccurate input data, the results may only be accurate to 1 or 2 places. The consumer, computer designer, and programmer must constantly be on the alert to avoid this trap. Also one must avoid becoming a lazy thinker by letting the computer produce a volume of numerical results for a particular problem when a little physical reasoning and an equation or two would provide a more complete and concise solution.

However, digital information processing does possess a number of characteristics which, when properly applied, can be very powerful. Some of the most important of these will now be considered:

1. *Problem-oriented languages.* Problem-oriented languages allow instructions to be given to the computer in the language of the problem, not that of the machine. This permits people in all fields (from theoretical physicists to high school students) to have direct access to the computer without working through an imtermediary (programmer). Two of the most well-known languages are FORTRAN and BASIC; however, there are many other languages which are even more useful and efficient in special applications. The language ECAP (Electronic Circuit Analysis Program), for example, permits the engineer to easily obtain solutions to very complicated electrical network problems without having to go to the trouble of writing mesh or loop equations. (Additional details concerning this subject are presented at the end of Chapter 3.)

2. *Versatility.* The general-purpose computer is truly a versatile device in that a program can rapidly create an information processing structure to solve virtually

any problem. Moreover, new programs can be inserted in a very short time so that an entirely different processing structure is established. No other device can be so easily adapted to such a wide variety of tasks.

3. *Accuracy.* Results to 10 or more significant decimal places is an achievement not even approached by most other systems. When this is compared with the two significant figures obtainable with most analog devices and the time required to manipulate large numbers manually, the achievement is all the more remarkable.

4. *Automation.* The ability to specify a list of tasks (arithmetic, logical, or other operations) and have them completed without human intervention is a key to the automation of a wide variety of processes, including the automated milling machine shown in Figure 1.9. Actually more can be accomplished than just slavishly producing a fixed sequence of commands. There is no trouble in accepting information from the process and using this to change program parameters or even to switch to an entirely new program. Mechanical and electromechanical systems previously used for such tasks were limited in their ability to accept parameter changes. They required an expensive and time-consuming hardware change whenever the process was modified, and they lacked the arithmetic and logical capabilities of digital systems.

Figure 1.9 PDP-8 Computer Controlling a Large Milling Machine for Manufacture of Precision Aircraft Components. (Courtesy of Digital Equipment Corporation.)

5. *Speed.* Individual steps of an information-processing job are accomplished in a few microseconds or less. This means that extensive computational tasks can be performed concurrently with the operations of other systems (in real time), e.g., during the reaction time of a machine operator.
6. *Large memory.* Another aspect of computer power is storage capacity for millions of words (numbers). Semiconductor, core, disk, and magnetic tape are all popular storage media in the current generation of machines. The fastest of these, semiconductor, allows any word in the memory to be selected in less than 100 nanoseconds. However, it is also the most expensive memory type, being orders of magnitude more costly than magnetic tape.
7. *Ease of communication.* Programs (data too) are easily and inexpensively communicated. This is accomplished via telephone channels, magnetic tape, decks of cards, etc. In most cases, a FORTRAN or similar program from one installation will run on another computer with little or no modifications. When this is compared with the difficulty of duplicating electrical or mechanical hardware using drawings produced by another organization, the real beauty of digital systems becomes evident.
8. *Simulation.* A digital computer is capable of *simulating* (or *modeling*) any process that can be described in quantitative terms. In recent years, great progress has been made in biology, economics, engineering, and other fields through the use of computer programs to model situations which would be impossible to study through ordinary physical experiments because of cost, the time required, or other limitations.

An interesting facet of computer simulation was recently brought out by Dr. Ruth M. Davis formerly of the National Bureau of Standards. She made the following observation at an International Federation of Information Processing Societies meeting:

> World War I was fought with chemistry. World War II was fought with physics. World War III is being fought with Computer Science. The first battles of World War III may well have occurred when mathematical formulations of strategies and counter-strategies of realistic proportions were able to be tried out as war games on computers.
>
> With realistic wars being able to be fought in 20 minutes or 20 hours on computers, decisions to engage in such encounters have been nil. No statistical correlations are needed to validate the fact that no major encounters have occurred between "large computer possessing" nations.

1.5 PROLOGUE TO THE REMAINDER OF THE TEXT

Having very briefly examined some of the historical, statistical, and general background for computer technology, we should now be prepared to "dive" into the engineering details. The reader should now examine the table of contents and notice that the text is divided into three major parts: Principles of Digital Systems, Modern Structured Logic Design, and Advanced Topics. As the titles imply, Part I establishes the fundamentals, and Part II builds on these with emphasis on practical logic circuit design for systems ranging from simple counters to large digital computers. Finally, more depth and breadth concerning topics from several earlier chapters is added in Part III. These advanced chapters also require slightly more sophistication in electronics and in mathematics than the earlier ones. For many, Part II will be the main focus of the book; its material was selected to promote the continued development of Structured Hardware Design. This concept has already reached prominence in software engineering (see Chapter 3), but the nucleus of the idea has been applied in hardware development for considerably longer. Much more will be said about Structured Design elsewhere, particularly in Chapters 7 through 9. For the present, however, we concisely describe it as a systematic technique which emphasizes careful consideration of the alternatives before a commitment is made to detailed design. The finished system is very regular, not random, in all its aspects. The method is frequently associated with the term "top-down design" mentioned in the first paragraph of this chapter; it will be given detailed consideration in Chapter 7.

The first topic presented, in Chapter 2, is number systems. Part of this material will be a review for some readers, but the information is a prerequisite for the machine language programming in Chapter 3, study of the arithmetic unit in Chapter 6, and the tools to be employed in several other chapters.

In Chapter 3, computer software will be considered. There is a very intimate relationship between the areas of *hardware* (the equipment) and *software* (the step-by-step directions for solving a problem); thus, a person cannot be a good designer in one area without some knowledge of the other. This chapter also serves as an introduction to machine organization and the programming of a practical microprocessor.

In Chapters 4 and 5, Digital Logic and Sequential Circuits, we begin to reach the heart of our subject, since it is here that the basic tools and physical understanding employed for the remainder of the text are developed. Chapter 6, Computer Architecture, is the highest-level structure employed by the designer in planning any digital system. Computer organization is considered here along with introductions to semiconductor memories and the arithmetic subsystem. The design of digital logic circuits with modern semiconductor components is presented in Chapter 7. Then, similar designs with microprocessors are taken up in Chapter 8. Also, in that chapter, comparisons are made for a circuit designed first with components and then with a microprocessor. Chapter 9 is an extension to microprogrammable logic, which is an expensive but more powerful type of microprocessor. Much of the last three chapters consists of topics that are a direct extension of a section of an earlier chapter; e.g., Chapter 10 on Arithmetic is an extension of Section 6.6. On the other hand, certain topics complement ones appearing earlier; e.g., Chapter 12, Digital Electronics, develops the electrical properties of logic components and illustrates how they are constructed from transistors.

1.6 BIBLIOGRAPHY

1. Abshire, G. M. (ed.), *The Impact of Computers on Society and Ethics: A Bibliography.* Morristown, N.J.: Creative Computing Press, 1980.
2. Arden, B. W. (ed.), *What Can Be Automated?* Cambridge, Mass.: The MIT Press, 1980.

3. Denning, P. J., "Third Generation Computer System." *ACM Computing Surveys*, 1971, pp. 175–216.
4. Goldstine, H. H., *The Computer from Pascal to von Neumann*. Princeton, N.J.: Princeton University Press, 1972.
5. Huskey, H. D., and V. R. Huskey, "Chronology of Computing Devices." *IEEE Transactions on Computers*, Vol. C-25, 1976, pp. 1190–99.
6. Metropolis, N., J. Howlett, and G.-C. Rota (eds.), *A History of Computing in the Twentieth Century: A Collection of Essays*. New York: Academic Press, Inc., 1980.
7. Ralston, A., and C. L. Meek (eds.), *Encyclopedia of Computer Science*. New York: Petrocelli/Charter Publishers, Inc., 1976.
8. Randell, B., *The Origins of Digital Computers-Selected Papers*. Berlin: Springer-Verlag, 1973.
9. Richards, R. K., *Electronic Digital Systems*. New York: John Wiley & Sons, Inc., 1966 (especially Chapter 1).
10. Rosen, S., "Electronic Computers: A Historical Survey." *ACM Computing Surveys*, Vol. 1, 1969, pp. 7–36.
11. Weyrick, R. C., *Fundamentals of Analog Computers*. Englewood Cliffs, N.J.: Prentice-Hall, Inc., 1969.
12. Wise, K. D., K. Chen, and R. E. Yokely, *Microcomputers: A Technology Forecast and Assessment to the Year 2000*. New York: John Wiley & Sons, Inc., 1980.

1.7 TOPICS FOR FURTHER STUDY AND STUDENT REPORTS

1.1 Consult recent issues of *Scientific American* for articles concerning various facets of computers, and write a critical review of one of them.

1.2 Report on some of the nonnumerical applications of computers found in References 2 and 7 of the bibliography.

1.3 Using References 4 and 8, prepare a brief history of digital computers.

1.4 Consult *Fortune* or a similar magazine, and report on important current industrial and business applications of computers.

1.5 Using References 2 and 7 write a paper on the impact of computers on our world.

1.6 Discuss the strengths and limitations of computers; References 2 and 7 may be helpful.

1.7 There has been much discussion concerning whether computers can think.
   (a) Make a list of pros and cons on the subject. (The article on artificial intelligence in Reference 7 may be helpful.)
   (b) Propose an objective way to decide whether a particular system can think.

1.8 Find information on the analog computer and write a two-page summary on its principles and major applications.

1.9 Repeat Exercise 1.8 for the field of computer graphics. (Reference 7 is a possible source of information.)

1.10 Repeat Exercise 1.8 for the field of pattern recognition.

1.11 Write a paper on your estimate of the technological future of computers. (Reference 12 may be helpful.)

Chapter 2

Number Systems

Knowledge of number systems forms a basic tool which is useful in all phases of computer technology. It is important to the understanding of all subsequent chapters of this text, and it is crucial for our work on microprocessor programming in Chapter 3.

Have you ever tried to do mathematical calculations with Roman numerals? A comparison of the capabilities of the Roman with the Arabic (or positional) number system emphasizes the necessity of a well-developed notation for satisfactory progress in mathematics and engineering. We frequently take for granted such innovations as negative numbers, the concept of zero, and the ease with which fractions can be expressed in the positional system. However, these were not trivial or accidental advancements but required brilliant insight and centuries of evolutionary development.[4] Moreover, it is only our provincialism that makes the decimal system seem so unique. Other civilizations have successfully applied alternative number systems, and there is even a group today, the Duodecimal Society of America, which advocates changing from base 10 to base 12. Although it is highly unlikely that the revolution advocated will ever occur, there are several advantages that would accrue from such a modification.

Although we are all guilty, at times, of taking the power of modern mathematical notation for granted, the situation may be even worse when calculating machines are concerned. With the convenience of pocket calculators and personal computers, it is easy to fail to appreciate fully the rapid progress made in data processing machines, say, even since the relatively crude tabulating machine of Herman Hollerith, shown in Figure 2.1. (The Smithsonian Institute in Washington, D. C., contains a large display devoted to early computing devices of this type and the reader will find

Figure 2.1 An Early (circa 1900) Tabulating Machine Designed by Herman Hollerith. (Courtesy of International Business Machines Corporation.)

it an enlightening place to visit.) Although the present chapter is devoted exclusively to computer mathematics, the circuits required for its implementation will be considered in several later parts of this text, most notably in Chapters 4, 6, and 10.

2.1 ARABIC NOTATION[2]

Consider a positional number such 591.3, which is a concise representation for $5 \times 10^2 + 9 \times 10^1 + 1 \times 10^0 + 3 \times 10^{-1}$. The beauty of this notation is that it may be extended without limit in both directions from the decimal point to represent both larger numbers and ones of greater precision. Moreover, it holds for all other bases, not just the base 10. In general terms then, a number, N_b, in any base,* b, may be represented by the equation

$$N_b = a_n b^n + a_{n-1} b^{n-1} + \ldots + a_1 b^1 + a_0 b^0$$
$$+ a_{-1} b^{-1} + a_{-2} b^{-2} + \ldots + a_{-m} b^{-m}$$
$$= \sum_{-m}^{n} a_i b^i$$

where a_i is any of the b digits of the notation; the integers n and m control the size of the number and its precision. Thus, in representing $N_{10} = 591.3$, $b = 10$, $n = 2$, $m = 1$, $a_2 = 5$, $a_1 = 9$, $a_0 = 1$, and $a_{-1} = 3$. Special note should be taken concerning the use of the subscript following a number to represent the base used, e.g., 473.6_8 indicates that the given number is in base 8. (Of course, the number 591.3_8 is meaningless since a number in the base 8 system cannot possess the digit 9; it is only allowed 8 digits, 0 through 7.)

The three bases most frequently used in computer technology are described in Table 2.1.

Radix is another common term which is used interchangeably with *base*.

TABLE 2.1 COMPARISON OF BASES MOST FREQUENTLY USED

Name	Base	Digits	Main Use
Binary	2	0, 1	Internal representation of numbers
Octal	8	0 through 7	Convenient form for input and output
Hexadecimal	16	0 through 9, A, B, C, D, E, F	

2.2 CONVERSION OF INTEGERS

Since most data found in engineering, science, and business are in the decimal system and since most computers operate in other systems, it is important for the computer designer to be skilled in converting between various bases.

Example: Binary-to-Decimal Conversion

Convert 1101_2 to decimal. From the definition in Section 2.1:

$$1101_2 = 1 \times 2^3 + 1 \times 2^2 + 0 \times 2^1 + 1 \times 2^0 = 8 + 4 + 0 + 1 = 13_{10}$$

The above method is satisfactory as long as there are only a few places in the binary number. But computers frequently use numbers with more than 36 places; thus, the faster technique shown below is recommended. This method is described recursively by the equation

$$S_{i-1} = 2S_i + a_{i-1}$$

where a_{i-1} is defined in Section 2.1, S represents the sum, $S_n = a_n$, and S_0 (the last sum term) yields the decimal equivalent of the original binary number.

A convenient three-line form for the conversion is produced by writing the binary number on the first line, calculating the $2S_i$ terms on the second line, and calcualting the S_{i-1} terms on the third line:

Binary number	1	1	0	1
Product terms ($\times 2$)	—	2	6	12
Sum terms	1	3	6	13

Note how the calculation proceeds from left to right and how the sum in one column, when multiplied by 2, yields the product term in the next column to the right, which when added to the binary digit* in that column yields the new sum.

Example: Octal-to-Decimal Conversion

Convert 3767_8 to decimal. Here the recursive equation becomes $S_{i-1} = 8S_i + a_{i-1}$, and the calculation is completed as follows:

Octal number	3	7	6	7
Product terms ($\times 8$)	—	24	248	2032
Sum	3	31	254	2039

Thus $3767_8 = 2039_{10}$.

*The contracted form *bit* will also be employed.

Number Systems Chap. 2

It now appears that the general recursive equation to convert an integer in any base to base 10 is

$$S_{i-1} = bS_i + a_{i-1}$$

Proof: Given $S_n = a_n$. Using the recursive equation:

$$S_{n-1} = bS_n + a_{n-1} = ba_n + a_{n-1}$$

Continuing,

$$S_{n-2} = bS_{n-1} + a_{n-2} = b(ba_n + a_{n-1}) + a_{n-2} = b^2a_n + ba_{n-1} + a_{n-2}$$
$$S_{n-3} = bS_{n-2} + a_{n-3} = b(b^2a_n + ba_{n-1} + a_{n-2}) + a_{n-3}$$
$$= b^3a_n + b^2a_{n-1} + ba_{n-2} + a_{n-3}$$

Finally,

$$S_0 = b^n a_n + b^{n-1} a_{n-1} + \ldots + b^{n-n} a_{n-n} = \sum_{i=0}^{n} a_i b^i$$

which is the integer definition for the position notation ($m = 0$ for integers) which we had in Section 2.1.

There is also an easy way to convert from decimal to any other base, but first let us consider conversion to binary. The principle of the method involves successive division by 2, with the remainders forming the binary number and the final remainder becoming the *most significant digit*, MSD. Table 2.2 demonstrates the process.

TABLE 2.2 DECIMAL-TO-BINARY CONVERSION

Quotients	Remainders
13	—
6	1
3	0
1	1
0	1 (MSD)

Thus $13_{10} = 1101_2$.

The only change required in the method to go from decimal to any base, b, is division by b rather than division by 2. (Proof of the above and several other proofs are left as exercises at the end of the chapter.)

Let us demonstrate by converting 3809_{10} to octal (Table 2.3).

TABLE 2.3 DECIMAL-TO-OCTAL CONVERSION

Quotients	Remainders
3809	—
476	1
59	4
7	3
0	7 (MSD)

Thus $3809_{10} = 7341_8$.

Usually in converting between two bases, neither of which are decimal, it is easiest to convert the original number first to decimal and then change from decimal to the desired final base. However, conversion between binary and bases which are a power of 2 is particularly simple as the following illustration shows.

Due to the fact that $8 = 2^3$, there is a simple relation between octal and binary which permits conversion between the two systems by inspection. In converting from octal to binary, we merely change the octal digits individually to binary. For example, convert 7341_8 to binary:

$$7341_8 = 111 \quad 011 \quad 100 \quad 001 = 111011100001_2$$

To accomplish the reverse conversion, we break the binary number into groups of three digits and then convert each group to the equivalent octal digit. For example, convert 110101010111_2 to octal:

$$110101010111_2 = 110|101|010|111 = 6527_8$$

We may take advantage of the simple relation between binary and octal in converting between binary and decimal. First we convert the decimal number to octal by the division method, and then we use inspection to go to binary. For example, from the results of Table 2.3, $3809_{10} = 7341_8 = 111 \: 011 \: 100 \: 001_2$. To accomplish the direct conversion between decimal and binary, 12 divisions by 2 would have been required; in this case, however, 4 divisions by 8 and inspection produced the same results.

‡2.3 CONVERSION OF FRACTIONS

Converting a decimal fraction to another base follows the same general principles as for integers except that it requires a multiplication process, while for integers it requires division. The following example (Table 2.4) shows how 0.6875_{10} is converted to binary by successive multiplication of the fractional part by 2, with the integer part after each step becoming the next lower digit in the binary fraction.

TABLE 2.4 DECIMAL-TO-BINARY CONVERSION

Multiplication Process	Integer Part
0.6875 × 2 = 1.3750	1 (MSD)
0.375 × 2 = 0.750	0
0.75 × 2 = 1.5	1
0.5 × 2 = 1.0	1

The result is $0.6875_{10} = 0.1011_2$.

‡This double-dagger symbol will be employed exclusively throughout the remainder of the text to indicate a section that may be omitted, or not studied in detail, without seriously affecting the understanding of later material. (The symbol refers to an *entire numbered section*; but if the section is unnumbered, the symbol refers only to the subsection itself and any lower-level subsections that it contains. For example, the current symbol applies up to the beginning of Section 2.4.)

Of course, the conversion from decimal to octal follows the same pattern, as the example in Table 2.5 shows.

TABLE 2.5 DECIMAL-TO-OCTAL CONVERSION

Multiplication Process	Integer Part
0.8632 × 8 = 6.9056	6 (MSD)
0.9056 × 8 = 7.2448	7
0.2448 × 8 = 1.9584	1
0.9584 × 8 = 7.6672	7

The result is $0.8632_{10} = 0.6717_8 \cdots$.

Since conversion of fractions from decimal to other bases requires multiplication, it is not surprising that going from other bases to decimal requires a division process. The exact procedure for converting a binary fraction to decimal starts with a division by 2 of the least significant binary digit. Then the next binary digit is added to the left side of the decimal point of the previous quotient, and the result is again divided by 2. The method terminates in the desired decimal equivalent after the most significant digit has been added to the quotient and the result is divided by 2. In the following example (Table 2.6), the decimal equivalent of 0.1011_2 is determined.

TABLE 2.6 BINARY-TO-DECIMAL CONVERSION

Binary Fraction	Division Process
1 →	2 ⌡ ①.0
1 →	2 ⌡ ①.5
0 →	2 ⌡ ⓪.75
(MSD) 1 →	2 ⌡ ①.375
	0.6875_{10} Decimal result

2.4 ADDITION AND SUBTRACTION

The four arithmetic operations (addition, subtraction, multiplication, and division) follow the same basic rules of carry, borrow, etc., whether they are accomplished in decimal or in some other base. The major changes are the actual entries in the addition and multiplication tables, e.g., one plus one in binary must yield 10 since a separate symbol for 2 does not exist. Similarly, $7_8 + 2_8 = 11_8$. Binary addition is concisely summarized by Table 2.7. (Table entries are for $A + B$.)

TABLE 2.7 BINARY ADDITION TABLE

A \ B	0	1
0	0	1
1	1	10

Use of Table 2.7 is demonstrated by the addition shown below.

$$\begin{array}{r} 1011.01 \\ +\ 0111.10 \\ \hline 10010.11 \end{array}$$

In this example, note that carries are handled in a way analogous to that done in decimal arithmetic. (As a check, the reader should convert all of the above numbers to decimal.)

An easy way to accomplish octal addition (a similar procedure can be employed in any base) without committing a full addition table to memory is to observe that the sum of two numbers is the same as in decimal unless the result is greater than or equal to 8. In that case, the sum is *modulo* 8 and a carry is produced. Modulo 8 means that at 8 the sum starts counting over from zero; for example, $7_8 + 4_8 = 13_8$. Here

$$7 + (1 + 3) = (7 + 1) + 3 = 10 + 3 = 13_8$$

There are two main ways to perform subtraction: with a table, as we all learned in elementary school, or through the *complement* method. Particularly in minicomputers, the complement method is preferred because it requires less circuitry and it is the method which will be emphasized here.

To get a better physical understanding of the complement method, examples using the decimal system will be presented before binary is discussed. The 10's complement of a number, N, is defined the following way:

$$10\text{'s complement} \equiv 10^n - N$$

where n is the number of digits contained in the computer's register. (A register is an electrical circuit which temporarily stores numbers in a computer.) If $N = 252$ and $n = 8$, the 10's complement becomes

$$\begin{array}{r} 100000000 \\ -\ \ \ \ \ \ \ \ 252 \\ \hline 99999748 \end{array}$$

To perform the operation $621 - 252$, we add the 10's complement of 252 to 621 as follows:

$$\begin{array}{r} 00000621 \\ +\ 99999748 \\ \hline \not{1}00000369 \end{array}$$

which is the correct result. Now you should be objecting to the fact that the leftmost 1 was crossed off. However, recall that the 10's complement is really $10^n = 10^8$ too large; thus, 10^8 should be subtracted from the final answer, which is exactly what we did when the 1 was dropped.

Another form of complement, called the *9's complement*, is defined in the following way:

$$9\text{'s complement} \equiv (10^n - 1) - N$$

If $n = 6$ and $N = 252$, the 9's complement becomes

$$\begin{array}{r} 999999 \\ -252 \\ \hline 999747 \end{array}$$

To perform the operation $621 - 252$, we add the 9's complement of 252 to 621 as follows:

$$\begin{array}{r} 000621 \\ +\ 999747 \\ \hline ①000368 \\ +1 \\ \hline 000369 \end{array}$$

Using reasoning similar to that for the 10's complement problem, we should not have been surprised that the leading 1 in the answer was removed. But why was it added to the least significant digit? The answer, of course, is found in the fact that the 9's complement is one less than the 10's complement. Incidentally, addition of the leading 1 to the least significant digit is known as *end-around carry*.

As mentioned previously, our real purpose here is to learn how to perform the complement method for binary numbers since this method must be finally implemented electronically in the computer. The 2's *complement* is analogous to the 10's complement and is defined in the following way:

$$\text{2's complement} \equiv 2^n - N$$

Rather than presenting an example for this method, let us immediately consider the *1's complement*, which is analogous to the 9's complement:

$$\text{1's complement} \equiv (2^n - 1) - N$$

If $n = 6$ and $N = 4_{10} = 100_2$, the 1's complement becomes

$$\begin{array}{r} 111111 \\ -000100 \\ \hline 111011 \end{array}$$

Here is where the real savings in the 1's complement method occurs. Note that the result of the above subtraction could have been produced by merely complementing the individual digits, i.e., making the exchange $0 \leftrightarrow 1$. Later we will see that this is very easily accomplished electronically.

Let us complete our example by using 1's complements to subtract $4_{10} = 100_2$ from $13_{10} = 1101_2$:

$$\begin{array}{r} 001101 \\ +\ 111011 \\ \hline ①001000 \\ +1 \\ \hline 001001 \end{array}$$ (The reader should check this result by converting final answer to decimal.)

Sec. 2.4 Addition and Subtraction

Note how the end-around carry was used in the above example much as it was in the 9's complement method.

If we subtract a large number from a smaller one, we will obtain a negative result. But with the complement method, how is this detected just by examining the result? The conventional way to solve the problem is to employ the MSD as a sign bit. (For emphasis, we also employ the symbol ⌋ to separate the sign bit from the others.) If the sign bit is 1, the number is negative and the magnitude bits are in complemented form. Thus, the 1's complement number 1⌋0110 is a negative number since the MSD = 1. To convert it to a number having a positive magnitude with an algebraic sign, we affix the sign and complement all bits, e.g.,

$$1{\rfloor}0110_2 = -(\overline{1{\rfloor}0110})_2 = -(0{\rfloor}1001)_2 = -9_{10}$$

where the bar over the bits is a mathematical notation for complement.

Now let us employ our complement method and a six-bit number (five bits for magnitude and one for sign) to determine $4_{10} - 13_{10}$. Now,

$$-13_{10} = -(0{\rfloor}01101) = \overline{0{\rfloor}01101} = 1{\rfloor}10010$$

Continuing with our problem:

$$\begin{array}{r} 0{\rfloor}00100 \\ +1{\rfloor}10010 \\ \hline 1{\rfloor}10110 \end{array}$$

Note that the MSD of both the augend and the result indicates that they are negative numbers. (The fact that there is no end-around carry is also an indication that the final answer is negative.) To get the result into the form of a magnitude plus its sign, we merely complement the original result and affix the sign:

$$-(\overline{1{\rfloor}10110}) = -(0{\rfloor}01001) = -9$$

as expected. (Some additional facts concerning the complement method are brought out in Exercise 2.18 of Section 2.8.)

2.5 MULTIPLICATION AND DIVISION

Since multiply and divide operations will be considered in detail in Chapter 10, only a brief presentation will be given here with emphasis on the fact that the main difference between decimal operations and those in other bases is in the multiplication tables.

TABLE 2.8 BINARY MULTIPLICATION TABLE

A \ B	0	1
0	0	0
1	0	1

Binary multiplication is concisely summarized by Table 2.8. (Table entries are for $A \times B$.)

Use of Table 2.8 is demonstrated by the multiplication shown below.

$$\begin{array}{r} 1011.01 \\ \times 110.1 \\ \hline 101101 \\ 1011010 \\ 101101 \\ \hline 1001001.001 \end{array}$$

(The reader should check the result by converting both the problem and final answer to decimal, performing the decimal multiplication, and comparing the answers.)

Division is also a simple process as the following example shows:

$$\begin{array}{r} 10.1 \\ 1100 \overline{)11110.0} \\ 1100 \\ \hline 1100 \\ 1100 \\ \hline 0000 \end{array}$$

It is really not necessary to memorize the complete multiplication table to solve an octal division problem. This is demonstrated by the next example.

Perform the division $5362_8 \div 65_8$ and round the answer to three places to the right of the octal point. The key to the method is to find first the 2 to 7 multiples of 65_8 by adding 65_8 to itself (see Table 2.9).

TABLE 2.9 MULTIPLES OF 65_8

Multiplier	Results
2	152
3	237
4	324
5	411
6	476
7	563

Since the needed multiples of 65_8 are now handy, it is a fairly simple matter to complete the division process as shown below. Of course, we must be careful to carry out octal rather than decimal subtraction, but this is fairly simple if the previously mentioned modulo concept is kept in mind.

$$\begin{array}{r} 64.674_8 \\ 65_8 \overline{)5362.000_8} \\ \underline{476} \\ 402 \\ \underline{324} \\ 560 \\ \underline{476} \\ 620 \\ \underline{563} \\ 350 \\ \underline{324} \\ 24 \end{array}$$

Therefore, $5362_8/65_8 = 64.674_8$.

2.6 CODES[3,5,6]

Suppose that we want to place the number 14_{10} into a computer. We could write 1110_2, or the individual digits of the decimal number could be coded separately as 0001 0100. The second method is called *binary-coded decimal*, BCD, and has been used in some commercial computers because of its similarity to the decimal system.

One should carefully distinguish between the terms *conversion* and *encoding*. As is obvious from the above example, conversion is the more drastic process in that the structure of the number itself is changed, e.g., a different base is used. Encoding leaves the basic structure of the number the same, but the individual digits are represented by different symbols.

One of the first questions that must be answered is "How many bits are needed to code the ten symbols 0 through 9?" A solution is desired for the equation $2^n = 10$, where n is the required number of bits. Using a calculator, we find that $n = 3.32$; but since n must be an integer, four bits are needed. However, $2^4 = 16$; thus, there are $16 - 10 = 6$ excess symbols. The next question that may be asked is "By using 10 of 16 symbols, how many different codes can we construct?" We are thus asking for the number of permutations of 16 things taken 10 at a time, $_{16}P_{10}$. Now,

$$_{16}P_{10} = \frac{16!}{6!} \approx 3 \times 10^{10}$$

which certainly is a staggering number. Fortunately, these codes can be broken down into a few classes and, except for very special applications, the computer engineer need only become familiar with a few members of each class. (Actually, we are speaking here of only a small part of the subject of coding, which has become a separate discipline of its own within the field of communication theory.)

Some typical BCD codes are shown in Table 2.10. The (8, 4, 2, 1) code is the common binary numbering system employed at the beginning of the chapter. The name "excess-3" is an apt description of the next code since it is formed by the addition

TABLE 2.10 TYPICAL BCD CODES

Decimal Digit	Common Binary (8, 4, 2, 1)	Excess-3	Gray*	Code X
0	0000	0011	0000	0100
1	0001	0100	0001	0010
2	0010	0101	0011	0111
3	0011	0110	0010	1110
4	0100	0111	0110	0011
5	0101	1000	1110	1100
6	0110	1001	1010	0001
7	0111	1010	1011	1000
8	1000	1011	1001	1101
9	1001	1100	1000	1011

*The most common form of Gray code containing 16 members is shown in Table 5.16; but for convenience, variations containing fewer members are considered here and in other chapters.

of three (0011) to the corresponding common binary entries. The remaining codes, Gray and code X, will be considered shortly.

Weighted Codes

Consider the common binary code:

$$1001 = 1 \times 2^3 + 0 \times 2^2 + 0 \times 2^1 + 1 \times 2^0$$
$$= 1 \times 8 + 0 \times 4 + 0 \times 2 + 1 \times 1$$

Note that, starting at the MSD, the places of the number are weighted in decreasing powers of 2. Because of this, it is called an 8, 4, 2, 1 *weighted* code. (Codes may also have negative weights, e.g., 2, 3, 8, −4.) Weighted codes are useful because of the concise way they can be described, the ease of converting them to decimal numbers, and the circuit economy realized when they are employed for certain tasks, e.g., analog-to-digital conversion.

For a weighted code, the relation between the code and the corresponding decimal number is given by the equation

$$N = \sum_{i=0}^{n-1} a_i w_i + B \qquad (2.1)$$

where N is the decimal equivalent, a_i is the code coefficient, w_i is the weight, n is the number of bits in the code, and B is the positive (or negative) bias expressed in decimal.

Some authors omit the bias term in Equation (2.1). This results in a less general definition and considerably reduces the number of possible weighted codes, e.g, under these conditions, excess-3 is not a weighted code. (We will always employ the more general definition, except where specifically indicated.)

Now let us determine whether code Y, shown in Table 2.11, is a weighted code by using Equation (2.1) to generate the following five equations:

TABLE 2.11 ELEMENTARY CODES TO BE TESTED FOR THE WEIGHTED PROPERTY

Decimal Digit	Code Y	Code Z
0	001	011
1	000	110
2	011	111
3	010	000
4	100	101

$$0 = 0 \times w_2 + 0 \times w_1 + 1 \times w_0 + B \tag{2.2}$$
$$1 = 0 \times w_2 + 0 \times w_1 + 0 \times w_0 + B \tag{2.3}$$
$$2 = 0 \times w_2 + 1 \times w_1 + 1 \times w_0 + B \tag{2.4}$$
$$3 = 0 \times w_2 + 1 \times w_1 + 0 \times w_0 + B \tag{2.5}$$
$$4 = 1 \times w_2 + 0 \times w_1 + 0 \times w_0 + B \tag{2.6}$$

In order to solve a set of equations such as these, it is helpful to note that many of the a_i coefficients are zeros and the remainder are ones; thus, certain unknowns may be solved either by inspection or with very little effort, as the following equations show:

By inspection of Equation (2.3), $B = 1$

From Equation (2.2), $w_0 = -B = -1$

From Equation (2.5), $w_1 = 3 - B = 2$

From Equation (2.6), $w_2 = 4 - B = 3$

Checking the remaining equation, (2.4), with the above values, we get $2 = 0 + 2 - 1 + 1$, which is consistent and proves that code Y is the weighted code 3, 2, −1; 1. Note that the bias is separated from the weights by means of a semicolon. When the semicolon is absent, the bias is assumed to be zero.

As a second example, consider code Z in Table 2.11. Again, use of Equation (2.1) generates five equations.

$$0 = 0 \times w_2 + 1 \times w_1 + 1 \times w_0 + B \tag{2.7}$$
$$1 = 1 \times w_2 + 1 \times w_1 + 0 \times w_0 + B \tag{2.8}$$
$$2 = 1 \times w_2 + 1 \times w_1 + 1 \times w_0 + B \tag{2.9}$$
$$3 = 0 \times w_2 + 0 \times w_1 + 0 \times w_0 + B \tag{2.10}$$
$$4 = 1 \times w_2 + 0 \times w_1 + 1 \times w_0 + B \tag{2.11}$$

By inspection of Equation (2.10), $B = 3$. Only three of the four remaining equations are required to solve for the three unknown weights; this leaves one equation to check for consistency. Using Equations (2.7) through (2.9) and the known bias value, we can write the following new equations:

$$w_1 + w_0 = -3 \tag{2.12}$$

$$w_2 + w_1 = -2 \qquad (2.13)$$
$$w_2 + w_1 + w_0 = -1 \qquad (2.14)$$

These may be solved to give $w_2 = 2$, $w_1 = -4$, and $w_0 = 1$. ($B = 3$ was previously established.) When these are substituted into the remaining equation, (2.11), the result is $4 = 2 + 0 + 1 + 3$, $4 \neq 6$; therefore, Equation (2.11) is not consistent with the other equations, which, of course, means that code Z is not a weighted code.

Although any convenient method can be employed to solve equations such as the above, it is helpful to note that frequently two equations contain exactly the same variables, with one exception. Then, subtraction will immediately yield the value of the odd variable. This process can be applied between Equations (2.12) and (2.14), yielding $w_2 = 2$, and between Equations (2.13) and (2.14), yielding $w_0 = 1$.

It is left as an exercise for the reader to show that, in Table 2.10, excess-3 is a weighted code while the Gray code and code X are not.

‡Self-Complementing Codes

Another useful way of classifying a code concerns whether or not it is *self-complementing*. This means that the complement of the binary code leads to a number which is the 9's complement of the original number. Table 2.12 illustrates this situation for the excess-3 code. For example, decimal 4 has the excess-3 code 0111; the complement $\overline{0111} = 1000$, which is the excess-3 code for 5, which in turn is the 9's complement of 4. Self-complementing codes are very useful in BCD computers because of the simple electronic circuits needed to perform subtraction. It is left as an exercise for the reader to show that, for the other codes of Table 2.10, code X is self-complementing while the (8, 4, 2, 1) code and the Gray code are not.

TABLE 2.12 SELF-COMPLEMENTING PROPERTY OF EXCESS-3 CODE

Decimal Digit	Excess-3 Code	Complement of Excess-3
0	0011	1100
1	0100	1011
2	0101	1010
3	0110	1001
4	0111	1000
5	1000	0111
6	1001	0110
7	1010	0101
8	1011	0100
9	1100	0011

Table 2.13 is a summary of the properties of the four codes we have just discussed. Note that all four possible combinations of the two properties are shown by the codes.

TABLE 2.13 PROPERTIES OF THE FOUR CODES SHOWN IN TABLE 2.10

Code	Self-Complementing?	Weighted?
8, 4, 2, 1	No	Yes
Excess-3	Yes	Yes
Gray	No	No
Code X	Yes	No

‡Unit-Distance Codes

Consider an analog voltage being sampled by a digital voltmeter and the results transmitted to a digital computer. If the (8, 4, 2, 1) BDC code shown in Table 2.10 is employed by the voltmeter, as the analog voltage changes from 3(0011) to 4(0100) volts, bit 2 may change first, resulting in a temporary reading of 0111, which is 7. Thus, a considerable error will exist. (This type of error can occur whenever a mechanical or electrical analog quantity is converted to a digital representation.) The error, however, can be eliminated by the use of a code in which only one bit changes at a time. The class of codes with a one-bit change for adjacent integers is given the descriptive name: *unit-distance codes*.

One code of this class is the Gray code shown in Table 2.14. Note that not only does one bit change at a time in the sequence 0 through 9 but the 9-to-0 transition also has only a one-bit change. The latter fact is important in BCD applications in that errors are eliminated when the count advances from 9 to 10 (i.e., 0000 1000 to 0001 0000). The Gray code of Table 2.14 has the property that it is *reflective*. This means that, except for bit 3, the second half of the table is a mirror image of the first half.

In order to study unit-distance codes more fully, we now consider a graphic system for representing them. In Table 2.14, note that the first four lines form a separate two-bit Gray code. (In particular, the 3-to-0 transition is a one-bit change.) Using the two-bit Gray code for both axes, the discrete graph (or map) shown in

TABLE 2.14 THE GRAY CODE

Decimal Digit	Gray Code $y_3 y_2 y_1 y_0$
0	0000
1	0001
2	0011
3	0010
4	0110
5	1110
6	1010
7	1011
8	1001
9	1000

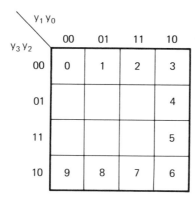

Figure 2.2 Map for the Gray Code.

Figure 2.2 is constructed for the four-bit Gray code. The reader should study the following features of the map:

1. Bits $y_3 y_2$ determine the vertical coordinate and bits $y_1 y_0$ determine the horizontal coordinate.
2. The binary digits associated with each row (column) are in correspondence with the $y_3 y_2 (y_1 y_0)$ labels, e.g., $y_3 y_2 = 10$ means $y_3 = 1, y_2 = 0$.
3. The map entry at the intersection of a row and column is the decimal equivalent of the row and column binary labels, e.g., $5 = y_3 y_2 y_1 y_0 = 1110$, which, of course, agrees with Table 2.14. (Observe that all the other entries also agree with the table.)
4. Since both coordinates change one bit at a time, adjoining squares in the vertical or horizontal direction are separated by only one bit in their binary descriptions. This is true even for the transition from the end of a row (column), e.g., the 9-to-0 transition is 1000 to 0000.

From the above discussion, it can be seen that the rules for forming a four-bit unit-distance code are:

1. Write the increasing decimal sequence for the desired code, within adjoining squares on the map.
2. Be sure that the last decimal entry in the sequence is effectively in an adjoining square to the first entry.

To demonstrate how the map method can be employed in constructing codes, assume that we desire to construct a unit-distance BCD code which does not exhibit the reflective property. From Figure 2.2, note that the lower two rows form a mirror image of the upper rows. This is a requirement for a reflected code. The requirement for a unit code that is not reflected can now be solved if we satisfy the unit-distance rules but avoid a mirror image. Figure 2.3 is the map for the new code. As an exercise, construct the corresponding table for this code to further verify that it displays both the unit-distance and the nonreflected property. Another useful exercise is to place each of the codes from Table 2.10 on a map and show that, except for the Gray code, they do not exhibit the unit-distance property.

The map method just discussed will be considerably amplified in Chapter 4

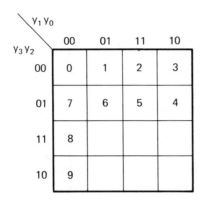

Figure 2.3 Unit-Distance BCD Nonreflected Code.

where it will be employed as a major tool in simplifying logical functions. At that point, it will be shown that the map is not limited to just four variables (bits), but that larger versions are easily constructed; thus, the method may be applied to unit-distance codes and other situations where more than four bits are required.

Error-Detecting Codes

For many computer applications, such as reading magnetic disk, paper tape, or telephone-transmitted data, it is important to detect any errors in the received information. When the probability of errors is not too large, a system called *parity checking* can be very effective. This method employs an extra bit, called the *parity bit*, whose value is selected to make the number of ones in the word even (odd) for an even (odd) parity code. For example, if even parity is employed and the parity bit, y_3, is to be added for the octal number $y_2 y_1 y_0 = 010$, the result will be $y_3 y_2 y_1 y_0 = 1010$. Here $y_3 = 1$ to yield the even number of ones required by even parity.

It is readily seen that a single-bit error in a word, including the parity bit, will always change the parity, which is easily detected. In the above even parity word, $y_3 y_2 y_1 y_0 = 1010$, if the noise changes bit y_2, the result will be $y_3 y_2 y_1 y_0 = 1110$, which is detected as a change to odd parity. On the other hand, if noise changes bit y_3, the result will be $y_3 y_2 y_1 y_0 = 0010$, which again is detected as odd parity.

Even and odd parity codes for octal numbers are shown in Table 2.15. In each

TABLE 2.15 OCTAL PARITY CODES

Decimal Digit	Even Parity $y_3 y_2 y_1 y_0$	Odd Parity $y_3 y_2 y_1 y_0$
0	0 0 0 0	1 0 0 0
1	1 0 0 1	0 0 0 1
2	1 0 1 0	0 0 1 0
3	0 0 1 1	1 0 1 1
4	1 1 0 0	0 1 0 0
5	0 1 0 1	1 1 0 1
6	0 1 1 0	1 1 1 0
7	1 1 1 1	0 1 1 1

case, bit y_3 carries the parity information. The reader should check the accuracy of the parity bits in each code and note that bits y_2 through y_0 are the standard octal numbers.

Here again the map can give us additional insight into the nature of a code. Figure 2.4 is a map representation of the even parity code of Table 2.15. The fact that each map entry is separated from its neighbors by two squares means that noise could change *any single bit* and the new word would not be a valid even parity character. Thus all single errors can be detected. Study of Figure 2.4 reveals that, in general, if all single errors are to be detected, each character must be separated from its nearest neighbor by at least two squares. Thus, a code such as that in Figure 2.4 is called a *minimum-distance-of-two code*. Figure 2.5 demonstrates that the parity property is not a necessary condition for an error-detecting code with a minimum distance of two. (The code for 1 has odd parity, and all the others have even.)

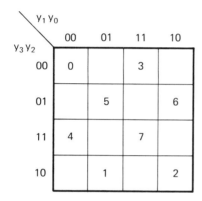

Figure 2.4 Even-Parity Code.

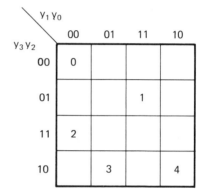

Figure 2.5 Nonparity Code with a Minimum Distance of Two.

The main advantage of the parity code of Figure 2.4 is that it is easy to construct a logic circuit to check the parity of a character. If it is not even parity, the circuit produces a signal which causes the data to be reread, or it can generate a warning signal. The nonparity code of Figure 2.5 does not possess this advantage.

Not only is it possible to detect errors, but with the proper code it is even possible to correct them. Error-correcting codes are beyond the scope of this text, but the interested reader is encouraged to consult References 3 and 5 in the Bibliography.

‡Special Codes

In addition to the basic codes just discussed, there is a special code class with which the computer engineer should become familiar. One major division of the class is the *alphameric codes*. These emphasize alphabetic and numeric information, but punctuation and control characters are also often included. Another major division of the special codes is the *display codes;* these control the formation of characters from segment or dot primitives for use in data output displays for devices such as digital voltmeters and pocket calculators.

Alphameric codes. Teletypewriters, keyboards, paper tape readers, magnetic disk units, and punched-card readers are typical devices which employ alphameric codes. The American Standard Code for Information Interchange, ASCII, shown in Table 2.16, is a popular code of this type. Since the version shown contains seven bits, it encompasses $2^7 = 128$ characters. Note that the characters represented include uppercase letters, lowercase letters, numbers, punctuation marks, special symbols, and control characters. The word standard associated with ASCII is somewhat misleading since there is not just one standard alphameric code. On the contrary, EBCDIC, Extended Binary-Coded Decimal Interchange Code, an eight-bit code containing most of the same characters as ASCII, is widely employed, particularly in the IBM Systems 360 and 370. (The coding of ASCII and EBCDIC are much different, e.g., the number 2 has the hexadecimal representation F2 in EBCDIC and 32 in ASCII.) Moreover, even ASCII has six- and eight-bit versions which have different information-carrying capacities than the seven-bit version. (Of course, the coding is similar in the three versions.)

TABLE 2.16 OUTLINE OF THE ASCII SEVEN-BIT CODE

Seven-Bit Code (Hexadecimal)	Characters		Seven-Bit Code (Hexadecimal)	Characters	
00	Null	Special control characters	41	A	Uppercase letters
...	...		42	B	
20	Space		...	...	
21	!	Punctuation and special symbols	5A	Z	
...	...		5B	[	Special symbols
2F	/		...	...	
30	0	Numbers	60	`	
31	1		61	a	Lowercase letters
32	2		62	b	
...	...		...	...	
39	9		7A	z	
3A	:	Punctuation and special symbols	7B	{	Special symbols
...	...		...	...	
3F	?		7E	~	
40	@		7F	Delete	

Bit arrangement for ASCII-7 code $\equiv \underbrace{b_6 b_5 b_4}_{\text{MSD}} \underbrace{b_3 b_2 b_1 b_0}_{\text{LSD}}$

Display codes. Again, there is not just one code employed to display information on cathode ray tubes, digital voltmeters, and other types of readouts. On the contrary, many schemes are in use, and no doubt many more will soon be invented. The basic principles of these are brought out by Figure 2.6, which shows a seven-segment display for decimal numbers. This type of display is extensively used in pocket calculators. Generally, numbers in a calculator or similar device are processed in another number system, such as BCD; but before they are displayed, there is a logic network which converts from the main code to the segment code in order to

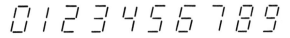

(a) Sample Numbers

(b) Seven-Segment Labeling

Decimal	BCD	Segments $S_6 S_5 S_4 S_3 S_2 S_1 S_0$
0	0000	0 1 1 1 1 1 1
1	0001	0 0 0 0 1 1 0
2	0010	1 0 1 1 0 1 1
3	0011	1 0 0 1 1 1 1
4	0100	1 1 0 0 1 1 0
5	0101	1 1 0 1 1 0 1
6	0110	1 1 1 1 1 0 1
7	0111	0 0 0 0 1 1 1
8	1000	1 1 1 1 1 1 1
9	1001	1 1 0 1 1 1 1

(c) Segment Coding

Figure 2.6 Numeric Display Code.

produce the display. As an exercise, the reader should verify that the codes in the table of Figure 2.6 do indeed generate the sample numbers shown.

To control errors, a parity bit can be added to any of the special codes. Because they are usually transmitted over longer distances and through noisier environments, however, the alphameric codes are more likely to require parity protection than are display codes.

In later chapters, we will consider the implementation aspects of coding. For example, in Chapters 4 and 7, methods for designing logic to convert one code to another will be presented; in Chapter 6, a character generator will be described which employs a 4×6 array of dots to produce an alphameric display.

2.7 BIBLIOGRAPHY

1. Bartee, T. C., *Digital Computer Fundamentals* (4th ed.). New York: McGraw-Hill Book Company, 1977.

2. Dudley, U., *Elementary Number Theory*. San Francisco: W. H. Freeman and Company, Publishers, 1969.

3. Hamming, R. W., *Coding and Information Theory*. Englewood Cliffs, N.J.: Prentice-Hall, Inc., 1980.

4. Newman, J. R., *The World of Mathematics*, Vol. 1. New York: Simon and Schuster, 1956 (especially pp. 430–520).

5. Peterson, W. W., "Error Correcting Codes." *Scientific American*, Vol. 215, No. 3, 1962, pp. 96–110.
6. Rhyne, V. T., *Fundamentals of Digital Systems Design*. Englewood Cliffs, N.J.: Prentice-Hall, Inc., 1973.
7. Richards, R. K., *Arithmetic Operations in Digital Computers*. New York: Van Nostrand Reinhold Company, 1956.
8. Walter, T. M., and W. W. Cotterman, *An Introduction to Computer Science and Algorithmic Processes*. Boston: Allyn and Bacon, Inc., 1970.

2.8 EXERCISES

2.1 (a) Count in base five from 0 to 30_5.
(b) Count in octal from 0 to 20_8.

2.2 Construct a table containing a hexadecimal column next to one with the equivalent decimal number and show how counting is accomplished, between 0 and 40_{16}, in the hexadecimal system. Every number need not be expressed, but it is important to show entries in the table at all possible points of confusion, e.g., at A_{16}, F_{16}, 10_{16}, $1A_{16}$, etc.

2.3 Perform the indicated conversion.
(a) $(111101)_2 = (\quad)_8$
(b) $(1001101)_2 = (\quad)_8$
(c) $(110111.01101)_2 = (\quad)_8$
(d) $(763)_8 = (\quad)_2$
(e) $(3715.27)_8 = (\quad)_2$
(f) $(615)_8 = (\quad)_{10}$
(g) $(915)_{10} = (\quad)_8$
(h) $(612)_7 = (\quad)_{10}$

2.4 Perform the indicated conversion and, where applicable, round the result to three places right of the radix point.
(a) $(101111)_2 = (\quad)_{10}$
(b) $(11011.011)_2 = (\quad)_{10}$
(c) $(793)_{10} = (\quad)_2$
(d) $(0.831)_{10} = (\quad)_8$
(e) $(3124)_5 = (\quad)_2$
(f) $(111011)_2 = (\quad)_6$
(g) $(839.41)_{10} = (\quad)_8$
(h) $(371.426)_8 = (\quad)_{10}$
(i) $(4163)_7 = (\quad)_4$
(j) $(0.3541)_6 = (\quad)_5$

2.5 Convert the following hexadecimal numbers to their binary equivalents.
(a) C7F
(b) 0.AF9
(c) 8EA.DA2

2.6 Convert the following hexadecimal numbers to their decimal equivalents.
(a) 3A8
(b) 0.F2A
(c) D9.7E

2.7 Convert the following decimal numbers to their hexadecimal equivalents.
  (a) 42
  (b) 93
  (c) 39.725
  (d) 46791

2.8 Considering numbers in all bases through 10, develop a general process (sequence of operations) for a nonprogrammable calculator that performs the following conversions.
  (a) From an integer in any base to the equivalent base 10 number
  (b) From a number less than 1 in any base to the equivalent in base 10
  (c) From a base 10 number less than 1 to the equivalent number in any base
  (d) From a base 10 integer to the equivalent number in any base

2.9 Repeat all parts of Exercise 2.8 under the following circumstances:
  (a) The other base besides decimal is hexadecimal.
  (b) All bases greater than 10 are allowed.
  (c) A complete program is to be written for a programmable calculator.

2.10 Prove the algorithm for decimal-to-binary conversion given in Section 2.2.

2.11 Construct a complete addition table for octal analogous to Table 2.7.

2.12 Construct a complete multiplication table for octal analogous to Table 2.8.

2.13 Determine the following sums (binary numbers except where indicated).
  (a) $11001.11 + 1011.01$
  (b) $10111.101 + 1010.01$
  (c) $65.27_8 + 54.67_8$
  (d) $AE.6F_{16} + C3.94_{16}$

2.14 Perform the following operations (binary numbers except where indicated).
  (a) $1011.11 - 110.01$
  (b) $11011.10 - 1101.01$
  (c) $(1110.01)(101.1)$
  (d) $(1101.1)(110.01)$
  (e) $(1101) \div (101)$
  (f) $(1110.1) \div (1101)$
  (g) $(6473_8) \div (462_8)$

2.15 Perform the following operations entirely within the base indicated.
  (a) $7346_8 + 5627_8$
  (b) $6512_8 - 2735_8$
  (c) $AB8_{16} + C35_{16}$
  (d) $E1D_{16} - 3A4_{16}$
  (e) $EH19_{19} + 38GC_{19}$
  (f) $6510.436_7 - 215.5432_7$

2.16 Perform the following operations entirely within the base indicated, where applicable, and round the result to one place right of the radix point.
  (a) $A350A3_{11} \div 26_{11}$
  (b) $32341_5 \times 42_5$
  (c) $31213_4 \times 12133_4$

2.17 If the accumulator for a new computer is to contain nine bits and perform arithmetic in the 2's complement system, what is the *full* range of numbers that can be expressed? Present your results in octal.

2.18 (a) Using the 1's complement method, with six-bit numbers, calculate $(6 - 10)_{10}$. Check the result by converting it back to a base-10 magnitude plus a sign.

(b) Starting with the definition of 2's complement, prove a conversion method, based on complementing individual bits, similar to the 1's complement method.

(c) Repeat part (a) but use the 2's complement method.

(d) Using first the 1's complement method and then the 2's complement method, calculate $-5_{10} + 5_{10}$. Comment on the results.

2.19 Construct a table similar to Table 2.10 for the (5, 4, 2, 1) and the (6, 4, 2, 1) codes.

2.20 How many punch positions (possible holes) exist per column on an IBM card? What is the maximum number of characters that could be represented by these punch positions in a single column? Using up to the maximum number of holes commonly punched per column, how many characters can be coded?

TABLE 2.17 CODES TO BE ANALYZED

Decimal Digit	Code A	Code B	Code C	Code D	Code E	Code F
0	0000	0001	0000	0000	0000	0101
1	0001	0010	1101	0001	0011	0110
2	0010	0011	1000	0101	0101	0111
3	0011	0100	0100	?	0110	1000
4	0100	0101	0011	1111	1111	1001
5	1011	0110	1100	?	1000	1010
6	1100	1000	1011	?	*	1011
7	1101	1001	0111	?	*	1100
8	1110	1010	0010	1001	*	1101
9	1111	1100	1111	?	*	1110

*Not required.

Analyze the codes shown in Table 2.17 according to the following directions:

2.21 Is code A a weighted code? If so, what are the weights?

2.22 Repeat Exercise 2.21 for code B.

2.23 If code D is to have the unit-distance property, fill in the missing members.

2.24 Does code E possess an error-detection capability? Explain.

2.25 Add an even parity check bit to code A.

2.26 Repeat Exercise 2.21 for code C.

2.27 Which of the codes possess the self-complementing property?

2.28 Repeat Exercise 2.21 for code F.

2.29 Which of the codes in Table 2.10 show the self-complementing property?

2.30 Is the excess-3 code of Table 2.10 a weighted code? If so, what are the weights?

2.31 In the ASCII seven-bit code, what is the hexadecimal representation for the following symbols: D, e, 5, and 7? If an even parity bit were added as a new MSB to the code, what would the binary representation of the following symbols be: @, 1, 2, and z?

2.32 Using the segment labeling of Figure 2.5, what would the codes be for the following alphabetic characters: A, C, F, H, and S?

2.33 Can an 8, 4, 3, 2; weighted code ever be employed to represent BCD numbers? Explain.

2.34 Is this table a weighted code? If so, give the weights. If not, prove it by showing the weights that satisfy certain members of the code fail for at least one other member.

Decimal Digit	Code
0	0001
1	1000
2	1010
3	1001
4	1011
5	0100
6	0110
7	0101
8	1101
9	1110

2.35 Is this table an acceptable weighted code for BCD numbers? If not, explain specifically where it fails. List all weights that you determined. (Note: $* \equiv$ proper code to be determined by you.)

Decimal Digit	Code	Decimal Digit	Code
0	0010	5	1100
1	0100	6	*
2	1000	7	0001
3	0110	8	*
4	1010	9	*

Chapter 3

Software

Hardware refers to the electronic circuits, as well as the other electrical and mechanical components, which constitute the physical portions of a digital computer. By contrast, software refers to the programs and *algorithms* which are used by the hardware to solve problems and process data. (An algorithm is a mathematical procedure, a detailed set of directions, for the solution of a particular class of problems. The manual method for forming a quotient by long division is a type of elementary algorithm.) The main subject of this book is hardware design, but it is necessary to discuss various aspects of software, both in this chapter and in Chapters 8 and 9, so that the reader can more fully appreciate how hardware and software work together during a data-processing task. Until recently, these two facets of computer technology were developed relatively independently of each other during the design of a new machine. Engineers now realize, however, that a much more powerful system can be produced if hardware and software are developed in close relationship with each other. With a knowledge of software, the engineer can make better design decisions in all phases of hardware development. Moreover, an understanding of programming is a necessity in testing computers during development and in the field. We are introducing programming concepts early in the text in order to give a clear and complete picture of the nature of a digital computer, which should accelerate understanding of the material in all later chapters.

If you have completed a college-level course employing a modern higher-level language like PL/I or PASCAL you are probably familiar with a disciplined method of software design such as *Structured Programming*.* This tool has the capabilities of

---

*Although we only consider Structured Programming[11] here, several variations exist, such as Top-Down Development, the Jackson Methodology, and Metastepwise Refinement.[10]

producing excellent software by insisting on a single, nonlooping control sequence between all program sections, modules. To facilitate this, GOTO statements are strongly discouraged; these tend to result in multiple module entries and to feedback from one module to another, making it extremely difficult to verify that logic errors do not exist. The philosophy at the heart of Structured Programming is that all modules (or subroutines) must contain *only* one entry and one exit, but within modules the following two structures are permitted:

1. IF–THEN–ELSE statements to facilitate tests and branching (nesting is allowed)
2. A statement type which permits looping

Structured Programming with a high-level language is appropriate for most problem solutions, especially those involving conventional numerical calculations. It leads to software that is more reliable, easier to understand, simpler to debug, more maintainable, and in short is more satisfactory than older methods. (See Reference 11 for more details.) Although a lengthy discussion of Structured Programming is beyond the scope of our text, Structured Hardware Design is one of our main topics and considerable details will be presented in Chapters 7 through 9. Moreover, it is not surprising that both areas have the same basic goal—that of a highly reliable system which is cost effective.

The software presented in the current chapter (also that in Chapter 9) is motivated both by a desire to complement the software knowledge possessed by the reader and to gain a deep physical understanding of the principles of computer logic and architecture. Our specific needs are to:

1. Study a computer's own unique internal instruction set because it yields information about the actual structure and operation of the processor itself. (Thus, machine-independent instructions are not appropriate here. It is important to understand that these internal instructions, called machine language, implement very elementary operations. They are discussed at length in Sections 3.2 and 3.3.)
2. Learn how to write special nonnumerical programs, such as the bit-manipulating routines often required at a microprocessor's input and output ports for interfacing with peripherals. (These tasks are often more efficiently accomplished in machine or assembly* language.)
3. Provide a medium that allows freedom for experimentation with various methods of writing and modifying machine instructions.
4. Gain insight into the nature of machine language so that we can more fully understand its use, its advantages, and its disadvantages.

The above reasons lead us to adopt classical machine language as a *major tool in studying computer hardware*, yet, of course, when writing applications software we demand a high-level language where feasible. Since the formal method of Structured Programming is not employed, one must not jump to conclusions concerning the way

---

*Machine language and assembly language are closely related, but their exact differences will be clearly described in Section 3.3.

programs will be written in this text. We still insist on carefully thought out structure (note the lowercase "s" in structure) and the goal is to make programs as systematic, reliable, and as easy to debug as possible. On the other hand, for purposes of learning, demonstrating, and experimenting, we will occasionally employ a technique that is not recommended in normal programming practice. (The one or two times this is done, care will be taken to point it out.)

What is a program, anyway? It is the directions, or "recipe," that the hardware follows in processing the data it receives. You may already be familiar with a language like FORTRAN or PL/I. Experience with a higher-level language gives one an appreciation of the practical value of the computer and an understanding of its external operation. It does not, however, give the programmer much insight into machine organization, logic, or circuitry.

We begin this chapter with a discussion of organization, introduce machine language instructions, present a brief discussion of programming languages, and end with microprocessor programming.

3.1 BASIC MACHINE ORGANIZATION

Later, a whole chapter will be devoted to machine organization, and several alternative configurations will be indicated; but at this point it is necessary to present enough fundamentals concerning one approach so that the reader will understand what portions of the computer are being controlled by particular machine language instructions. Also, it is important to keep the computer's structure in mind as the logic design techniques are developed in the next several chapters.

Figure 3.1(a) shows the basic block diagram for a digital computer, and Figure 3.1(b) shows an exploded view of the various sections in a modern minicomputer—the Hewlett-Packard 21MX. Note the correspondence between the actual hardware and the block diagram items in the two parts of the figure. The small size and the modularity of the 21MX system is typical of modern, well-designed minicomputers. This type of construction facilitates tailoring the computer to a wide variety of applications.

The input device in Figure 3.1 typically receives information from a keyboard punched cards or magnetic tape, converts it to electrical signals, and sends it to the *Control Unit* within the *Central Processing Unit*, CPU. From there the information is usually stored in semiconductor integrated circuits within the *Memory*. The Memory of a computer is divided into many individual storage locations, each containing one word (number). Every location is given a numerical address in a string configuration much like the address employed to locate a particular house on a long street. After the program and data have been transferred to the Memory, the Control Unit begins allowing one program instruction at a time, in sequence, to be executed. This is accomplished by initializing the *Program Counter*, within the control logic, to the starting address of the program and incrementing the counter by one each time an instruction is completed. Frequently an instruction will call for data to be removed from storage, processed in the Arithmetic Unit, and returned to storage. The address of the word being loaded or stored is determined by the *Memory Address Register*,

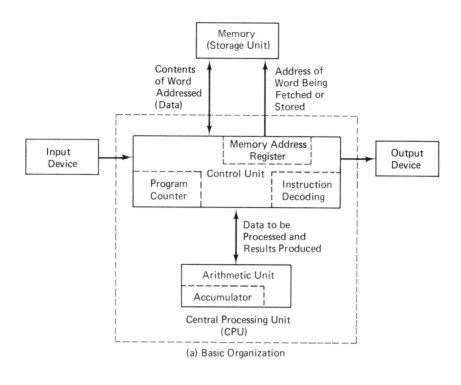
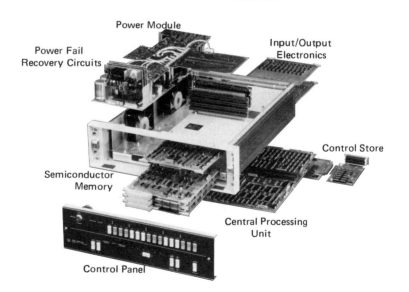

Figure 3.1 Digital Computer System-Block Diagram and Hardware. (Photo courtesy of Hewlett Packard Data Systems.)

MA, which in turn receives its information either from the Program Counter or from the operand portion of the instruction being executed. The *Accumulator*, the main register in the arithmetic unit, always contains one of the numbers during execution of one or two operand arithmetic instructions. After the data have been processed, the results are read from Memory, under the supervision of the Control Unit, and presented to the user in a convenient form by the output device. This peripheral can be a teletypewriter, a cathode ray display, or one of a large number of other units currently available.

As is evident from the block diagram, the Control Unit is at the heart of the computer. It provides pulses (not indicated on the diagram) to synchronize the operations of the other units; it decodes the program instructions; and it generally supervises the operation of all the other units. (A detailed discussion of machine organization is given in Chapter 6.)

3.2 MACHINE LANGUAGE PROGRAMMING

Machine language instructions directly control the various functions of a digital computer, and these may be divided into the following four classes:

1. *Transfer*. This class includes instructions both for storing and fetching information from memory.
2. *Arithmetic and logic*. Instructions include add, multiply, AND, and OR.
3. *Branch and control*. Included here are instruction sequence control functions such as jumps from one section of a program to another, halt the CPU, etc.
4. *Input/output*. This class includes operation of a teletypewriter, line printer, card reader, and cathode ray tube display.

Although it is theoretically possible to code any solvable problem through the use of a very small number of instructions, a desire to make programming easy and convenient may cause a designer to include more than 200 machine language instructions in the repertoire of a modern computer. This section is intended to provide an elementary understanding of machine language and to form an appreciation of the programmer's task, but not to actually develop expert programmers. Thus, only a limited number of instructions will be studied. In order to make our work more realistic, the instruction set presented here is taken from the well-known LINC minicomputer family,* but the machine language operation of many other computers is quite similar.

The LINC is a second-generation computer, but the very simplicity and freedom from extraneous details that this implies makes it an ideal processor for learning the first principles of both machine language programming and computer architecture. These principles are still employed in today's most modern computers and their use will likely continue far into the future. On the other hand, our later (in this chapter)

---

*The classic LINC, Laboratory INstrument Computer, developed by Wesley Clark and his associates, and its descendent, the Digital Equipment Corporation PDP-12, are well-known members of this group.

study of a standard microprocessor, the 6800 family, completes the fundamentals by presenting the more complex features which have recently been invented. One should be clear about the fact that the LINC is *not* employed for its historic value, but its basic nature and realism makes it a good learning tool. Moreover, the microprocessor is our primary computer and considerable details will be presented in this and several later chapters. Incidentally, the LINC is a 12-bit machine, which means that all the information it processes and stores in its memory will be represented by 12-bit numbers. (The leftmost bit is for the sign, and the other 11 bits form the magnitude in a 1's complement system.)

The *add* instruction is very frequently needed in computer programs, and it can appear in one of several forms, e.g., ADD 235. The meaning of this instruction is: "add the number found in memory location 235 to the number currently in the main arithmetic register, called the accumulator, and leave the result in the accumulator." This is a so-called single-address instruction in that only one memory location is explicitly mentioned. Since addition requires at least two operands, the second operand must be implied, and this is where the accumulator becomes necessary. (Computers have been built using two, three, and even four addresses.)

Let us consider a simple program for summing the numbers in locations 26, 27, and 30 and leaving the result in location 31. Table 3.1 shows the sequence of instructions starting at memory address 20.

TABLE 3.1 PROGRAM TO ADD THREE NUMBERS

Memory Address	Memory Contents
(Start) 20	CLR
21	ADD 26
22	ADD 27
23	ADD 30
24	STC 31
25	HLT
26 through 31	(Data and result)

The first instruction, CLR, clears the accumulator (sets it to zero). Then the numbers in locations 26, 27, and 30 are added in sequence, and the result appears in the accumulator after the instruction in 23 has been executed. Next the STC instruction $\underline{ST}$ores* the sum in location 31 and $\underline{C}$lears the accumulator. Finally, the computer is halted by means of the HLT instruction.

Several points should be noted concerning Table 3.1:

1. Certain instructions, such as ADD and STC, have both an operation part and an address part.
2. Other instructions, such as CLR and HLT, have only an operation part.
3. The three-letter abbreviations for the instructions, called *mnemonics*, form a convenient shorthand.

*The underlined uppercase letters give the key to the name of the instruction.

Since all symbols and numbers in a digital computer are presented internally by means of binary numbers, the memory contents of Table 3.1 must be coded before it can be executed by the computer. As a convenience, these codes will be represented in octal, e.g., the code for STC 31 is 4031. Again, the number 4031 must code both the operator and the address. It is probably apparent that the first digit, 4, represents the operation, and the next three digits, 031, represent the address.

A brief list of LINC machine language instructions and their octal codes are given in Table 3.2. (For tutorial reasons, only a small fraction of the actual LINC instructions are listed here.)

Observe the following facts concerning the table:

1. The underlined uppercase letters in the last column give the key to the mnemonics.

TABLE 3.2 BRIEF LIST OF PSEUDO-LINC* INSTRUCTIONS

Mnemonic	Machine Code Number (Octal)	Meaning
ADD XXX	2XXX	ADD the contents of the accumulator to the contents of location XXX and leave the result in the accumulator, using 1's complement addition.†
APO	0451	Skip the next instruction if the Accumulator is POsitive, i.e., the sign bit equals zero.
CLR	0011	CLeaR accumulator.
COM	0017	COMplement the contents of the accumulator.
HLT	0000	The machine is ordered to HaLT.
JMP XXX	6XXX	JuMP to location XXX to obtain the next instruction.
MUL XXX	5XXX	MULtiply the contents of the accumulator by the contents of location XXX and leave the least significant 11 bits in the accumulator.†
NOP	0016 or YYYY	No OPeration. This instruction merely provides a delay before the next instruction is executed. The alternate code YYYY is any undefined code and has the same meaning as the standard code.
ROR n	0300 + n	ROtate Right. The contents of the accumulator are moved around to the right n places, with the least significant bit connecting into the most significant bit as shown:

[12-bit accumulator diagram]

(move right n bits)

| STC XXX | 4XXX | STore the contents of the accumulator into location XXX and then Clear the accumulator. |

*These are called pseudo-LINC instructions because there are some minor variations between these instructions and those actually available on the LINC.
†The contents of location XXX are not changed.

2. The JMP and APO are branch instructions which furnish considerable programming power by allowing the normal sequence of operations to be modified. The APO instruction is particularly valuable in that it permits different actions to be taken, based on the sign of the accumulator. (Note that APO does not skip on -0.)
3. Because of the emphasis on fundamentals in this chapter, instructions such as *print*, which are rather specialized to a particular computer, are not presented. (Input/output functions are considered in Chapter 11.)
4. When arithmetic operations, such as ADD and MUL, employ one or more negative operands, the algebraic sign for the result is automatically placed in the leftmost bit. (This includes operations with ± 0.)

Example: Evaluating Expressions

Evaluate the expression $R = Y - X^2$. Assume that the number representing X is in location 100, and Y is in 101. We desire to have the result, $R = Y - X^2$, in location 102. The required program is shown in Table 3.3, with the results for $X = 3$, $Y = 4$, in the last column. In the program, note how subtraction is accomplished through the use of the complement operation, and that all numbers are given in octal. One should also recognize the fact that six instructions were required here (HLT is not counted) to do a task requiring a single statement in a higher-level language such as PASCAL.

TABLE 3.3 PROGRAM TO CALCULATE $R = Y - X^2$

Memory* Address	Memory Contents Mnemonic	Machine Code	Accumulator Contents (After Execution of Instruction), and Remarks
(Start) 20	CLR	0011	0
21	ADD 100	2100	$X = 0003$
22	MUL 100	5100	$X^2 = 0011$
23	COM	0017	$-X^2 = 7766$
24	ADD 101	2101	$Y - X^2 = 7772 = -5$
25	STC 102	4102	0
26	HLT	0000	0
...	...	...	...
100	X	0003	
101	Y	0004	
102	R	7772	(Result after program runs)

*Note that each memory location contains a 12-bit word (see machine code column) and that the addresses are numbered in octal.

‡*Example: Modification of Instructions*

As an exercise in concise programming, we now produce a routine which modifies an instruction while the program runs. This technique is not normally recommended because it can be difficult to debug, especially in longer programs. It is interesting, however, and does furnish some insight into the nature of machine language software. Assume that it is desired to test which of the numbers, 1 through 7, is present in register 100. If a 1 is present, a 1 is to be placed in location 201; if a 2 is present, a 1 is to be placed in location 202; etc. Finally, if a 7 is present, a 1 is to be placed in location 207. (The locations 201 through 207 are assumed to initially contain zeros.) The solution to this problem is given in Table 3.4, and you should study the program by verifying the accumulator contents (column 4) after each instruction is executed. In particular, observe how the instruction in address 25 is modified by the data in location 100 so that a 1 is stored in the proper location.

TABLE 3.4 INSTRUCTION MODIFICATION PROGRAM

Memory Address	Memory Contents Mnemonic	Machine Code	Accumulator Contents (After Execution of Instruction), and Remarks
(Start) 20	CLR	0011	0
21	ADD 100	2100	$X = 0003$
22	ADD 25	2025	$4200 + X = 4203$
23	STC 25	4025	0000
24	ADD 101	2101	0001
25	STC 200	4200	0000 (After instruction in 23 is executed, location 25 contains 4203)
26	HLT	0000	0000
...	...	...	...
100	X	0003	(Number being tested)
101	1	0001	(A constant)
...	...	...	
201	0	0000	
202	0	0000	
203	0	0000	Results (After instruction in 25 is executed, location 203 contains 0001)
204	0	0000	
205	0	0000	
206	0	0000	
207	0	0000	

In the program, the contents of location 25 act both as data and as an instruction, but to the computer hardware it merely looks like a number in both cases. This dual ability need not concern us, and as a matter of fact, it is very convenient. However, the inability of the computer to distinguish between instructions and data means that we must be careful to avoid accidental mixing of the two; otherwise, the computer might try to execute a piece of data in place of an instruction, resulting in a *program bug*.

‡*Example: Summing the Numbers in a Group of Registers*

Frequently it is necessary to perform a repetitive task or to iterate a particular mathematical operation. A very short program can usually be written in such cases by having the program loop through a particular group of instructions a specified number of times. (A DO LOOP in compiler language accomplishes the same thing.)

Assume that it is desired to sum the numbers in memory locations 200 through

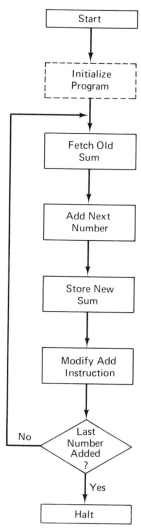

Figure 3.2 Algorithm for Summing Numbers.

225 and to place the result in location 300. Figure 3.2 shows a flow diagram for the desired iterative algorithm; note how concisely the desired processor action may be described.

The resulting machine language program is given in Table 3.5, with the Remarks column included as an aid to understanding the process. The trickiest part of a program like this is determining the correct value of the constant in location 302 so that the program comes out of the loop after the proper number of iterations. In the current problem, after the 25th loop has finished, the number in location 225 has not yet been added to the sum; thus, a 26th loop is required. During that loop the accumulator first becomes positive at the APO instruction (location 31), causing the JMP 20 instruction to be skipped. This should be compared with loop 25, where the accumulator contains 7777. Since 7777 is negative zero, the skip does not occur.

Approximately twice as many instructions would have been required if the

TABLE 3.5 PROGRAM TO SUM THE NUMBERS IN LOCATIONS 200 THROUGH 225

Memory Address	Memory Contents Mnemonic	Memory Contents Machine Code	Remarks	Accumulator after Execution of Instruction 1st Time	2nd Time	25th Time	26th Time
Start							
20	CLR	0011	⎫	0	0	0	0
21	ADD 200	2200	⎬ Sum and	$C(200)$*	$C(201)$	$C(224)$	$C(225)$
22	ADD 300	2300	⎭ store	$C(200)$	$C(200) + C(201)$	$\sum_{200}^{224} C(i)$	$\sum_{200}^{225} C(i)$
23	STC 300	4300		0	0	0	0
24	ADD 21	2021	⎫ Instruction	2200	2201	2224	2225
25	ADD 301	2301	⎬ modifi-	2201	2202	2225	2226
26	STC 21	4021	⎭ cation	0	0	0	0
27	ADD 21	2021	⎫	2201	2202	2225	2226
30	ADD 302	2302	⎬ Loop	7753	7754	7777	0001
31	APO	0451	⎬ counter	7753	7754	7777	0001
32	JMP 20	6020		7753	7754	7777	—
33	HLT	0000	⎭	—	—	—	0001
...							
200–225			Data (numbers to be summed)				
...							
300	0	0000	Result (initial value shown)				
301	+1	0001	⎫ Constants				
302	−(2225)	5552	⎭				

*The notation $C(Y)$ is frequently employed in computer technology and means the *contents of location Y*.

brute-force method of writing a sequence of ADD instructions for each of the locations 200 through 225 had been used. Moreover, the brute-force program grows linearly as the numbers to be summed increase, while the size of the Table 3.4 program is constant. In Section 3.4, we will find that by introducing an index instruction it is possible to make the program in Table 3.5 even more concise.

Often a program is loaded once into memory, and then several sets of data are processed, with the results being removed and new data inserted after each run. One difficulty, however, will be observed in the case of Table 3.5: the contents of memory addresses 21 and 300 are changed by the program. Consequently, these locations must be initialized before each run. The dashed section in Figure 3.2 shows how initialization fits into the flow diagram, and the added instructions are shown in Table 3.6. With the described improvement, each time the program is started at location 14, the two memory addresses in question are first initialized. Thus, the new program may be

TABLE 3.6 INITIALIZING ROUTINE FOR USE WITH PROGRAM IN TABLE 3.5

	Memory Address	Memory Contents Mnemonic	Machine Code	Remarks
	13	ADD 200	2200	
(Start)	14	CLR	0011	
	15	STC 300	4300	Initializing steps
	16	ADD 13	2013	
	17	STC 21	4021	
	20	...	...	Program from Table 3.5

executed as many times as desired, and the correct instructions will always be present after the operation in location 17 is completed.

3.3 PROGRAMMING LANGUAGES

There is a wide variety of programming languages in general use (actually hundreds), each of which is adapted to solving a particular group of problems. These programming languages can be arranged into a few classes, and this section presents a brief sketch of the classes and their basic properties. One scheme for partitioning languages into four main divisions is presented below:

Microprogram
Machine
Assembler
Complier
    General-purpose or procedure-oriented
    Special-purpose or problem-oriented

Microprogram. This is the lowest-level programming language. It is used to compose conventional machine language instructions from more primitive operations, e.g., a multiply instruction can be constructed from a number of shift and add operations. Some contemporary machines, e.g., the PDP-8, are very flexible because they can be microprogrammed by the user, but in most instances this level of programming is either not available or is reserved for the expert programmer. (More details will be given in Chapter 9.)

Machine language. Programming in machine code was discussed at length in Section 3.2. It is the lowest-level language available to the programmer on most machines.

Assembler. Assembly language is sometimes called *symbolic machine language* which provides a key to its nature. Machine language denotes the numerical codes that the computer executes in accomplishing a data-processing task. They are listed in the third column of most of the programs presented in Section 3.2 (see Table 3.3). Assembly language allows the programmer to enter on a card or other input medium the mnemonic ADD 100 instead of the machine code 2100. The symbols ADD 100, however, are not actually entered directly into the program to be run. Rather, the assembler interprets ADD 100 as 2100 and this is the number which is inserted in the proper memory location. The replacement of machine code by a mnemonic may not seem like much of an advance, but when we recall that the complete machine language may include 100 instructions and these may each have several variations, there actually is quite a benefit to the user. Moreover, assembly language provides a large number of other conveniences such as:

1. Symbolic referencing
2. Relative addressing
3. Editing of program manuscripts*
4. Maintaining an index of programs which can be loaded by name*
5. Error detection for illegal mnemonics, undefined symbols, etc.
6. Several other useful features such as printing a program listing which shows each memory address and the memory contents in mnemonics as well as in machine code

Except for the first two, items in this list should be self-explanatory. *Symbolic referencing* means the use of a group of letters or a name to represent the contents of a particular memory location, instead of giving the actual address. The program in Table 3.7 shows how the calculation $R = X + Y$ is accomplished by referring to locations 27, 30, and 31 not by their actual addresses, but by their preassigned names, R, X, and Y. When the program is submitted to the assembler, the R in location 27 is recognized by the system as the convention for a symbolic reference.

Relative addressing is another way to avoid giving an absolute address in a program. Here, each instruction is considered to exist at p, the present address, and

TABLE 3.7 EXAMPLE OF SYMBOLIC REFERENCING

Memory Address	Memory Contents
(Start) 20	CLR
21	ADD X
22	ADD Y
23	STC R
24	HLT
25	...
26	...
27	R 0000
30	X 1053
31	Y 0261

*These features will often be part of other sections of the software system.

TABLE 3.8 EXAMPLE OF RELATIVE ADDRESSING

Memory Address	Memory Contents
(Start) 20	CLR
21	ADD $p + 7$
22	ADD $p + 7$
23	STC $p + 4$
24	HLT
25	...
26	...
27	0000
30	1053
31	0261

all other addresses are relative to p, e.g., $p - 2$ refers to the location two addresses before the current instruction. Table 3.8 shows the program of Table 3.7, but using relative addressing.

The advantage of both symbolic and relative addressing is that the same program may be located anywhere in memory without the need for changing the address references in the instructions. Also, they save time and trouble when instructions must be added or removed from an existing program.

Compiler. A *compiler* is a program which translates concise statements in the source language to machine code. "Concise" implies that usually a single statement in the source language can replace many machine instructions. The source language should be very close to the notation of the problem and it should be machine-independent. For example, a PASCAL program written for one machine should run, with few changes, on another machine; but the compiler, which provides the interface between the source language and the hardware, will be significantly different for each machine. A *procedure-oriented compiler* is one that allows the user to write programs in a language applicable to a rather broad class of problems, e.g., the language of mathematics. Typical compilers of this class are PL/I, FORTRAN IV, PASCAL, ADA, ALGOL, and BASIC.

With a well-designed *problem-oriented compiler*, it is easy to write a program for a fairly narrow class of problems. Thus, the trade-off between simplicity and generality is strongly weighed in favor of simple statements for a specialized class of problems, with a loss of ability to solve other problem classes. ECAP[9] (Electronic Circuit Analysis Program) is a typical* example of a problem-oriented language. This system will analyze very complicated dc, ac, and transient networks and print out the results. The topology of the network and the values of the circuit elements are input to the computer by means of a simple notation. The person using ECAP needs to know almost nothing about computer programming, and the mathematical knowledge required is no higher than a first course in algebra. Engineers and scientists in many

---

*More modern circuit analysis programs exist, but the fundamental and complete resources available for ECAP, e.g., Bibliography reference 9, make it excellent for tutorial purposes.

fields are now making considerable practical use of problem-orientated languages.[2,4] In contrast with its network capabilities, ECAP is not designed to handle programs requiring direct arithmetic calculations. On the other hand, PASCAL can be used to make network calculations and a wide variety of other types of calculations. The solution of complicated electrical networks with PASCAL, however, requires considerable mathematical as well as programming knowledge.

3.4 MICROPROCESSOR PROGRAMMING

Instructions for the LINC with its 12-bit words were discussed earlier in this chapter. There a typical single instruction word, e.g., ADD 21, was composed of one octal digit for the instruction code and three digits for the address of the operand; yet many variations of this format are possible. In the IBM 360/370 family a 32-bit instruction is typically broken into three sections: the operation code, the address of the first operand (which later serves as storage for the result), and the address of the second operand. Other large computer systems employ instructions 64 or more bits long containing the following five sections: operation code, address of first operand, address of the second operand, address for the result, and finally the address of the next instruction. These large machines make efficient use of their large instruction words by being able to specify more than one operand and, in other cases, by being able to address directly extensive sections of memory.

At the other extreme from the large computers is the *microprocessor*, which, as the name implies, is the CPU of a very small computer. The photograph of a common microprocessor, the Intel 8086, is shown in Figure 3.2. The microprocessor circuit itself typically comes in a package approximately 5 by 2 by 1.5 centimeters (with the latter dimension including the pins, or terminals, that form the circuit connections). Typical popular devices in this category are the 6800, 8080, 8085, and Z8000.[6] Networks consisting of 100 or more conventional integrated circuits are currently being replaced by a single microprocessor for control and logic functions in automobile ignition systems, smart electrical instruments, home appliances, calculators, and electronic games. Most of these applications are characterized by relatively small programs and a requirement for a low-cost system. Efficiency is achieved here by using small instruction words—often eight bits. In order to permit a fairly rich variety of instructions and to permit a reasonable memory addressing range, however, it is necessary to employ multiple word instructions in these systems. The first word of the instruction is reserved for the instruction code; thus, for eight-bit words, $2^8 = 256$ instruction codes are available. The operand must then be placed in the second and sometimes also in the third word. Where two words are allowed for the operand address $2^{8+8} = 65,536$ memory locations may be directly addressed.

Microprocessor Instruction Modes

In the application of multiple-word instructions, the desire for good computing power and the need to conserve memory has led to the invention of a variety of instruction modes. The five modes that we will be using for microprocessor programming are found in Table 3.9. For most cases, the names of modes themselves provide a good

TABLE 3.9 MICROPROCESSOR INSTRUCTION MODES

Mode (Line) Number	Name	Bytes Required	Sample Instruction	Description
1	Implied	1	CLRA	Clear accumulator. The accumulator is specified (implied) by the mnemonic CLRA itself; therefore, no explicit operator is needed.
2A	Extended	3	ADD $32B	Add the number in memory location $32B_{16}$ to the current accumulator contents.
2B	Extended	3	ADDA BASE	Add the number which is in the memory location labeled with the name BASE to the accumulator.
3	Immediate	2	ADD#$3A	Add the number $3A_{16}$ to the accumulator. The operand 3A is stored in the second byte of the instruction itself. This mode is identified in assembly language instructions by the symbol "#."
4	Conditional or (relative)	2	BEQ START	Branch if equals zero: If the present value of the condition code register bit Z=1, branch to (take the next instruction from) the location previously tagged START; otherwise take the next instruction in sequence.
5A	Indexed	2	ADDA 2,X	The number 2 is added to the contents of the index register and this number becomes the address of the operand, which is added to the accumulator. The actual contents of the index register are not changed by the ADDA 2,X instruction. For assembly language instructions this mode is identified by the "X" symbol.
5B	Indexed	2	ADDA BASE,X	Same as line 5A except that the numerical value corresponding to the symbol (label) whose name is BASE is added to the contents of the index register and the result is the address of the operand. (Notice that $0 \leq BASE \leq FF$.)

indication of their nature. For example, the Implied Mode contains only single-byte instructions; they do not require a second or third byte since the single operand, usually the accumulator, is automatically specified (implied) by the name of the instruction itself.

The advantage of the Extended Mode is that it allows instructions to reach all 65K memory locations. In order to accomplish this, it of course requires a total of three bytes. Modes 2A and B demonstrate two completely different ways to specify the address of the operand. The Mode 2A example is straightforward since the address is directly specified by the hex number following the mnemonic. Note that in specifying the operand for an instruction, we employ the symbol "$" to indicate that the number is in hex; otherwise decimal is assumed. Thus $10 = 10_{16}; $10 = 10_{10}$; $1A = 1A_{16} = 26_{10}$; and 1A = undefined. By contrast the 2B example is less obvious in that the operand address is specified by a name (a label or tag) which appears elsewhere in the program to indicate the memory location of interest.

The Immediate Mode is a convenient way to store a constant adjacent to the instruction employing it. On line 3 observe that the mnemonic ADDA is followed by the special symbol "#" to distinguish the Immediate from the Extended Mode and thereby to separate the case where 3A is the address of an operand from that where 3A is the operand itself.

Conditional branching instructions (see line 4 of Table 3.9) are extremely useful because they allow a numerical or logic test to control which of two instructions is executed next in the program. The way our system remembers information needed for these instructions is to employ a Condition Code Register, CCR. After each instruction is executed, pertinent bits of the register are updated representing the conditions produced by the last operation. The following information is stored in individual bits of the CCR:

1. The C-bit is set to one if an operation produces a result with a carry out from either Bit-0 or Bit-7; otherwise it is cleared. (The C-bit is labeled Bit-0 in the CCR register.)
2. The V-bit is set to one if an operation produces a 2's complement overflow condition; otherwise it is cleared. (It is CCR Bit-1.) This situation occurs when the accumulator is too small to contain the results of an operation and information is lost at either end of the data word. It can happen on the left end of the word when the sign bit inadvertently changes during an operation; e.g., two large positive numbers are added and the result is so large that a carry is produced into the sign bit making it appear negative. It can happen on the right end of the word when precision is lost during an operation; e.g., if the accumulator containing 01_{16} is shifted right by one bit (divided by 2), the result is 00, which is an error in precision because the correct result should be one-half. More information concerning overflow is contained in Chapter 10.
3. The Z-bit is set to one if an operation produces a result with *all* bits containing zero; otherwise it is cleared. (It is CCR Bit-2.)
4. The N-bit is set to one if an operation produces a result with Bit-7 (the sign bit) equal to one; otherwise it is cleared. (It is CCR Bit-3.)

*B*ranch if *EQ*ual Zero, BEQ, is a typical conditional instruction; here a true result from the test causes the next instruction to be taken from the location whose address is the algebraic sum of the second byte of the conditional instruction and the value of the program counter. Use of the algebraic sum is the reason these are often called relative instructions. (The address is plus or minus relative to the PC.) The BEQ instruction performs its test on the Z-bit of the Condition Code Register.

Often there is a program requirement to process numbers from a sequential list of memory addresses. The Indexed Mode is well-suited to sequential tasks of this nature because the operand employed by the instruction is based on the contents of a special 16-bit register whose value may be conveniently incremented (indexed) by another instruction. Examples of this instruction are found on lines 5A and 5B of Table 3.9. One should note the similarities and differences between the Extended and Indexed families of instructions. They both work on the same group of two-operand instructions with the accumulator being the second operand. The Indexed Mode employs two-byte instructions, meaning that only the second byte is available for the operand. The second byte, however, is treated as a positive eight-bit number; thus, these instructions can reach forward up to 255 locations from the contents of the index register itself, which is somewhat limited compared to the Extended Mode's capabilities. On the other hand, the index register does contain 16 bits and carry does propagate into the higher-order bits when the operand is added to it. Therefore, one way the index instructions can reach all contiguous portions of the 65K possible memory locations is by setting the operand to zero, setting the register itself to the data starting address, and incrementing the register each time the next sequential data word is required.

Another convenient programming feature contained in the 6800 system is the Stack. This is a push-down section of memory which can store a list of data in a *Last In First Out*, LIFO, fashion much as plates are stored in many cafeterias. In our case, the memory is not physically pushed, but a reference address stored in a register called the Stack Pointer, SP, actually keeps track of the 16-bit address for the top of the stack. The 6800 system is arranged so that the SP decrements, moves down to lower memory addresses, when information is pushed onto it and increments when information is pulled (or popped) out of it. Several instructions will soon be presented which facilitate manipulating the stack.

Microprocessor Instructions and Examples

The computer industry has developed many convenient microprocessor instruction sets, and these are continually being expanded. The abbreviated set that we have adopted here is based on the very popular 6800 microprocessor, and an added benefit of this instruction set is that many readily available commercial microcomputers employ it so that the reader can easily get "hands-on" programming experience with our instructions. Our main motivation, however, is not in making 6800 experts; nor is it to expose the reader to a large variety of popular instruction sets. On the contrary, our goal is to develop real depth of knowledge in programming so that the reader can quickly understand and use new microprocessors as they become available. A summary of information about several popular microprocessor languages, including details not covered here on the 6800, is found in Reference 6; and an extensive treatment of programming, including some language features beyond the scope of the present text, is found in References 1, 3, 8, 12, and 13.

The list of instructions we will employ in this text is presented in Tables 3.10 through 3.12. Note that the mnemonic, verbal description, machine code, and functional description are given for each instruction. In addition, all except the Conditional

TABLE 3.10 IMMEDIATE, INDEX, AND EXTENDED MICROPROCESSOR INSTRUCTIONS

Mnemonic	Verbal Description	Instruction Code (Hex) Immed. (2 Bytes)	Index (2 Bytes)	Extended (3 Bytes)	Functional Description*	Condition Code Reg. 3 N	2 Z	1 V	0 C
ADDA	Add Accumulator	8B	AB	BB	$A \leftarrow A + M$				
ANDA	AND Accumulator	84	A4	B4	$A \leftarrow A \cdot M$			0	—
CLR	Clear Memory Location	—	6F	7F	$M \leftarrow 0$	0	1	0	0
CMPA	Compare Accumulator	81	A1	B1	$A - M$				
CPX	Compare Index Register	8C†	AC	BC	$X_H - M$, $X_L - (M+1)$				—
DEC	Decrement Memory Contents	—	6A	7A	$M \leftarrow M - 1$				—
INC	Increment Memory Contents	—	6C	7C	$M \leftarrow M + 1$				—
JMP	Jump for Next Instruction	—	6E	7E	$PC \leftarrow M$	—	—	—	—
JSR	Jump to Subroutine	—	AD	BD	See the text	—	—	—	—
LDAA	Load Accumulator from Memory	86	A6	B6	$A \leftarrow M$			0	—
LDS	Load Stack Pointer	8E†	AE	BE	$\begin{cases} SP_H \leftarrow M \\ SP_L \leftarrow (M+1) \end{cases}$			0	—
LDX	Load Index Register	CE†	EE	FE	$\begin{cases} X_H \leftarrow M \\ X_L \leftarrow (M+1) \end{cases}$			0	—
ORAA	OR with Accumulator (inclusive)	8A	AA	BA	$A \leftarrow A + M$			0	—
STAA	Store Accumulator	—	A7	B7	$M \leftarrow A$			0	—
STS	Store Stack Pointer	—	AF	BF	$\begin{cases} M \leftarrow SP_H \\ (M+1) \leftarrow SP_L \end{cases}$			0	—
STX	Store Index Register	—	EF	FF	$\begin{cases} M \leftarrow X_H \\ (M+1) \leftarrow X_L \end{cases}$			0	—
SUBA	Subtract Accumulator	80	A0	B0	$A \leftarrow A - M$				

*A, accumulator; M, content of memory address; PC, program counter; SP_H, higher byte of stack pointer; SP_L, lower byte of stack pointer; X_H, higher byte of index register; X_L, lower byte of index register; —, CCR bit is not affected by this instruction.

†This is a three-byte instruction.

Branch Instructions effect the CCR and these actions are also given in the tables. In the Functional Description column, the symbol "←" should be read "replaced by." Thus, for the ADDA instruction of Table 3.10, $A \leftarrow A + M$ means the contents of the accumulator are replaced by the original accumulator contents plus the contents of memory location M. This functional notation will be extended considerably in Chapter 6, where it will be given the formal name Register Transfer Logic.

As a typical instruction, consider the one with the mnemonic LDAA in Table 3.10. The code for this instruction is 86_{16} in the immediate, A6 in the indexed, and $B6_{16}$ in the extended mode. As expected, the symbol " $\updownarrow$ " in the CCR section indicates

TABLE 3.11 IMPLIED MICROPROCESSOR INSTRUCTIONS

Mnemonic	Verbal Description	Implied Instr. Code (Hex) (1 byte)	Function Description	Conditional Code Reg.			
				3 N	2 Z	1 V	0 C
ASLA	Arith. Shift Left Acc.	48	$C \leftarrow [A_7 \leftarrow A_0] \leftarrow 0$	↕	↕	↕	↕
ASRA	Arith. Shift Right Acc.	47	$[A_7 \rightarrow A_0] \rightarrow C$	↕	↕	↕	↕
CLRA	Clear Accumulator	4F	$A \leftarrow 0$	0	1	0	0
COMA	1's Complement Accum.	43	$A \leftarrow \bar{A}$	↕	↕	0	1
DEX	Decrement Index Reg.	09	$X \leftarrow X - 1$	—	↕	—	—
INCA	Increment Accumulator	4C	$A \leftarrow A + 1$	↕	↕	↕	—
INX	Increment Index Reg.	08	$X \leftarrow X + 1$	—	↕	—	—
LSRA	Logic Shift Right Acc.	44	$0 \rightarrow [A_7 \rightarrow A_0] \rightarrow C$	0	↕	↕	↕
NOP	No Operation	01	$PC \leftarrow PC + 1$	—	—	—	—
PSHA	Push A onto Stack	36	$M_{sp} \leftarrow A; SP \leftarrow SP - 1$	—	—	—	—
PULA	Pull from Stack to A	32	$SP \leftarrow SP + 1; A \leftarrow M_{sp}$	—	—	—	—
ROLA	Rotate Acc. Left Acc.	49	$C \leftarrow [A_7 \leftarrow A_0] \leftarrow$	↕	↕	↕	↕
RORA	Rotate Acc. Right Acc.	46	$\rightarrow [A_7 \rightarrow A_0] \rightarrow C$	↕	↕	↕	↕
RTI	Return from Interrupt	3B	Restore accumulator, index reg. Cond. Code Reg., etc., to contents before interrupt	↕	↕	↕	↕
RTS	Return from Subroutine	39	See text (this section)	See text			
SEC	Set Carry of CCR	0D	$C \leftarrow 1$	—	—	—	1
TAP	Transfer Acc. to CCR	06	$CCR \leftarrow A$	↕	↕	↕	↕
TPA	Transfer CCR to Acc.	07	$A \leftarrow CCR$	—	—	—	—
WAI	Wait for Interrupt	3E	Instruction execution is suspended, and microprocessor is temporarily isolated from the memory and other units until an interrupt is received.	—	—	—	—

*See Table 3.10 for an explanation of abbreviations.

that the N and Z bits are free to assume values determined by the number being loaded from memory. The overflow bit, V, becomes a logical zero since overflow cannot occur during a load operation; and the carry bit is undisturbed; thus, the symbol "—" appears in the C column. One important fact that should be observed about all our microprocessor instructions is that *two's complement arithmetic* is employed, which is

TABLE 3.12 CONDITIONAL BRANCHING MICROPROCESSOR INSTRUCTIONS*

Mnemonic	Verbal Description	Conditional Instr. Code (Hex) (2 Bytes)	Branch Test†
BCC	Branch if Carry Clear	24	$C = 0$
BCS	Branch if Carry Set	25	$C = 1$
BEQ	Branch if Equal Zero	27	$Z = 1$
BMI	Branch if Minus	2B	$N = 1$
BNE	Branch if Not Equal Zero	26	$Z = 0$
BPL	Branch if Plus	2A	$N = 0$
BRA	Branch Always	20	None
BVS	Branch if Overflow Set	29	$V = 1$

*Condition Code Register is not affected by these instructions.
†Condition Code Register bits:

$C = 1$: There was a carry out of bit 7.
$N = 1$: Result was negative.
$V = 1$: Overflow occurred.
$Z = 1$: All result bits are zero.
$C, N, V, Z = 0$: The opposite condition.

in contrast to the LINC programming earlier where one's complement arithmetic was employed.

Determination of the N, Z, and C bit should be relatively easy from the description of the instructions given in Tables 3.10 through 3.12. The V-bit, however, is somewhat more complicated; but it is always set when the data processed by an instruction represent a potentially serious error. A few pages earlier we described this bit and gave a simple example for the addition of two positive numbers that cause the sign bit to change. A somewhat similar situation occurs with the ASLA instruction in Table 3.11, which basically implements multiplication by two. If $C(A) = 43_{16}$ and the ASLA instruction is execution, the result is $C(A) = 86$ with the new CCR being NZVC = 1010. Thus, multiplication of a positive number, seven-bit magnitude plus sign bit, led to a negative result with the consequence that the overflow bit is set; $V = 1$. For the above instruction as well as the other members of the shift family of instructions (ASLA, ASRA, LSRA, ROLA, and RORA), the function *always* implemented is: Set V if the contents of the N and C bits are opposite after instruction execution, i.e., $N = 1, C = 0$, or $N = 0, C = 1$. (This is the most complicated V function employed by any of our instructions.)

Our examples until now have illustrated very serious mistakes, where $V = 1$ indicated that a fatal error was made. On the other hand, the overflow bit is also set for less serious situations. For example, starting with $C(A) = 25_{16}$, execution of ASRA, basically a divide by two operation, results in $C(A) = 12_{16}$. Since 25_{16} does not divide evenly by two, the presence of a remainder is indicated by the V-bit being set. (This value, of course, comes directly from the V-function for the shift family.) Therefore, the CCR becomes NZVC = 0011.

A result that may surprise you is the V-bit value for the instruction ASRA when the accumulator starts with an even negative number. Consider the starting $C(A) = FE$; after the ASRA instruction, the result is $C(A) = FF$. Thus, the operation correctly performed is $-2/2 = -1$. Applying the rule for setting V in the shift family, we obtain: $N = 1$, $C = 0$ resulting in $V = 1$, which was unexpected in the case of an even number. Moreover, with $C(A) = FD$ and executing the ASRA instruction, the result is $C(A) = FE$ and $NZVC = 1001$. Here $-3/2 = -1$. Since $V = 0$, the remainder is not indicated by the V-bit. This behavior is *consistently obtained* when both ASRA and RORA act on negative numbers. Thus, as long as we understand the V-function implemented, we can properly allow for it when writing and checking programs. On the other hand, if we need to detect the remainder when performing divide with the ASRA instruction, only checking the C-bit is required since it will always contain the remainder.

When a microprocessor is interfaced to other equipment, frequently new data need to be communicated on a fairly tight time schedule. For this purpose a special feature called an *interrupt* is included, which allows the microprocessor to be stopped in the middle of a task (but at the end of the instruction already in progress) to receive (or transmit) new data through the interface, and to return to the original task without disturbing the program or partial results already obtained. In our system, the interrupt signal is received through a separate terminal, a pin, on the microprocessor chip, causing the CPU to start executing an interrupt service program whose first address is specified (previously stored) in memory locations $FFF8_{16}$ and $FFF9_{16}$.

Since the interrupt service program could write into some or all of the microprocessors registers, they must be stored before servicing begins. The stack is the solution to this problem in that the contents of all of the registers, including the program counter, are placed on the stack in the order shown in Figure 3.3. Observe that the usual notation is employed for contents of the registers stored in the stack; e.g., PC_L is the lower eight bits of the program counter. When the interrupt is received the register information is pushed onto the stack starting at the current SP address, which in this case is location m. Then the stack pointer is decremented and the PC_H is stored at m-1. Finally, after six such cycles, the CCR is stored at m-6 and the SP decremented one final time. Next, the microprocessor executed the service routine as it would any other portion of a program until the return from interrupt, RTI, instruction is encountered. This instruction tells the processor that it is time to restore the original registers. Since our stack follows the LIFO order, the SP is incremented, the original CCR information pulled from the stack, and the cycle repeated until finally the PC_L is restored. Two key points here concerning the sequence of events are:

For PUSH: 1. First the data are entered into the stack.
              2. Then the SP is decremented.

For PULL: 1. First the SP is incremented.
              2. Then the data are removed from the stack.

Tables 3.10 and 3.11 list several instructions which permit the programmer to control stack operations. For example: LDS, load the stack pointer from memory, allows the stack pointer to be initialized near the top of memory. Then, as data are

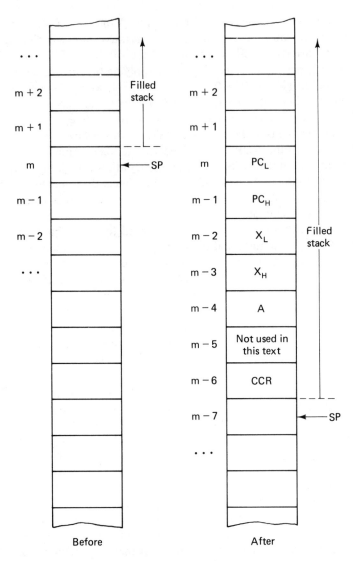

Figure 3.3 Register Status Saved on the Memory Stack.

Notes: 1. m is the memory address before the interrupt.
2. The contents of each memory address is 8 bits.

pushed on the stack, there will be a minimum of interference with user programs, which usually start in the lower part of memory. PSHA and PULA allow data to be entered into and removed from the stack independent of the interrupt operation. This stack feature can be extremely useful in advanced programming, particularly where lists of data are being processed. The wait for interrupt instruction, WAI, also facilitates interrupt servicing. If the WAI instruction is executed, the CPU is forced to wait for an interrupt to occur; and when it does the same process as decribed above is followed. Both the RTI and WAI instructions are listed in Table 3.11.

The *Subroutine* is another valuable programming feature that relies on the stack. Here an isolated section of program that executes a frequently required calculation is arranged in a form that allows it to be *called*, its instructions executed, at any desired point in the main program. With a single copy of the instructions, the subroutine, serving in place of several copies at various places in a large program, efficiency is promoted and a better structured program results. The JSR, jump to subroutine, instruction of Table 3.10 causes the program to jump to a new memory address where the subroutine begins. The instructions for the calculation itself are written like any other routine; but the last instruction must be an RTS, return from subroutine, in order to get back to the original position in the main program. The stack comes into play as a storage medium for the PC value required for a proper return to the main program. Stack operation details can be visualized by referring again to Figure 3.3. The stack before the JSR instruction is, of course, the same as for the interrupt operations. The After part of the figure is also the same, except that in the case of the JSR instruction, only PC_L and PC_H are placed on the stack,* which means that the SP would only go as far as m-2 and the other register information would not be pushed on the stack. The RTS instruction follows the usual stack rules for pull, but it is just the opposite of the JSR; i.e., the PC_H and PC_L data are pulled, with the SP finally residing at memory location m when the process is complete.

The unconditional branch instruction JMP has already been presented in Table 3.10, but there are many times that an algorithm requires a branch based on one of the CCR bits. The Conditional Branching instructions listed in Table 3.12 satisfy these requirements and a typical member BEQ has already been considered in Table 3.9.

In order to assist the reader to understand fully the rather large number of details in the above tables, the isolated instructions in Table 3.13 were prepared. These should be studied carefully before going to the set of example programs that follow.

Example: Branch Instruction Program

To demonstrate several of our microprocessor instructions, the following example is presented: A number at address $1FF_{16}$ is to be tested. If the number is zero, $A5_{16}$ is to be loaded into the accumulator and the program is to continue at address 220; and if the number is not zero, $C8_{16}$ is to be loaded into the accumulator and the program is again to continue at address 220. The instructions are to start at address 200_{16}.

The program to satisfy the above requirements is shown in Table 3.14, with the dual assembly machine language format allowing easy comparison between the two instruction forms at the same address. The labels NUMB, START, etc., are employed to mark major reference addresses. Another important point to consider is the conditional branching at address 203 and its displacement 05. Observe that the displacement is added to the address (PC value) of the next instruction to obtain the branch address, which is 020A and is given the label PLACE. Notice that the Z-bit tested by the BEQ instruction was established by the LDAA instruction at address 0200. The

---

*Note that the programmer is responsible for saving any other current CCR information needed later by the main program.

TABLE 3.13 EXAMPLE OF OPERATIONS FOR TYPICAL MICROPROCESSOR INSTRUCTIONS

Mne-monic†	Address (Hex)	Memory Contents Instr. Code (Hex)	Accumulator Contents, A Before* (Hex)	Accumulator Contents, A After (Hex)	Condition Code Register Before* (NZVC)	Condition Code Register After (NZVC)	Next Instr. Address (Hex)	Remarks
ADDA #$15	0100 0101	8B 15	F2	07	1000	0001	0102	$F2 + 15 = 107_{16}$. Thus, after: $A = 07$, $N = 0$, $C = 1$. Also $V = 0$, since only the addition of numbers with the same sign can result in overflow into the sign bit.
TAP	0100	06	16	16	0000	0110	0101	The accumulator's most significant digit, i.e., 1, had no effect on CCR.
ROLA	0100	49	84	08	1000	0011	0101	CCR after-condition is $C \neq N$; therefore, N (Bit 7) has changed signs, which means that $V = 1$.
BMI 7	0100 0101	2B 07	F3	F3	1000	1000	0109	CCR: $N = 1$ is consistent with accumulator having its most significant bit = 1. Next instruction $= 102 + 7 = 109_{16}$.
BVS $FC	0100 0101 0005	29 FC A2	F3	F3	1010	1010	00FE	Next instruction $= 0102 + FFFC = 00FE_{16}$.
INC 5	0100 0101 0102	7C 00 05	25	25	0000	1000	0103	After the instruction, memory address 5 contains A3. Since the MSB of A3 is 1, CCR bit $N = 1$. Note that the CCR agrees with the operation just done at address 5, and it no longer agrees with the accumulator.
LDAA $105	0100 0101 0102 ... 0105	B6 01 05 ... F2	52	F2	0011	1001	0103	Location 105 is not affected by the instruction.
CPX #$50	0100 0101 0102	8C 00 50	3A	3A	0000	0100	0103	If $C(X) = 0050_{16}$, then after the instruction $Z = 1$ is expected since the operand, 0050, is equal to the contents of X.
STAA 5,X	0100 0101 ... 0151	A7 05 ... 25 Before instruction C1 After instruction	C1	C1	1011	1001	0102	The content of the index register is 14C both before and after the instruction is executed. Notice that the after CCR bits are as expected.

*It is assumed that all these bits were affected by the last accumulator instruction.
†These are independent instructions, not members of a program.

TABLE 3.14 PROGRAM DEMONSTRATING THE BEQ INSTRUCTION

Assembly Language (Symbolic) Program		Machine Language Program		Remarks
Label	Instruction	Address	Code (Hex)	
NUMB	(Any one-byte number)	01FF	(Any one-byte number)	
START	LDAA NUMB	0200	B6	⎫ LDAA must be an extended instruction since two bytes are required for the address 01FF. (The N and Z bits are controlled by the contents of location NUMB.)
—	—	0201	01	
—	—	0202	FF	⎭
—	BEQ PLACE	0203	27	
—	—	0204	05	205 + 5 = 20A, which is the address with the PLACE label.
—	LDAA #$C8	0205	86	NUMB ≠ 0 ⎱ Immediate instruction
—	—	0206	C8	
—	JMP CONT	0207	7E	
—	—	0208	02	
—	—	0209	20	
PLACE	LDAA #$A5	020A	86	NUMB = 0 ⎱ Immediate instruction
—	—	020B	A5	
—	JMP $220	020C	7E	⎫ The operand is the numerical address 220_{16} rather than the symbolic address CONT.
—	—	020D	02	
—	—	020E	20	⎭
...	...	...	...	
CONT	(Continuation of remainder of program)	0220	(Continuation of remainder of program)	

Remarks column of Table 3.14 should provide a guide to the detailed operation of these and other facets of the program. If PLACE had been before address 203, say at location 195, the relative address (displacement) would be negative: Displacement $= 0195 - 0205 = $ FF90. For a check we obtain: (Address of next instruction) $+$ (Displacement) $= 0205 + $ FF90 $= 0195 = $ (Branch address), as expected. Several points should be noted concerning the negative displacement:

1. The numbers used in the calculations contain four hexadecimal digits because our microprocessor employs a 16-bit address.
2. Since only eight bits are available to contain the displacement at the second byte of the branch instruction, the number 90 would actually be stored instead of FF90.
3. When the microprocessor executes the branch instruction and calculates the address of the next instruction, however, it interprets 90 as FF90. Thus, it effectively performs the 16-bit addition shown above, yielding the address 0195 as the correct result.
4. The carry out of 1 from the address calculation in (3) is, of course, dropped.

Example: Servicing an Interrupt

Now consider the following problem: At a certain point in a program information is needed from an external data source; and the program waits until an interrupt arrives indicating that the information is available. After the interrupt, the information is read from the external source through an address called DATA and it is then stored at an address called STORE. (The location DATA may not be employed for semipermanent storage since it is a fairly rapidly changing source.) Action returns to the main program and the information in STORE is used but details are not given.

The reader must carefully study the resulting program in the two forms shown in Table 3.15. The remarks in particular should be helpful in understanding some of the less obvious details of the program.

TABLE 3.15 INTERRUPT SERVICE PROGRAM

Assembly Language (Symbolic) Program		Machine Language Program		Remarks*	
Label	Instruction	Address	Code (Hex)		
...	...	...	...	At this point the main program waits for the interrupt before going to address 350.	MAIN
—	WAI	205	3E		
—	—	206	—	Program returns here after RTI at address 355.	PROGRAM
...	...	...	...		
DATA	LDAA #0 (External data)	350 351	86 00	The 0 is a dummy to be replaced by the external source as the program runs.	INTERRUPT
	STAA STORE — —	352 353 354	B7 03 56	STAA is not available as an immediate instruction or else two memory locations could be saved.	SERVICE
	RTI	355	3B	This instruction causes a return to the main program at address 206.	ROUTINE
STORE	00	356	00	The 00 is a dummy to be replaced as the program runs.	
...	...	...	...		
(INTERRUPT ADDRESS)	03 50	FFF8 FFF9	03 50	After the interrupt, this causes the service routine to start at address 350.	

*Recall that the interrupt process was considered a few pages back in this section. One major point required from that discussion is the fact that the interrupt causes an unconditional branch to the Interrupt Service Routine whose starting address ia always placed in memory locations $FFF8_{16}$ and $FFF9_{16}$.

Example: Indexed Instruction Programs

The Indexed Mode forms a very valuable set of instructions for sequential operations. Although its real power is most evident in fairly complicated programs, the fundamentals can be demonstrated by means of the following simple problem,

TABLE 3.16 NUMBER SUMMING PROGRAM

Assembly Language (Symbolic) Program		Machine Language Program		Remarks
Label	Instruction	Address	Code (Hex)	
START	LDX# $120	100	CE	This establishes the starting point for the sum.
		101	01	
		102	20	
	CLRA	103	4F	
TOP	ADDA 0,X	104	AB	Observe how the loop is formed between the BNE instruction and this one. (See Problem 3.20 concerning an assumption made here.)
		105	00	
	INX	106	08	
	CPX# $151	107	8C	$C(X)$ must equal 151 before the CCR bit will be set to $Z = 1$ at this point.
		108	01	
		109	51	
	BNE TOP	10A	26	
		10B	F8	Relative address = 104 − 10C = FB
	STAA $151	10C	B7	
		10D	01	
		10E	51	
END	JMP END	10F	7E	Jumping back to itself is a convenient way to produce a halt.
		110	01	
		111	0F	
		...	...	
		120	—	Numbers to be summed
		...	...	
		150	—	Storage for result
		151	—	

which involves the calculation of a sum. It is desired to add the numbers in locations 120_{16} through 150_{16} and place the result in location 151_{16}. The program is to start at address 100. Table 3.16 is the solution to this problem.

The solution to the above should be compared with a similar LINC program shown in Table 3.5 and its initialization routine shown in Table 3.6. The comparison reveals that our new solution, with the index operation, requires fewer instructions, does not need initialization, and is generally a better structured program.

Conclusions

Our treatment of microprocessor programming in this section has necessarily been limited to the most critical features—those important to our tutorial objectives; thus, even many valuable aspects of the 6800 have not been covered. Moreover, there are some strict format rules, e.g., concerning allowed length of labels, that must be observed in the commercial 6800 assembly language. We have taken an informal approach here by relaxing these rules.

At this point, the reader has sufficient information to fully understand the interaction between hardware and software design to be discussed in later chapters, particularly the use of microprocessors in logical design presented in Chapter 8. Moreover, you should now be at the point where manufacturers' manuals on microprocessor programming, including the more advanced features of the 6800 family, can be much more easily understood.

3.5 BIBLIOGRAPHY

1. Bishop, R., *Basic Microprocessors and the 6800*. Rochelle Park, N.J.: Hayden Book Company, Inc., 1979.
2. Bowers, J. C., and S. R. Sedore, *SCEPTRE: A Computer System for Circuit and Systems Analysis*. Englewood Cliffs, N.J.: Prentice-Hall, Inc., 1971.
3. Bruce, R., *Software Debugging for Microcomputers*. Reston, Va.: Reston Publishing Company, Inc., 1980.
4. Calahan, D. A., et al. (eds.), "Special Issue on Computers in Design." *Proceedings of the IEEE*, Vol. 60, 1972, pp. 1–124.
5. Donovan, J. J., *Systems Programming*. New York: McGraw-Hill Book Company, 1972.
6. Duncan F. G., *Microprocessor Programming and Software Development*. Englewood Cliffs, N.J.: Prentice-Hall Inc., 1979.
7. Haupt, F., *Elementary Assembler Language Programming*. Columbus, Ohio: Charles E. Merrill Publishing Company, 1972.
8. Jensen, R., and C. C. Tonies, *Software Engineering*. Englewood Cliffs, N.J.: Prentice-Hall, Inc., 1979.
9. Levin, H., *Introduction to Computer Analysis: ECAP for Electronic Technicians and Engineers*. Englewood Cliffs, N.J.: Prentice-Hall, Inc., 1970.
10. Peters, L. J., and L. L. Tripp, "Comparing Software Design Methodologies." *Datamation*, Vol. 23, November 1977, pp. 89–94.
11. Rohl, J. S., and H. J. Barrett, *Programming via PASCAL*. Cambridge: Cambridge University Press, 1980.
12. Wakerly, J. F., *Microcomputer Architecture and Programming*, Vol. 1. New York: John Wiley & Sons, Inc., 1981.
13. Zelkowitz, M. V., *Principles of Software Engineering and Design*. Englewood Cliffs, N.J.: Prentice-Hall, Inc., 1979.

3.6 EXERCISES

3.1 Classify the LINC instructions in Table 3.2 according to the categories listed at the beginning of Section 3.2.

3.2 (a) List the sequence of instructions required to produce skip on negative accumulator.
(b) Repeat part (a) for skip on zero accumulator. In parts (a) and (b), use only the versions of the LINC instructions given in Table 3.2.

3.3 After the machine language program shown has run, determine the following quantities using *octal* numbers.
(a) The mnemonics corresponding to the LINC codes and the magnitude and sign of any data contained in this program
(b) The memory address where the program halts

(c) The number in the accumulator when it halts

The program starts in location 24, and all locations except those listed initially contain zero.

Memory Address	LINC Code	Mnemonics
24	0011	
25	2051	
26	0017	
27	2050	
30	4031	
31	6052	
32	0451	
33	0000	
34	6040	
...		
50	2327	
51	0277	

3.4 Write a program which searches memory locations 100 through 500 to try to find a number greater than the one stored in location 70. If a suitable number is found, halt the computer at the address immediately following that number. If no suitable number is located, halt the computer at address 501.

3.5 After the LINC program shown has run, determine the following quantities using *octal* numbers.
 (a) The memory address where the program halts
 (b) The number of times the instruction in location 22 is executed
 (c) The address of any locations changed by the program and the final contents of those locations

The program starts at address 20 and all locations except those shown initially contain zero.

Memory Address	LINC Code
17	7776
20	0011
21	2024
22	2017
23	4024
24	4040
25	2024
26	0451
27	6020
30	6050

3.6 After the LINC program shown has run, determine the following quantities using *octal* numbers.
 (a) The memory address where the program halts

(b) The number in the accumulator when it halts

The program starts in location 23 and all locations except those listed initially contain zero. The starting contents of the accumulator is somewhere within ± 3777.

Memory Address	LINC Code
23	4070
24	4033
25	2070
26	0017
27	2070
30	0451
31	6033
32	6040
33	6041

3.7 Write a program which converts 8, 4, 2, 1 code to excess-3 code. The original number is in location 100 and the result is to be placed in 200. The program should check that the content of 100 is a positive binary number between 0000_2 and 1001_2. If it is, halt the program at 350; otherwise, halt at 300.

3.8 Write a program which accepts 8, 4, 2, 1 code and converts it to code X, shown in Table 2.10. The number to be converted is found in location 100 and the result is to be placed in 200. (The content of 100 is a proper single BCD digit.)

3.9 With the fewest possible instructions, modify the program in Table 3.4 so that it is self-initializing with respect to the data output locations as well as the program itself.

3.10 Using the fewest possible pseudo-LINC instructions, write a program to add the contents of locations 101 through 104. Normally, halt at 40 and leave the results in the accumulator; but if overflow occurs (the capacity of the accumulator is exceeded), halt at 50 and clear the accumulator. Assume that all numbers are positive.

3.11 Using the fewest possible pseudo-LINC instructions, write a program to determine whether the number initially in the accumulator is odd. If it is odd, halt at 100; otherwise, halt at 101.

3.12 After the program (employing pseudo-LINC instructions *only*) runs, determine the following quantities using octal numbers.
(a) Does the program halt? If so, at what memory address?
(b) The number of times the instruction in 20 is executed.
The program starts at 20, and all locations, except those shown, initially contain zero.

Memory Address	LINC Code
20	0011
21	0405
22	0451
23	0000
24	2021
25	2020
26	4021
27	6020
30	6100

3.13 After the machine language program shown has run on the LINC, determine the following quantities:

(a) The LINC mnemonics corresponding to the machine codes and the magnitude and sign of any data. (Fill in table below.)

(b) The memory address where the machine halts.

(c) The number of times the instruction in location 20 is executed. (Give results in octal.)

The program starts in location 20, and all locations except those listed initially contain zero.

Memory Address	LINC Code	Mnemonics
20	0011	
21	2200	
22	2201	
23	4200	
24	2033	
25	2201	
26	4033	
27	2200	
30	2202	
31	0451	
32	6020	
33	6040	
...	...	
200	2300	
201	0001	
202	5437	
...	...	
300	0001	
301	0005	
...	...	

3.14 Using the fewest possible pseudo-LINC instructions (data locations also count), write a program to determine whether the number initially in the accumulator satisfies $1000_8 \leq A \leq 1777_8$ or $3000_8 \leq A \leq 3777_8$, where A is the accumulator contents. If it satisfies the above relation, halt the computer at 201; otherwise halt at 200. Place a comment next to each group of one to three instructions to document your program.

3.15 Repeat Exercise 3.5 for the program shown. Use only the instructions from Table 3.2; and add: (d) Determine the number of times the instruction in location 27 is executed. If the program does not halt, explain why.

Memory Address	LINC Code	Memory Address	LINC Code	Memory Address	LINC Code	Memory Address	LINC Code
20	0017	24	0310	30	2023	34	1072
21	0011	25	4032	31	0455	35	0000
22	2032	26	0011	32	6601	36	6026
23	7677	27	2035	33	3615	37	1072

Sec. 3.6 Exercises

The following exercises are to be worked with the 6800 microprocessor instruction set:

3.16 For the isolated instructions (no program) given below, prepare a complete table similar to Table 3.13. Assume that above 0300_{16} all odd-numbered memory locations contain $E6_{16}$ and all even ones contain 07. If any instruction is illegal, state why in detail.

Problem Part	Mnemonic	Starting Address (Hex)	Accumulator Contents Before (Hex)	CCR Before (NZVC)
(a)	ADDA #$23	0100	93	1000
(b)	ADDA #$81	0100	93	1000
(c)	BRA $F3	0100	13	0011
(d)	BNE $3C	01A4	A3	1011
(e)	BEQ $4E	01A4	16	0011
(f)	CMPA $3A1	0101	C3	1001
(g)	ASRA	0100	75	0000
(h)	ASLA	0100	75	0001

3.17 Repeat Exercise 3.16 for the isolated instructions below, but add the following two columns to your table: stack pointer after and index register after. Include the following additional assumptions, which are valid before each instruction: $SP_H = E8$, $SP_L = 34$; $X_H = 03$, $X_L = 05$; starting address 0100_{16}.

Problem Part	Mnemonic	Accumulator Contents Before (Hex)	CCR Before (NZVC)
(a)	ADDA $23,X	53	0011
(b)	ADDA $24,X	53	0010
(c)	LDX $391	53	0011
(d)	LDX $91,X	53	1011
(e)	LDX $391,X	53	1011
(f)	PULA	00	0111

3.18 In both parts of this exercise, the following constraints hold.

1. The code 00 is defined to be a halt (HLT) instruction and execution stops at the address containing that code; the CCR is left unchanged.

2. In addition to 01, all other codes not defined by Tables 3.10 through 3.12 are defined to be NOP instructions.

3. All memory locations not explicitly listed in the program contain 00.
   (a) The program shown has the initial values CCR = 7, $C(X) = 3$, $C(120) = 5$.

Address (Hex)	Mnemonics
00FA	ADDA $120
	SEC
	BCC $55
(Execution starts here)	LDAA #$85
	DEX
	BNE $F6
	BRA $15

Determine the following:
(1) The number of times the instruction at location 100 is executed.
(2) Does the program halt? Where?
(3) At halt what is the contents of the CCR and the accumulator? (Give all values in hex.)

(b) The program shown has the initial value CCR = 5.

Address (Hex)	Mnemonics
00F6	CLRA
	STAA $113
	ROLA
	BPL $20
(Execution starts here)	LDAA $113
	RORA
	BVS $F4
...	...
0112	STAA $FC63

Determine the following:
(1) The number of times the instructions at locations $F6_{16}$ and 100_{16} are executed.
(2) Does the program halt? Where?
(3) At halt, what is the contents of the CCR and the accumulator? (Give all values in hex.)

3.19 Using the smallest number of total memory locations, write a sequence of instructions to accomplish each task described below. Start each sequence at address 200_{16} and prepare your sequences in a form similar to the program in Table 3.14.
(a) The Z-bit is to be set and we do not care what happens to the other bits of the CCR.
(b) The Z-bit of the CCR is to be set and the other bits must be the same as their original values.
(c) Search the memory until the number F3 is found and store its address at location 0100_{16}. Exactly one F3 is guaranteed to be found between 00 and FF. When the task is done, take the next instruction from location 5691_{16}.
(d) Branch if overflow is not set.
(e) In addresses 300_{16} to 400_{16}, alternately store 02 in the even memory locations and

01 in the odd locations. When the task is done, take the next instruction from location AF10.

(f) Repeat (e), but fill only locations 300 through 308. In addition, assume that your primary goal is to write a program which is as easy as possible for another person to understand and that using the smallest number of locations is secondary.

3.20 What major assumption is made about the calculated result in the program of Table 3.16? What instructions should be added to remove this assumption?

3.21 Write a program, employing the fewest possible instructions, which reads the locations (given in hex) 1013, 521, 71A, 1F59, 1E72, A27, 111, and 9AF. If the contents of any location is even, change it to the next more positive odd value and store in back in the original location. Otherwise do not change the original memory contents.

3.22 Each time an interrupt occurs, a number is available to be read from a data source at location 741_{16}. After all the data words have been read, they are to be returned to a receiver at location $8A7_{16}$ in LIFO order. (Again, the location is ready each time an interrupt is received.) The number zero never appears in the data; thus, it may be employed as a sentinel (marker) whenever desired. More specifically, a zero is used by the original data source to mark the end of the data. No more than 30_{10}K numbers are to be received.

3.23 Apply the constraints from Exercise 3.18 to this problem. However, employ the following new initial values: CCR = B, $C(A) = 07$, $C(SP) = 0283_{16}$, $C(X) = 09$. The first byte of the LDAA instruction shown is in address 00F7 and program execution starts with the BMI instruction.

Mnemonics

LDAA $010F
STAA $010A
LDS $0109
INX
INCA
PSHA
CMPA #$15
NOP
BMI $F0
JMP $0196

Determine the following:
(a) Does the program halt? Where?
(b) The number of times the instruction at location 100 is executed.
(c) At halt determine the contents of the CCR, the accumulator, the stack pointer, and the index register.

Chapter 4

Logical Design

To ensure reliability and to satisfy other practical requirements, the signals in a digital computer consist only of two values e.g., -1.5 volts and $+4.0$ volts. These voltages have logical significance, with one value representing the existence of a particular condition and the other representing the absence of that condition. There is a branch of mathematics called *Boolean algebra* which is especially qualified to represent such situations. It was first developed by George Boole,[2] around 1850, in his analytical studies of the thought process. Although other philosophers and mathematicians employed Boole's work in the intervening years, it was not until 1937 that Claude Shannon[10] made a major application of it in engineering to describe digital circuits.

In this chapter we will develop the theorems of Boolean algebra and will show how they can be employed in the design of digital systems. Also of major interest will be the use of map methods for the simplification of logic circuits. Finally, we will present the techniques for realizing logical expressions with standard electronic circuits, called *gates*.

It is important to obtain a thorough understanding of this chapter because the fundamentals developed here will be applied in *all* later chapters.

4.1 BASIC LOGICAL OPERATIONS

Digital computer logic can be expressed through combinations of three operations: AND, OR, and NOT.

AND. The AND operation requires that all input conditions are simultaneously true before the output is true. Thus, the circuit in Figure 4.1 requires that both switch A AND switch B be operated before the light comes on.

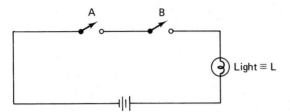

Figure 4.1 AND Operation.

In Boolean algebra, the following notation is employed for any variable, X:

Input X operated is represented by $X =$ true or $X = 1$.
Input X unoperated is represented by $X =$ false or $X = 0$.

Thus, the binary values 0 and 1 represent a convenient shorthand notation for the logical state of a variable.

Another way to express the AND circuit of Figure 4.1 is by means of a truth table such as that shown below. Note that this table presents the outputs for all possible values of the inputs, A and B.

TABLE 4.1 TRUTH TABLE FOR THE AND OPERATION

A	B	L
0	0	0
0	1	0
1	0	0
1	1	1

Boolean algebra allows us to represent the AND operation for Table 4.1 in a very concise way by the equation

$$L = A \cdot B = AB$$

Note that the symbol "·" is read as AND and that it is the same as the multiply operation of ordinary algebra. Moreover, the electronic circuit, or gate, which implements the AND operation is symbolized below:

OR. The OR operation is developed in a parallel way to that used for AND. The output of an OR circuit, however, is on if at least one of the switches is operated. This is represented by switches in parallel in Figure 4.2.

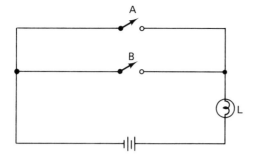

Figure 4.2 OR Operation.

TABLE 4.2 TRUTH TABLE FOR THE OR OPERATION

A	B	L
0	0	0
0	1	1
1	0	1
1	1	1

The truth table corresponding to Figure 4.2 is shown in Table 4.2. Again Boolean algebra allows the operation to be represented by a concise equation:

$$L = A + B$$

This time the symbol "+" is read as OR and it corresponds to the addition operation of ordinary algebra.* In Boolean algebra, however, it should be observed that $1 + 1 = 1$ (not $1 + 1 = 2$), i.e., a true variable ORed with a true variable results in the output being true.

The electronic gate symbol for OR is shown below:

NOT. The last of the three fundamental operations, which is variously called NOT, negation, complementing, or inversion, is really simpler than the other two. Its switching circuit representation is shown in Figure 4.3. Note that the switch A is normally closed. When the switch is unoperated ($A = 0$), its contacts are closed, causing the light to come on; when it is operated ($A = 1$), the contacts are open, causing the light to be off.

The very simple truth table corresponding to this circuit is shown in Table 4.3, and the resulting equation is

$$L = \bar{A}$$

*We are using engineering notation for these operations, but the corresponding mathematical symbols are:
$$\text{OR} \equiv \vee, \quad \text{AND} \equiv \wedge, \quad \text{and} \quad \text{NOT} \equiv \sim$$

Sec. 4.1 Basic Logical Operations

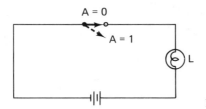

Figure 4.3 NOT Operation.

TABLE 4.3 TRUTH TABLE
FOR THE NOT OPERATION

A	L
0	1
1	0

Note the use of the bar over the *A* to represent the NOT operation; this is the general engineering notation for taking the complement whether a single variable or a large expression is being considered. The electronic gate that performs the NOT operation is called an *inverter* and is represented by the symbol

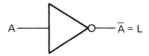

The triangular portion is the symbol for an amplifier, but the small circle is the general representation for inversion.* It will be used later in conjunction with AND and OR gates to form NAND and NOR gates, which are very popular in integrated-circuit technology.

4.2 THEOREMS[3,5,7]

In the design of digital systems, Boolean algebra provides a way of describing various complex operations in equation form by means of the AND, OR, and NOT parameters. Moreover, there is a group of theorems in the algebra which allow the equations to be manipulated into forms that are more advantageous for circuit mechanization. Frequently, the theorems also allow an original expression to be simplified by eliminating the occurrence of a variable or a group of variables. A convenient feature about this new algebra is that most of the laws of ordinary algebra can be shown to be valid; e.g., the commutative, associative, and distributive laws hold.

Table 4.4 contains a group of postulates from which all the theorems of Boolean algebra can be derived. (Using the previous definitions of the primitive operations, the reader should critically examine Table 4.4.)

*A wedge is also often employed:

A ──▷── $\bar{A} = L$

TABLE 4.4 POSTULATES OF BOOLEAN ALGEBRA

AND Operations:	$0 \cdot 0 = 0$ $0 \cdot 1 = 1 \cdot 0 = 0$ $1 \cdot 1 = 1$
OR Operations:	$0 + 0 = 0$ $0 + 1 = 1 + 0 = 1$ $1 + 1 = 1$
NOT Operations:	$\overline{1} = 0$ $\overline{0} = 1$

Applying the postulates listed in Table 4.4 and the fact that any variable, e.g., A, can only take on the two values, 0 and 1, it is easy to prove the elementary theorems shown in Table 4.5.

TABLE 4.5 ELEMENTARY THEOREMS OF BOOLEAN ALGEBRA

AND Operations:	$A \cdot 1 = A$ $A \cdot 0 = 0$ $A \cdot A = A$ $A \cdot \overline{A} = 0$
OR Operations:	$A + 0 = A$ $A + 1 = 1$ $A + A = A$ $A + \overline{A} = 1$
NOT Operations:	$\overline{\overline{A}} = A$

In the theorems listed in Table 4.5, as well as in the others we shall develop, the replacement can go in either direction, e.g., occasionally it is convenient to replace A by $A + A$.

Perfect induction. We will now discuss a general though frequently brute-force way to prove theorems. First recall that, by definition, a theorem is true only if both sides always assume the same value for a particular combination of variables. Usually in mathematics, any variable can take on an infinite number of values, but in Boolean algebra it can take on only two values, 0 and 1. Thus, it is feasible here to prove theorems by comparing both sides for all combinations of the variables, and this method is known as *perfect induction*. Suppose we desire to prove what will be called theorem 1, which is $A + AB = A$. We can do this conveniently with perfect induction by placing the information in the configuration shown in Table 4.6.

We complete the proof by simply checking for agreement between entries in the individual lines of the first column and the corresponding lines in the last column.

TABLE 4.6 PROOF BY PERFECT INDUCTION

A	B	A + AB
0	0	0 + (0)(0) = 0
0	1	0 + (0)(1) = 0
1	0	1 + (1)(0) = 1
1	1	1 + (1)(1) = 1

It is easy to show by perfect induction that the commutative, associative, and distributive laws are valid, and these are explicitly stated in Table 4.7. Note that the distributive law provides a means for transforming between *product-of-sums*, POS, and *sum-of-products*, SOP, expressions. These two basic forms lead to different but equivalent circuits, which will become obvious in Section 4.5.

TABLE 4.7 THREE FAMILIAR LAWS OF ALGEBRA

Law	AND Form	OR Form
Commutative	AB = BA	A + B = B + A
Associative	A(BC) = (AB)C	A + (B + C) = (A + B) + C
Distributive	(A + B)(A + C) = A + BC	AB + AC = A(B + C)

Once the distributive law is available, theorem 1 can be proved algebraically as follows:

$$A + AB = A \quad \text{(Statement of theorem 1)}$$
$$A(1 + B) = A \quad \text{(The distributive law)}$$
$$A(1) = A \quad \text{(Elementary theorem)}$$

Therefore,
$$A = A$$

Since any variable can represent a compound expression, a good way to remember theorem 1 is:

A quantity ORed with the product of that same quantity and another expression can be simply reduced to the original quantity.

Frequently electronic components can be saved, and generally a better design produced, if a given function can be simplified. *Simplification* is defined to be a process for the elimination of variables or terms from the original expression without changing the form of the expression, i.e., by staying either in SOP or POS form. (The reduction of functions by changing the form of the expression is discussed in Section 4.5.)

Example

Simplify $f = x + \bar{y} + (x + \bar{y})(\bar{w} + z)$ using theorem 1 with $x + \bar{y} = A$ and $(\bar{w} + z) = B$:

$$f = x + \bar{y}$$

By comparing the original expression with the final one, we see that considerable simplification has taken place. The most complex term, $(x + \bar{y})(\bar{w} + z)$, was completely eliminated and SOP form was maintained.

In addition to the elementary theorems and the three familiar laws of algebra, there is a set of four very useful theorems, occurring both in SOP and POS forms, which are not at all obvious. These are presented in Table 4.8. (Note that the theorem numbers for the POS form carry an asterisk to distinguish them from corresponding theorems in SOP form.) It is certainly not self-evident that the two forms of theorems with the same number are closely related, but this will be proved later in this section.

TABLE 4.8 MAIN THEOREMS OF BOOLEAN ALGEBRA

Sum-of-Products Form	Product-of-Sums Form
1. $A + AB = A$	1.* $A(A + B) = A$
2. $A + \bar{A}B = A + B$	2.* $A(\bar{A} + B) = AB$
3. $AB + A\bar{B} = A$	3.* $(A + B)(A + \bar{B}) = A$
4. $AB + \bar{A}C + BC = AB + \bar{A}C$	4.* $(A + B)(\bar{A} + C)(B + C) = (A + B)(\bar{A} + C)$
5. $\overline{A + B} = \bar{A} \cdot \bar{B}$	5.* $\overline{A \cdot B} = \bar{A} + \bar{B}$

Any of the theorems in Table 4.8 could be proved by perfect induction, but in order to develop facility with the mathematics, let us perform some algebraic proofs. Since theorem 1 has already been proven, consider theorem 2.

$A + \bar{A}B = A + B$ (Statement of theorem)

$A + \bar{A}B = (1)(A + B)$ (Elementary theorem)

$A + \bar{A}B = (A + \bar{A})(A + B)$ (Elementary theorem)

Therefore,

$A + \bar{A}B = A + \bar{A}B$ (Distributive)

The above work is typical in that there are few set ways to approach a proof, but imagination and experience often lead to very concise arguments.

De Morgan's theorem. The two forms of theorem 5 are known as *De Morgan's theorem*, and together they allow any complemented expression to be reduced to a form with, at most, individual variables complemented. (Theorem 5 can easily be proved by perfect induction.) As an example, consider $\overline{(\bar{X}Y + \bar{W})Z}$. Since this is basically a product-of-sums expression, we first apply theorem 5*, with the result $\overline{(\bar{X}Y + \bar{W})} + \bar{Z}$. The left term is now in sum-of-products form, so we apply theorem 5 to it: $\overline{(\bar{X}Y)}\overline{\bar{W}} + \bar{Z} = \overline{(\bar{X}Y)}W + \bar{Z}$. Now we apply theorem 5* to the bracketed term; $(\bar{\bar{X}} + \bar{Y})W + \bar{Z} = (X + \bar{Y})W + \bar{Z}$. Thus $\overline{(\bar{X}Y + \bar{W})Z} = (X + \bar{Y})W + \bar{Z}$. This same method of alternately using the two forms of theorem 5 can be applied to any complemented expression in order to simplify it.

The dual operation. The *dual* is similar to the complement except that the variables are left unchanged. (The operations $(+, \cdot)$ and the constants $(0, 1)$, however, are complemented.) For example:

$$\text{dual}\,[X\cdot 1 + (\bar{Y} + Z)\bar{W}] = [\text{dual}\,(X\cdot 1)][\text{dual}\,(\bar{Y} + Z)\bar{W}]$$
$$= [X + 0][\text{dual}\,(\bar{Y} + Z) + \bar{W}] = X[\bar{Y}Z + \bar{W}]$$

Consider theorem 3, which is

$$AB + A\bar{B} = A \qquad (4.1)$$

Since this is a general theorem, it also holds for the complemented variables, i.e., with the substitutions $\bar{A}$ for A, $\bar{B}$ for B, and B for $\bar{B}$.

$$\bar{A}\bar{B} + \bar{A}B = \bar{A}$$

Complementing both sides of the equation,

$$\overline{\bar{A}\bar{B} + \bar{A}B} = \bar{\bar{A}}$$

Therefore,

$$(A + B)(A + \bar{B}) = A \qquad (4.2)$$

Now we take the dual of Equation (4.1):

$$(A + B)(A + \bar{B}) = A \qquad (4.3)$$

Note that Equations (4.2) and (4.3) are exactly the same. It can be shown that, in general, substituting variables for their complements and then taking the complement of both sides of a theorem is equivalent to taking the dual of both sides of the theorem. Usually the dual is a much simpler operation.

Thus, once a theorem is proved, a second form of that theorem is immediately available through the dual operation. We now recognize that the SOP and POS forms of the theorems in Table 4.8 are duals of each other, and only one of them need be proven.

Proof and examples for theorem 4. Theorem 4 is very useful for simplifying Boolean expressions, as will soon be seen; but first let us prove the theorem.

$$\begin{aligned}
AB + \bar{A}C + BC &= AB + \bar{A}C &&\text{(Statement)} \\
&= (1)AB + AB + \bar{A}C &&\text{(Elementary theorem)} \\
&= (C + \bar{C})AB + AB + \bar{A}C &&\text{(Elementary theorem)} \\
&= ABC + \bar{A}C + AB\bar{C} + AB &&\text{(Distributive)} \\
&= C(AB + \bar{A}) + AB\bar{C} + AB &&\text{(Theorems 1 and 2)}
\end{aligned}$$

Therefore, $AB + \bar{A}C + BC = AB + \bar{A}C + BC$, which proves the theorems.

1. Suppose it is desired to simplify the expression:

$$\begin{aligned}
f &= \bar{X}Y + XZ\bar{Y} + YZ \\
&= \bar{X}Y + (X\bar{Y} + Y)Z &&\text{(Distributive followed by theorem 2)} \\
&= \bar{X}Y + XZ + YZ &&\text{(Theorem 4)}
\end{aligned}$$

Therefore, the simplified expression is

$$f = \bar{X}Y + XZ$$

2. Simplify

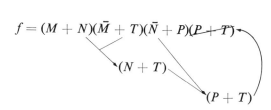

$$f = (M + N)(\bar{M} + T)(\bar{N} + P)(P + T)$$

$$(N + T)$$

$$(P + T)$$

Therefore,

$$f = (M + N)(\bar{M} + T)(\bar{N} + P)$$

In this example, we achieved simplification by applying theorem 4* backwards two times until the redundant term $(P + T)$ was discovered. Problems such as this demonstrate the importance of the right-to-left version of the theorems.

4.3 THE CANONICAL EXPANSION OF FUNCTIONS

In the field of digital computers, the term *canonical expansion* relates to a standard form of expansion, which we will soon see has both practical and theoretical significance. Consider the truth table for the OR function shown in Table 4.9. Note that the extra column labeled Minterms contains a product term which is the only combination of complemented and uncomplemented variables that equals 1 for the values of the variables in that row. Thus, reading the $f = 1$ minterms from the table, we obtain the following expression:

$$f = \bar{X}Y + X\bar{Y} + XY \tag{4.4}$$

TABLE 4.9 OR FUNCTIONS AND MINTERMS

X	Y	f	Minterms
0	0	0	$\bar{X}\bar{Y}$
0	1	1	$\bar{X}Y$
1	0	1	$X\bar{Y}$
1	1	1	XY

This is the conventional method for reading a function from a truth table in minterm expansion form. Equation (4.4) can be simplified to the usual expression for OR by the following procedure:

$$f = \bar{X}Y + X\bar{Y} + XY \quad \text{(Original expression)}$$
$$f = \bar{X}Y + XY + X\bar{Y} + XY \quad \text{(Elementary theorem)}$$
$$f = Y(\bar{X} + X) + X(\bar{Y} + Y) \quad \text{(Distributive)}$$
$$f = Y + X \quad \text{(Elementary theorem)}$$

By way of a formal definition, a *minterm* is a term in a sum-of-products form which contains all variables, either complemented or uncomplemented.

An alternate method of reading a function from a truth table is to select the minterms that go with $f = 0$. This, of course, results in the $\bar{f}$ expression. Let us use the above approach on the EXCLUSIVE-OR function shown in Table 4.10.

TABLE 4.10 EXCLUSIVE-OR FUNCTION

X	Y	f
0	0	0
0	1	1
1	0	1
1	1	0

Here $\bar{f} = XY + \bar{X}\bar{Y}$. We complement both sides in order to obtain f:

$$f = \bar{\bar{f}} = \overline{XY + \bar{X}\bar{Y}} = (\bar{X} + \bar{Y})(X + Y) \tag{4.5}$$

A term such as one of those in Equation (4.5), which contains all variables, either complemented or uncomplemented, in a product-of-sums expression is called a *maxterm*. (This definition is analogous to the minterm definition.) Of course, a sum-of-products expression in minterm form could have been obtained from Table 4.10 by reading the minterms with $f = 1$, but another method is to apply the distributive law to Equation (4.5), which results in $f = \bar{X}X + \bar{X}Y + \bar{Y}X + \bar{Y}Y$. Therefore, $f = \bar{X}Y + X\bar{Y}$ (elementary theorem). This sum-of-products form for the EXCLUSIVE OR function is probably more common than the product-of-sums form in Equation (4.5), but the minterm expansion and the maxterm expansion are equally valid. Incidentally, neither of these forms can be simplified.

Function of n Variables

Considerable time has already been devoted to discussing the common functions—AND and OR. Also some other two-variable functions such as the EXCLUSIVE-OR have been mentioned. We now investigate all the possible functions of two variables, as well as their nature. To show this, Table 4.11 has been constructed much like a usual truth table, except that all possible two-variable functions are represented. Note that the functions are arranged in order of increasing binary numbers, with the least significant bit being at the top. The subscript at the head of each column represents the decimal equivalent of the binary number. The most common operations are labeled with their names at the bottom of the table. For example, NOR is $F_1 = \bar{X}\bar{Y} = \overline{X + Y}$. This is OR followed by NOT, but the "N" is placed first for easy pronunciation. As would be expected, the gate symbol is

$$\overline{A + B} \equiv \text{NOR}$$

TABLE 4.11 LISTING OF ALL FUNCTIONS OF TWO VARIABLES

X	Y	F_0	F_1	F_2	F_3	F_4	F_5	F_6	F_7	F_8	F_9	F_{10}	F_{11}	F_{12}	F_{13}	F_{14}	F_{15}
0	0	0	1	0	1	0	1	0	1	0	1	0	1	0	1	0	1
0	1	0	0	1	1	0	0	1	1	0	0	1	1	0	0	1	1
1	0	0	0	0	0	1	1	1	1	0	0	0	0	1	1	1	1
1	1	0	0	0	0	0	0	0	0	1	1	1	1	1	1	1	1

Common names for functions: NULL, NOR, INHIBIT, $\bar{X}$, INHIBIT, $\bar{Y}$, EXOR (Exclusive OR), NAND, AND, EXNOR or EQUALITY (Exclusive NOR), Y, IMPLICATION, X, IMPLICATION, OR, IDENTITY

Similarly, the NAND is

$$\bar{F}_7 = XY$$

Therefore $F_7 = \bar{\bar{F}}_7 = \overline{XY}$. Here, the minterm associated with $\bar{F}$ rather than F was selected from the table because there is only one $F_7 = 0$ minterm, while there are three $F_7 = 1$ terms. However, if the three minterms had been selected, the resulting function could have been reduced to $F_7 = \overline{XY}$. The corresponding NAND gate symbol is

$$\overline{AB} \equiv \text{NAND}$$

Now let us determine the number of minterms there are for n variables. Since each variable in a minterm can be either complemented or not, there are two possibilities, but each variable is independent of the others. Thus, the total minterms are equal to

$$\underbrace{2 \cdot 2 \cdot 2 \ldots 2 \cdot 2}_{n \text{ factors}} = 2^n$$

Each of the 2^n minterms can either be in f or not; thus the total possible functions are equal to

$$\underbrace{2 \cdot 2 \cdot 2 \cdot \ldots \cdot 2 \cdot 2}_{2^n \text{ factors}} = 2^{2^n}$$

Although the equation for the number of functions looks innocent enough, it yields very large numbers even for fairly small values of n, as Table 4.12 shows.

TABLE 4.12 MINTERMS AND FUNCTIONS OF n VARIABLES

n	Number of Minterms	Functions of n Variables
1	2	4
2	4	16
3	8	256
4	16	65,536
5	32	4,294,967,296

4.4 THE KARNAUGH MAP[3,4,7]

The *Karnaugh* map* is one of the most important tools in digital system design. It is very useful in simplifying Boolean functions, and it forms an important part of many logic circuit synthesis techniques, such as those to be introduced in Chapter 5. We have already employed a rudimentary version in Chapter 2 for the analysis of codes.

The essence of the map method is that the 2^n possible minterms of n variables are represented by means of separate squares, or cells, on the map. Thus, it contains the same information found in a truth table. If a minterm is present in a particular function, the corresponding square contains a 1; otherwise, it contains a 0. Once all the minterms of the function have been placed on the map, there are some fairly simple rules for reading the map in terms of a simplified representation of the function.

Consider a four-variable situation ($n = 4$). Here $2^n = 2^4 = 16$; thus the map will contain 16 squares. Such a map is shown in Figure 4.4. Note how the bracket is

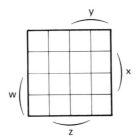

Figure 4.4 Four-Variable Karnaugh Map.

drawn adjacent to the last two rows to indicate the region where $w = 1$. (In turn, the top two rows represent the region of $w = 0$.) This, of course, is repeated in a similar way for the other variables. Another way to represent the regions on the map is to label each column and row separately. Both methods are shown in Figure 4.5, along with a representation of the function

$$f_1 = \bar{w}\bar{x}\bar{y}\bar{z} + wx + yz$$

*The name Karnaugh is used in honor of one of the map's inventors; however, the name *Veitch diagram* is also frequently used in recognition of its other inventor.

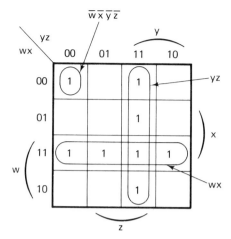

Figure 4.5 Two Methods for Labeling a Karnaugh Map. (Zeros on the map are frequently represented by blank squares.)

One of the requirements on Karnaugh map labeling is evident in Figure 4.5: the fact that adjacent columns (rows) must differ only in the value of one variable. This is done so that minterms which may be combined are contiguous on the map, e.g., note how the minterms $\bar{w}\bar{x}yz + \bar{w}xyz + wxyz + w\bar{x}yz$ combine to form the reduced term yz.

Plotting the 1-terms of a function on the map is a fairly simple task if we remember that the result of ANDing a group of variables on the map results in a term which is the intersection of the map regions represented by the individual variables. An example of this is the way the term wx is formed in Figure 4.5 from the intersection of the w and x regions. (If you are thinking that operations on a Karnaugh map are very similar to those with Venn diagrams, you are entirely correct.) Now return to Figure 4.5 and verify the labeling of regions $\bar{w}\bar{x}\bar{y}\bar{z}$ and yz.

Similarly, when several terms are combined through the OR operation, the resulting map coverage is the union of the various terms. The function, f_1, above is formed by such an operation.

Now that we can see how to plot a sum-of-products function on the map through the AND and OR operations, let us ask ourselves how a reduced function can be read from the map. First observe that when 2^m squares combine to form a single group, m variables are eliminated. An example of this is shown in Figure 4.5, where the four squares forming yz have a description with $m = 2$ variables missing. This, each coverage on a map must involve a number of squares which is equal to a power of 2.

The following sequence of rules for grouping 1-terms on the map should be observed:

1. First cover, circle, all isolated 1's. These will be described by all n variables and require no further attention.
2. Next, consider each remaining term separately. If it can be grouped in more than one way, go to another term; otherwise, include it in the *largest possible rectangular coverage that is a power of 2*.
3. Following exhausive application of the above rule, if at least one term remains

which can be grouped in more than one way, arbitrarily select one of the ways and return to rule 2.

4. The simplified function is complete as soon as all terms are covered; but in the process of making a group as large as possible, it is permissible to use previously covered terms.

Figure 4.6(a) shows an example of a reduced expression for the function, $f_2 = x\bar{y}z + \bar{w}yz + \bar{w}\bar{x}y\bar{z} + wx\bar{y}\bar{z}$, formed by using the above rules. Note that, according to rule 2, the squares containing the * symbol are the only valid places to begin the coverage—the other terms can be grouped in more than one way.

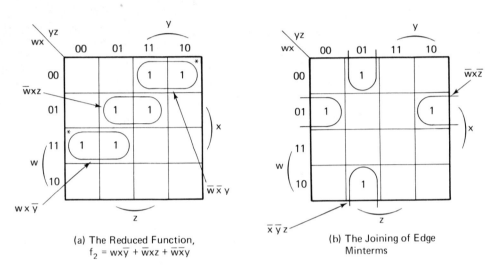

(a) The Reduced Function, $f_2 = wx\bar{y} + \bar{w}xz + \bar{w}xy$

(b) The Joining of Edge Minterms

Figure 4.6 Karnaugh Map Examples.

Examination of Figure 4.6(a) reveals that not only is there a single change of variables between adjacent columns (rows) but that the first and last columns (rows) also have only a single variable change. This leads us to suspect that the edges are effectively joined, and that minterms in the same row in both end columns are not isolated but join to form a common term. (A similar situation exists between the top and bottom rows.) Figure 4.6(b) is a representative example of the joining of edge terms.

One may be tempted to choose the largest possible coverage as one of the terms in a simplified function. This may be incorrect, however, if alternate coverages are possible, as Figure 4.7(a) shows. Here, according to rule 2, the squares containing the * symbol should be selected first, since they are the only minterms which may be grouped in only one way. Once these solid coverages are made, the largest (dashed) group has already been covered in an indirect way, and rule 4 indicates that the function is complete. (To add another term would clearly be introducing redundancy.)

In contrast, Figure 4.7(b) shows a function with only one minterm different from that in part (a) of the figure; but here the resulting function contains the group of four squares because the minterm in the lower right-hand section can be covered

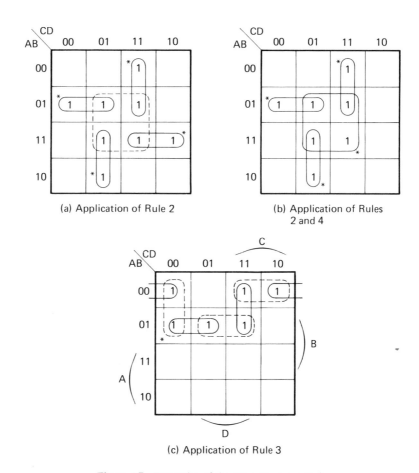

Figure 4.7 Examples of the Map Coverage Rules.

in only one way. Again the * symbol is employed to indicate which minterms may be considered first. Another interesting thing about Figure 4.7(b) is that it demonstrates rule 4 through the multiple use of minterms.

An application of rule 3 is demonstrated through Figure 4.7(c). Here, each of the minterms can be grouped in two ways; thus, one arbitrary selection must be made. If the selection of the solid coverage at the * minterm is made, all the other groups are fixed, and the function $f_3 = \bar{A}B\bar{C} + \bar{A}CD + \bar{A}B\bar{D}$ results. On the other hand, if the selection of the dashed coverage at the * minterm is made, a different fixed grouping is produced, resulting in the function $f_4 = \bar{A}\bar{C}\bar{D} + \bar{A}BD + \bar{A}\bar{B}C$. Now, functions f_3 and f_4 are obviously equal. They contain the same minterms, and they are both simplified representations of the original functions; yet not one term between the two functions is identical.

Of course, the map method is not limited to only four-variable functions. Problems with fewer than four-variables are a trivial variation of the four-variable case. Problems with more than four variables are handled by using sufficient four-variable maps to achieve the 2^n squares required by the n-variable function. In the groups of

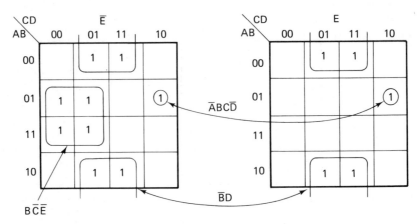

Figure 4.8 Simplified Function, $f_5 = B\bar{C}E + \bar{B}D + \bar{A}BC\bar{D}$.

four-variable maps representing an *n*-variable function, corresponding squares on adjacent maps may be combined to form a simpler term, just as the adjacent squares on a single map may be combined. Figure 4.8 shows how the five-variable function

$$f_5 = B\bar{C}\bar{D}\bar{E} + \bar{C}D\bar{E} + \bar{B}CD\bar{E} + \bar{A}BC\bar{D} + \bar{B}DE$$

map be simplified by using a pair of four-variable maps.

In the above five-variable problem, observe how the fifth variable is used to label the two maps. Also, the variable E is eliminated to create the term $\bar{A}BC\bar{D}$ formed between corresponding squares on the two maps. (The term $\bar{B}D$ is even a more interesting example of the same thing.)

Don't-Care Conditions

There are physical constraints in practice which either prevent a particular group of minterms (e.g., $\bar{w}xy$) from occurring or which make their occurrence in the function optional. These are called *don't-care* conditions. Constraints of this type can be very helpful in that the don't-care minterms can then be included with either the 0 or 1 Karnaugh map coverage, depending on which leads to the simplest function. To represent the above situation, a dash is entered into each don't-care Karnaugh map square. The map is then covered following the same rules as previously employed.

As an example, consider $f_6 = \bar{w}\bar{x}yz + wyz$, with the don't-care term $\bar{w}xy$. The resulting Karnaugh map is shown in Figure 4.9. Note that the simplified function $f_6 = yz$ can be obtained by grouping one of the don't-care squares with the 1's and allowing the other to be grouped with the 0's. We will make extensive use of don't-care terms when designing sequential circuits in Chapter 5.

The Complementary Approach

Sometimes it is more convenient to read the zero coverage from a Karnaugh map rather than reading the 1's. This is known as the *complementary approach*. It involves the same map rules as employed previously, but the resulting function is $\bar{f}$, and f is then obtained from De Morgan's theorem.

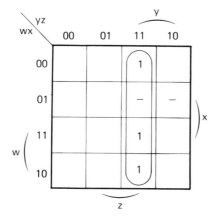

Figure 4.9 Simplification through the Use of a Don't-Care Term.

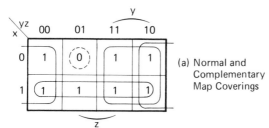

(a) Normal and Complementary Map Coverings

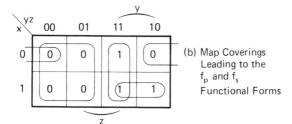

(b) Map Coverings Leading to the f_p and f_s Functional Forms

Figure 4.10 Comparison of the Normal and Complementary Methods.

Consider the three-variable map shown in Figure 4.10(a). The dashed coverage yields the function $\bar{f}_7 = \bar{x}\bar{y}z$; therefore, $f_7 = \bar{\bar{f}}_7 = \overline{\bar{x}\bar{y}z} = x + y + \bar{z}$. The solid coverage on the map leads to the same function, but it requires more effort. For larger Karnaugh maps, the complementary coverage of a single zero minterm, or a small group, leads to an even greater savings in effort.

Another facet of the complementary approach is the ability to place product-of-sums functions on the Karnaugh map. There are many ways of working with product-of-sums functions. The most straightforward method, however, is to find the complement of the function, $\bar{f}$, and place this as a zero coverage on the map. The following example demonstrates this approach.

Example

Simplify the following function:

$$f = (x + z)(x + y + \bar{z})(\bar{x} + y)$$

Sec. 4.4 The Karnaugh Map

Now $\bar{f} = \bar{x}\bar{z} + \bar{x}\bar{y}z + x\bar{y}$, which leads to the Karnaugh map shown in Figure 4.10(b).

Once the map is established, we obtain the sum-of-products function by covering the 1's, or the product-of-sums function by covering the 0's and then taking the complement.

The resulting sum-of-products function is

$$f_s = xy + yz$$

and the resulting product-of-sums function is

$$\bar{f}_p = \bar{y} + \bar{x}\bar{z}, \qquad f_p = \overline{(\bar{y} + \bar{x}\bar{z})} = y(x + z)$$

In this case, the product-of-sums form, f_p, can easily be expanded to the sum-of-products form, f_s.

$$f_p = y(x + z) = xy + yz = f_s$$

In some other cases, the expansion is not so obvious. It should be clear, however, that one can always obtain a product-of-sums function by reading $\bar{f}$ from the map and then applying De Morgan's theorem.

The presence of don't-care terms may lead to f_s and f_p functions from the same map which are actually unequal. In Figure 4.11(a), the don't-care minterm is included with the 1's, resulting in the function $f_s = \bar{B} + \bar{A}C + A\bar{C}$. On the other hand, part (b) of the figure shows the don't-care minterm being included with the 0's, resulting in the function

$$f_p = \overline{\bar{A}\bar{C} + ABC} = (A + C)(\bar{A} + \bar{B} + \bar{C}) = A\bar{B} + A\bar{C} + \bar{A}C + \bar{B}C$$
$$= A\bar{B} + \bar{A}C + A\bar{C}$$

where the $\bar{B}C$ term was eliminated by use of theorem 4. Both maps were properly covered following the rules, yet it is obvious that $f_s \neq f_p$. The reason for this is the fact that the don't-care minterm was included in both f_s and f_p. Thus, anytime the minterms grouped with the 1's include some terms also grouped with the 0's, there will be a difference between f_s and f_p.

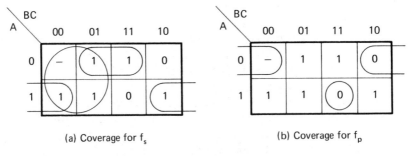

Figure 4.11 Example Where $f_z \neq f_p$.

We have already completed many important facets of the Karnaugh map method. More examples, however, and further developments of it as a design tool are covered in Sections 4.6, 4.8, 5.4, and 5.5.

Other Simplification Methods

Although the map method is quite popular, there are other good manual and computer techniques for simplifying Boolean functions. The most notable of these is the Quine-McCluskey method, explained in References 3, 5, and 7, which involves a tabular reduction of the original function into a simplified one somewhat similar to the Karnaugh map process. The tabular method has the advantage that it may be easily programmed, and the reader is encouraged to consult Rhyne's work in Reference 9 for a very interesting FORTRAN version.

4.5 LOGIC DIAGRAMS

Since in most cases pictures are a more efficient way to convey design information, the logic diagram is developed in this section as an alternate way to describe Boolean functions. We have already indicated that any function can be constructed with only AND, OR, and NOT primitives.

Consider the equation $f = \bar{A}B + C\bar{D}$. Since it is in sum-of-products form, the output gate will be an OR with AND gates feeding it, as Figure 4.12 shows.

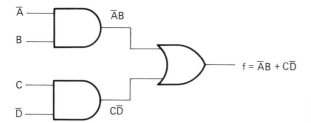

Figure 4.12 Logic Diagram for a Sum of Products.

One crude measure* of circuit complexity, and thus cost, is the number of gate inputs in the circuit. It is easy to count them directly from the logic diagram (e.g., there are six inputs in Figure 4.12). However, the same number can be obtained by inspection of the equation, and the reader should take note of this. Another fact that should be observed concerning Figure 4.12 is that each input passes through two *levels* of gating before the output is reached. The number of levels of gating in a digital circuit is an important parameter because it is a measure of electronic delay. Again observe that the number of levels required by the circuit can be obtained from the equation by inspection. For the above function, it was assumed that the complemented quantities, $\bar{A}$ and $\bar{D}$, were available. In the following example, however, inverters are required to generate the complements.

Example

Draw the logic diagram for the function

$$f = \bar{A}C\bar{D} + \bar{A}B\bar{D}\bar{G} + \bar{A}EF + \bar{A}B\bar{D}G \qquad (4.6)$$

---
*A better measure, frequently employed in modern logic design, is the number of integrated-circuit packages required. These practical measurers of complexity are considered in more detail in Chapter 7.

By inspection of Equation (4.6), we see that 3 inputs are required to invert A, D, and G; 14 inputs are required to generate the product terms; and 4 inputs are needed to combine them into the output f. Thus, as it stands, 21 inputs are required. Now before drawing the diagram, we check to see if f can be simplified. If we use theorem 3, the second and fourth terms in Equation (4.6) may be combined. Therefore,

$$f = \bar{A}C\bar{D} + \bar{A}B\bar{D} + \bar{A}EF \tag{4.7}$$

The result requires 14 inputs. By factoring Equation (4.7), however, we obtain the following new equation requiring only 12 inputs:

$$f = \bar{A}[\bar{D}(C + B) + EF] \tag{4.8}$$

The corresponding circuit representation is shown in Figure 4.13.

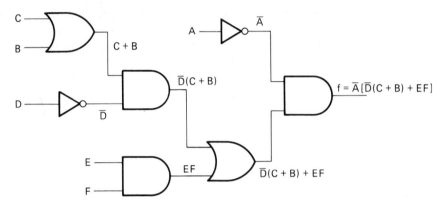

Figure 4.13 Mechanization of Equation (4.8).

In comparing Equations (4.7) and (4.8), we observe that the number of literals has been reduced from nine to six. (Every occurrence of a variable, whether complemented or uncomplemented, is defined as a *literal*.) However, the process of reducing the number of literals has changed the form of the expression in that Equation (4.8) is no longer a sum of products. This process of reducing the number of literals by changing the form of an equation should be contrasted with the process of simplification such as that used in reaching Equation (4.7). There, the form of the equation stayed the same, but the number of terms was reduced as was the number of variables. An elementary example of reducing the number of literals by changing the form of an equation is shown below:

$f = (A + B)(B + C)$ (Four literals in product-of-sums form)
$f = AC + B$ (Three literals in sum-of-products form after applying the distributive law)

4.6 NAND-NOR LOGIC

The basic *NAND* and *NOR* gates were mentioned in Section 4.3, but their current importance demands that design details be presented at this point.

First, study the two-level NOR circuit shown in Figure 4.14. Note that Figure

Figure 4.14 Two-Level NOR Circuit.

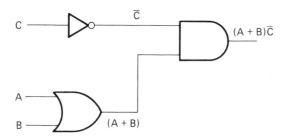

Figure 4.15 Two-Level OR-AND Circuit.

4.14 gives the same results as the two-level OR-AND circuit in Figure 4.15. Thus, two levels of NOR gating are equivalent to OR followed by AND, i.e., it realizes a product-of-sums function.

It should be observed that since the variable C in Figure 4.14 passes through only one level of gating, it is inverted in the output. Similarly, it is easy to show that a two-level NAND configuration is equivalent to AND followed by OR.

Example

Using either NANDs or NORs draw a logic diagram for the function $f = xy + \bar{x}\bar{z} + y\bar{z}$.

First we try to simplify the expression

$$f = xy + \bar{x}\bar{z} + y\bar{z} \quad \text{(By theorem 4)}$$

Therefore, $f = xy + \bar{x}\bar{z}$, and the resulting logic diagram is shown in Figure 4.16.

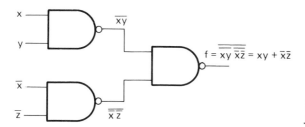

Figure 4.16 NAND Realization of $f = xy + \bar{x}\bar{z}$.

It is known that a sum-of-products equation can always be realized with a two-level NAND circuit; thus, the term written at the output of each gate is merely used for checking.

Since all functions can be reduced to sum-of-products and product-of-sums forms, the above method gives a straightforward, two-level circuit realization for a function with either NANDs or NORs. As was observed in Section 4.5, however, simpler circuits frequently result if neither a pure sum of products nor product of sums is used. Consider the function $f = AC + A\bar{D} + BE + BF$. As it stands, a circuit requiring 5 gates and 12 inputs is needed. Moreover, the function cannot be simplified, but the number of literals can be reduced if we change the form of the expression through factoring as follows:

$$f = A(C + \bar{D}) + B(E + F) \qquad (4.9)$$

This equation is not a pure sum of products; therefore, our previous design technique is not applicable. However, the expression may easily be mechanized by AND-OR gates as shown in Figure 4.17.

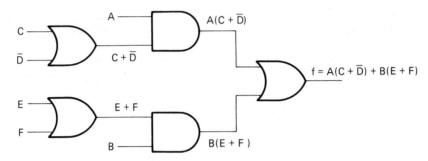

Figure 4.17 AND-OR Realization of Equation (4.9).

Consider the NAND gate, shown below, which realizes the function $f = \overline{AB}$.

Carrying out the complement operation in the expression, we get $f = \bar{A} + \bar{B}$, and this expression leads to the symbol

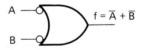

This is a second, though less familiar, form of the NAND symbol. Now let us replace the AND gates in Figure 4.17 by NAND gates of the first form, and replace the OR gates by NAND gates of the second form, with the results shown in Figure 4.18. Note that all complements between gates occur in pairs and are effectively canceled, but the circuit input variables must be complemented from those appearing in Figure 4.17 to account for the inverter present on each gate input. If desired, all gates in Figure 4.18 can now be replaced by the common NAND symbol (the first form), which results in Figure 4.19.

The reader should now show that the output of Figure 4.19 does indeed agree with Equation (4.9).

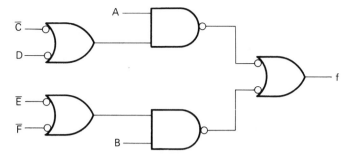

Figure 4.18 NAND Replacement of the Gates in Figure 4.17.

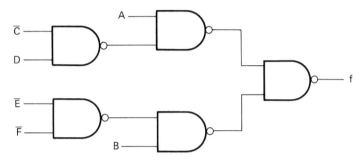

Figure 4.19 Realization of Equation (4.9) with Common NAND Gates.

Instead of NAND gates, suppose NOR gates had been required. The common form of NOR gate is given by $f = \overline{A + B}$:

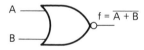

Again carrying out the complement operation, we get $f = \bar{A}\bar{B}$, which leads to the symbol

Now let us replace the OR gates in Figure 4.17 by the first NOR form above and replace the AND gates by the second NOR form. The resulting diagram is presented in Figure 4.20.

Again the complements between gates occur in pairs, so that they are canceled. The input variables A and B are complemented because of the inversions where they enter, and an inverter is inserted at the output gate so that f instead of $\bar{f}$ is produced. If desired, the NOR gates in Figure 4.20 can now be replaced by common NORs.

The above NAND (NOR) procedure is a general one that we will call the

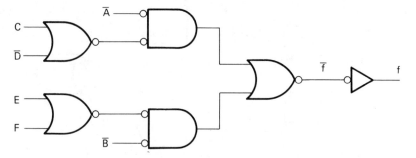

Figure 4.20 NOR Replacement of the Gates in Figure 4.17.

replacement method, and its steps are summarized as follows:

1. Simplify the given function.
2. Place the function in a form with minimum literals.
3. Draw the AND-OR block diagram.
4. Use the two forms of NAND (NOR) gates to replace the ANDs and ORs in the original block diagram. Make sure that any inverters on the input or output of the circuit are properly canceled.
5. If desired, replace all gates by those of the common form.

Example

Employing either all NOR or all NAND gates, synthesize a two-level logic circuit with fewest possible gate inputs for the function

$$f = (B + C + D)(\bar{A} + \bar{C})(\bar{B} + \bar{C} + \bar{D})(A + B + \bar{C} + D)$$
$$\times (A + \bar{B} + C + \bar{D})$$

(Assume complemented and uncomplemented input variables are available.)

Solution: Now

$$\bar{f} = \bar{B}\bar{C}\bar{D} + AC + BCD + \bar{A}\bar{B}C\bar{D} + \bar{A}B\bar{C}D$$

which results in the zero entries shown on the map of Figure 4.21.

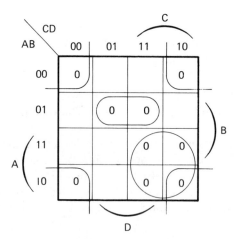

Figure 4.21 Optimum Zeros Coverage for the Function f.

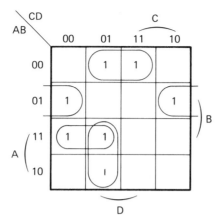

Figure 4.22 Optimum Ones Coverage for the Function f.

The same map with the optimum coverage of the 1's is shown in Figure 4.22, and the resultant two-level function is

$$f = \bar{A}\bar{B}D + \bar{A}B\bar{D} + A\bar{C}D + AB\bar{C}$$

Since a sum-of-products expression can be realized with two levels of NAND gates, and since each product term requires three gate inputs and the sum requires four gate inputs, a total of $T = 4 \times 3 + 4 = 16$ gate inputs are required.

As an alternate approach, we return to Figure 4.21 and read the optimum coverage shown there for $\bar{f}$:

$$\bar{f} = \bar{A}BD + \bar{B}\bar{D} + AC$$

Therefore, $f = (A + \bar{B} + \bar{D})(B + D)(\bar{A} + \bar{C})$. Here the total requirements for a two-level NOR network will be

$$T = 3 + 2 \times 2 + 3 = 10 \text{ gate inputs}$$

In yet another approach, the above product-of-sums function is realized in Figure 4.23 with NAND gates. Observe, however, that the NAND design necessitates

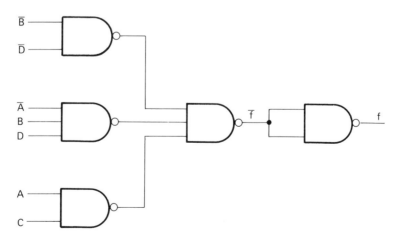

Figure 4.23 NAND Gate Realization of the Function f.

Sec. 4.6 NAND-NOR Logic

that one gate be employed as an inverter at the circuit output, which makes the circuit in Figure 4.23 a three-level circuit. Two additional inputs are also used, resulting in a total of 12 gate inputs required.

Thus, it appears that the best circuit design for the given function is the product-of-sums realization employing NOR gates. Incidentally, if the original function, without simplification, had been realized with a two-level NOR circuit, a total of 21 gate inputs would have been needed. Consequently, our design has realized more than a factor-of-2 savings in gate inputs.

This example points out the importance of understanding all facets of NAND-NOR design. It also emphasizes the value of considering alternate approaches before deciding on the final realization of a particular function.

‡4.7 EXCLUSIVE-OR LOGIC

Probably there will always be a need for lower-level *primitives** such as NANDs and NORs to implement certain logical functions. The birth of integrated circuits, however, has initiated a continuing trend toward a greater emphasis on higher-level primitives. This section and the next will explore this trend with a medium-level primitive, the EXCLUSIVE-OR gate, and then high-level primitives will be considered at length in Chapter 7.

Among the two variable functions listed in Table 4.11 are the EXCLUSIVE-OR EXOR, and its dual, the EXCLUSIVE-NOR, EXNOR. (The EXNOR function is often called the EQUALITY function because it is true when the two inputs match.) Although these functions are not as widely used as NAND and NOR, they are very handy in concisely performing certain arithmetic and coding operations, as we shall soon see in Section 4.8. Initially, it is our intention to define some new notation and then to present some of the properties of the EXOR function. (Although it will not be emphasized, the EXNOR has similar properties to the EXOR.)

The EXOR and the EXNOR have been assigned the following special symbols:

$$\text{EXOR} \equiv \oplus$$
$$\text{EXNOR} \equiv \odot$$

Thus, in place of writing $f = A\bar{B} + \bar{A}B$, it is more concise to write $f = A \oplus B$. (Similarly, $f = \bar{A}\bar{B} + AB$ can be more concisely written as $f = A \odot B$.) In addition, the following gate notation has been widely adopted:

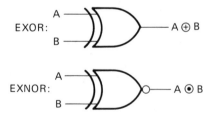

---

*The word primitive has a special connotation in computer technology. It specifies a basic building block which is employed in the development of more complicated structures.

The first thing to observe concerning the properties of EXOR is that it obeys the following laws of algebra:

Commutative: $A \oplus B = B \oplus A$

Associative: $(A \oplus B) \oplus C = A \oplus (B \oplus C) = A \oplus B \oplus C$

Distributive: $A(B \oplus C) = (AB) \oplus (AC)$

Now let us prove the distributive law for the EXOR function:

$A(B \oplus C) = (AB) \oplus (AC)$ (Original statement of the law)

$A(B \oplus C) = AB\overline{AC} + \overline{AB}AC$ (Definition of EXOR)

$A(B \oplus C) = AB(\bar{A} + \bar{C}) + (\bar{A} + \bar{B})(AC)$ (De Morgan's law)

$A(B \oplus C) = AB\bar{A} + AB\bar{C} + \bar{A}AC + \bar{B}AC$

$A(B \oplus C) = 0 + AB\bar{C} + 0 + A\bar{B}C$ (Elementary algebra)

$A(B \oplus C) = A(B\bar{C} + \bar{B}C)$

$A(B \oplus C) = A(B \oplus C)$ (Definition of EXOR)

(The other laws can be proved in a similar way.)

Table 4.13 lists a number of other important properties of the EXOR function. The proofs for these are straightforward and will be left as exercises.

TABLE 4.13 IMPORTANT PROPERTIES OF THE EXOR FUNCTION

Property Number	Statement
1	$1 \oplus 0 = 1$
2	$0 \oplus 0 = 1 \oplus 1 = 0$
3	$A \oplus 0 = A$
4	$A \oplus 1 = \bar{A}$
5	$A \oplus A = 0$
6	$A \oplus \bar{A} = 1$
7	$\bar{A} \oplus \bar{B} = A \oplus B$
8	$\overline{A \oplus B} = A \odot B$

(Each of the properties in Table 4.13 has a corresponding EXNOR form, but these will not be considered here.)

How does one quickly evaluate an EXOR function of more than two variables if the truth values are given? Consider the following function:

$f = A \oplus B \oplus C$ (With $A = 1, B = 0, C = 1$)

$f = 1 \oplus 0 \oplus 1$ (Values substituted)

$f = (1 \oplus 0) \oplus 1$ (Associative law)

Sec. 4.7 EXCLUSIVE-OR Logic

$$f = 1 \oplus 1 \qquad \text{(Property 1)}$$
$$f = 0 \qquad \text{(Property 2)}$$

Thus, an even number of 1's in this special case of a three-variable function has lead to $f = 0$. Actually, it can easily be shown that an EXOR function, with any number of variables, will always equal 0 if there are an even number of variables which are true; the function will equal 1 if there are an odd number of variables which are true. For these reasons, EXOR is described as a modulo-2 function, meaning its output is determined by counting the true variables in base 2 without carry. This is a formal way of saying the function is true if an odd number of the input variables are true.

The EXOR function has an interesting pattern on a Karnaugh map. For example, consider the function $f = A \oplus B \oplus C \oplus D$. By the modulo-2 property, this function will be true if one or three of the variables are true. Thus, it is easy to plot the function on a Karnaugh map by noting the number of 1's in the description of each square. For example, the square $AB = 01$, $CD = 11$ has three true variables; therefore, $f = 1$ for that square. The complete K-map for the function f is shown in Figure 4.24. Observe that the result looks like a checkerboard. This is characteristic of all EXOR functions. Previously, when such a map appeared, the only way to describe it was by laboriously listing the individual minterms. It can now be described very concisely by means of an EXOR function.

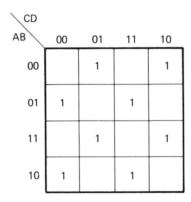

Figure 4.24 Map of the Function $f = A \oplus B \oplus C \oplus D$.

How may the gating circuit be drawn for a map like that in Figure 4.24? After recognizing it as an EXOR function, we may express it in the following ways:

$$f = (A \oplus B) \oplus (C \oplus D) \quad \text{or} \quad f = [(A \oplus B) \oplus C] \oplus D$$

where the brackets show how the function is assembled from two-variable gates. The resulting two circuits are shown in Figure 4.25. The tree-type circuit gets its name from its symmetrical branch-like appearance, which of course is even more obvious for a six- or eight-variable function. One advantage of the tree-type circuit is that it requires fewer levels of gating than the unbalanced circuit. In Chapter 12 we will see that an increase in the number of gate levels has the detrimental effect of slowing the circuit's response.

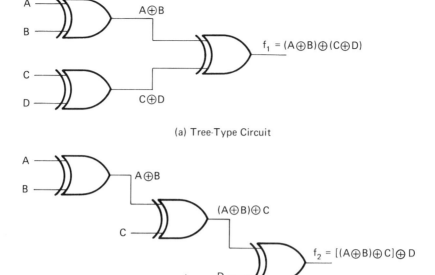

Figure 4.25 Two-Circuit Realization for $f = A \oplus B \oplus C \oplus D$.

‡4.8 IMPLEMENTATION OF COMPUTER CIRCUITS

In Section 2.6 we developed a number of codes and considered several algorithms useful in processing numbers. Now, as a practical application of EXOR gates, we will present the circuits for implementing some of these processes.

Conditional Complementor

When BCD numbers are processed, the 9's complement technique is frequently employed for subtraction, which means that a method of complementing the bits of the subtrahend is required. Since each bit is independent of the others, a conditional complementing circuit can be developed for a single bit and then duplicated as many times as required. Table 4.14 shows the logic requirements where B_i is one bit of the

TABLE 4.14 BIT LOGIC REQUIRED BY THE CONDITIONAL COMPLEMENTOR

B_i (Input)	C (Control)	R_i (Result)
0	0	0
0	1	1
1	0	1
1	1	0

number being complemented, C is the control signal ($C = 1$ for complement desired; $C = 0$ otherwise), and R_i is the resulting output signal. It is immediately evident that $R_i = \bar{B}_i C + B_i \bar{C}$, which, of course, is the EXOR function $R_i = B_i \oplus C$. The resulting circuit for an n-bit conditional complementor is shown in Figure 4.26. Note that when $C = 0$, the modulo-2 property requires that each input bit simply be transmitted through to the output without change; but when $C = 1$, each bit must be inverted.

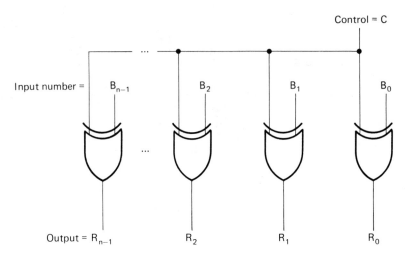

Figure 4.26 n-Bit Conditional Complementor.

Parity Circuits

When parity is employed, circuits are required to generate the parity bit from the source register and to check the parity at the receiving register. Figure 4.27 shows a system of this type employing EXOR gates. (The numbers next to the bit lines represent the logic values for a particular example. Note how the parity information is created in the parity generator and how it is processed at the receiving end by the parity checker circuit.) Assuming even parity, the sending circuit should detect any word whose data bits do not have an even number of ones and compensate by making the parity bit one. Otherwise, the output of the parity generation should be zero, causing the parity bit to be zero. Due to the modulo-2 property of the EXOR function, the sending circuit requires the parity (source) function to be

$$P_s \equiv B_0 \oplus B_1 \oplus B_2 \oplus B_3$$

where B_i is the ith bit of the source register. It is clear that an odd number of ones in the data word shown in Figure 4.27 results in $P_s = 1$. Now, if P_s is employed as a new bit, bit 4, added to the data word, the augmented word will have even parity. The opposite case of an even number of ones in the original data word results in $P_s = 0$, which maintains even parity in the augmented data word. When the signal arrives at the receiver, all five bits of the augmented word are checked with the parity (receiving) function, and

$$P_r \equiv B_0 \oplus B_1 \oplus B_2 \oplus B_3 \oplus B_4$$

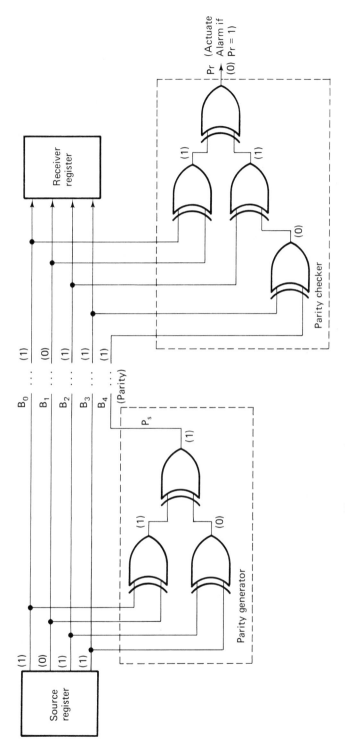

Figure 4.27 Small Even-Parity Error Detection System.

If no bits are in error, $P_r = 0$ (as it is in Figure 4.27). Should one (or any odd number) error occur, however, then $P_r = 1$, which can be used to sound an alarm so that corrective action can be taken.

As an exercise, the reader should now change one of the bits in the received signal, e.g., make $B_3 = 0$. This will change the outputs of some of the EXOR gates in the parity checker and will eventually result in $P_r = 1$, which indicates that the error has been properly detected.

Code-Conversion Logic

There are many situations where it is desired to convert from one code to another within a computer system. For example, the information originating in a digital-to-analog converter is often in Gray code; before it can be processed in the arithmetic unit, conversion to another code is required.

To better understand the code-conversion problem, consider a simple example—that of transforming octal code to Gray code. The relationship between these two codes is shown in Table 4.15, and it is evident that the conversion requires that each bit in the Gray code be considered a function of all three bits of an octal digit. Thus, we require three functions in the following form:

$$G_0(B_0, B_1, B_2), \quad G_1(B_0, B_1, B_2), \quad \text{and} \quad G_2(B_0, B_1, B_2)$$

TABLE 4.15 RELATION BETWEEN THE OCTAL AND GRAY CODES

Decimal	Octal B_2 B_1 B_0	Gray G_2 G_1 G_0
0	0 0 0	0 0 0
1	0 0 1	0 0 1
2	0 1 0	0 1 1
3	0 1 1	0 1 0
4	1 0 0	1 1 0
5	1 0 1	1 1 1
6	1 1 0	1 0 1
7	1 1 1	1 0 0

These functions can be easily determined if we place the Gray code truth values for each bit on a separate Karnaugh map. The three resulting independent maps are shown in Figure 4.28. Note that two of these functions require EXOR gates and the other is obtained directly from bit 2 of the octal code.

The circuit demanded by these functions is shown in Figure 4.29. (In this case, it just happens that a very concise result is produced with EXOR gates; usually other gate types will be employed.) Since a very simple example was just considered, a word of caution is in order. In general, the conversion from one code to another, usually four bits to four bits, requires a multiple-output logic circuit in which the optimum solution demands that some of the terms generating one function be employed to generate one or more of the other functions. Thus, the process of selecting the coverage

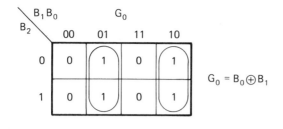

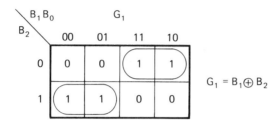

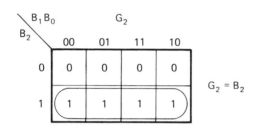

Figure 4.28 Octal-to-Gray Code Conversion Functions.

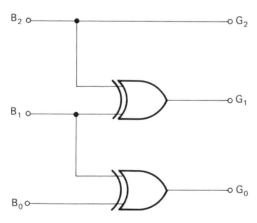

Figure 4.29 Octal-to-Gray Code Conversion Circuit.

for, say, four Karnaugh maps of four variables each is often a more complicated task than we have previously considered. The reader is referred to the bibliography—particularly Hill and Peterson's text[3]—for a formal method which guarantees an optimum solution. On the other hand, by employing our usual method of covering each Karnaugh map independently, we will obtain a reasonable suboptimum function. Moreover, once some tentative coverages have been placed on each map, it is often possible by inspection to make the final selections in such a way that an optimum or near-optimum set of functions result. In Chapter 7, we will return to the code-conversion problem and find that the advent of inexpensive semiconductor memories has in many cases greatly simplified the problem of code conversion.

4.9 BIBLIOGRAPHY

1. Becher, W. D., *Logic Design Using Integrated Circuits.* Rochelle Park, N.J.: Hayden Book Company, Inc., 1977.
2. Boole, G., *An Investigation of the Laws of Thought.* New York: Dover Publications, Inc., 1954.
3. Hill, F. J., and G. R. Peterson, *Introduction to Switching Theory and Logical Design* (3rd ed.). New York: John Wiley & Sons, Inc., 1981.
4. Karnaugh, M., "The Map Method for Synthesis of Combinational Logic Circuits." *Transactions of the AIEE, Communications and Electronics,* Vol. 72, Pt. 1, 1953, pp. 593–99.
5. Kohavi, Z., *Switching and Finite Automata Theory* (2nd ed.). New York: McGraw-Hill Book Company, 1978.
6. Lee, S. C., *Digital Circuits and Logic Design.* Englewood Cliffs, N.J.: Prentice-Hall, Inc., 1976.
7. Marcus, M., *Switching Circuits for Engineers* (3rd ed.). Englewood Cliffs, N.J.: Prentice-Hall, Inc., 1975.
8. Muroga, S., *Logic Design and Switching Theory.* New York: John Wiley & Sons, Inc., 1979.
9. Rhyne, V. T., *Fundamentals of Digital Systems Design.* Englewood Cliffs, N.J.: Prentice-Hall, Inc., 1973.
10. Shannon, C., "A Symbolic Analysis of Relay and Switching Circuits." *Transactions of the AIEE,* Vol. 57, 1939, pp. 713–23.

4.10 EXERCISES

Simplify the functions in Exercises 4.1 through 4.6 using the theorems.

4.1 $[E(B + C)(C + D)\bar{E} + A](A + B + \bar{A})(\overline{A\bar{D}})$

4.2 $(A + B)(\bar{A} + B)(\bar{A} + B + E)$

4.3 $DEH + \bar{E}G\bar{H} + \bar{H}E + HF\bar{E} + \bar{H}EJ$

4.4 $AB + \bar{A}C + \bar{B}CD$

4.5 $(\overline{V\bar{W}}) + \bar{Y} + \bar{W}XYZX + WY\bar{Z} + VZW\bar{V}$

4.6 $(A + B)(\bar{A} + C)(B + C + D)$

4.7 Draw a diagram similar to Figure 4.2 for the NOR operation. Employ the switch configuration shown in Figure E4.7.

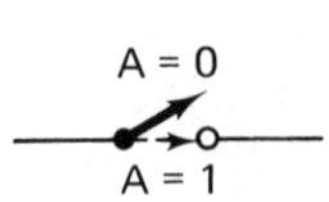

Figure E4.7

Simplify the functions in Exercises 4.8, 4.9, and 4.10 using the Karnaugh map.

4.8 (a) $A\bar{B}C + \bar{A}\bar{C}D + A\bar{C}$
(b) $B\bar{C} + \bar{A}B + BC\bar{D} + \bar{A}\bar{B}D + A\bar{B}\bar{C}D$
(c) $\bar{A}\bar{B}\bar{C}D + AC\bar{D}\bar{E} + ABC\bar{D}E + A\bar{B}C\bar{D}E + \bar{B}\bar{C}\bar{D}\bar{E}$
(d) $\bar{A}\bar{C} + AB\bar{C}\bar{D} + \bar{A}BD + \bar{A}\bar{B}C\bar{D} + A\bar{B}\bar{C}D$
(e) $\bar{A}\bar{B} + AB + A\bar{B}C + \bar{A}B\bar{C}$
(f) $(A + B)(\bar{A} + C)(B + C + D)$

4.9 (a) $B\bar{C} + \bar{A}D + \bar{A}B + A\bar{B}\bar{C}D$, in the f_p form.
(b) $\bar{A}\bar{B}C + \bar{A}\bar{B}\bar{C}\bar{D} + \bar{A}BC$, in f_s form with the don't care terms: $A\bar{B}\bar{C}\bar{D}$, $\bar{A}BC\bar{D}$, $A\bar{B}\bar{C}D$.
(c) Exercise 4.2 in f_s form.

4.10 $(C + \bar{D})(B + \bar{C} + D)(A\bar{B}C + \bar{B}\bar{C} + \bar{C}D)(A + \bar{B} + \bar{C})$

(In addition to Karnaugh maps, De Morgan's law may also be employed.)

4.11 Complement: $S[\bar{W} + I(T + \bar{C})] + H$.

4.12 Simplify using the theorems only:

$$(A + \bar{B} + C)(B + E)(A + \bar{B} + C + E + \bar{G})(B + C + \bar{D})(A + B\bar{C})(C + \bar{D} + \bar{E})$$

4.13 Using the fewest possible gates, realize the function $f = A(C + \bar{D}) + B(C + \bar{D})$ first with NORs, then with NANDs.

4.14 Using only NORs, synthesize a gate circuit with the fewest possible gate inputs for the functions listed below.
(a) $f = \bar{A}CD + \bar{A}BD\bar{G} + \bar{A}EF + \bar{A}BDG$
(b) $f = \bar{A}CD + BD + EF$
(c) $f = \bar{X}\bar{Y}\bar{Z} + XYZ$, $g = \bar{X}\bar{Z} + YZ$ (both functions with one circuit)
Both complemented and uncomplemented inputs are available.

4.15 Simplify using the theorems only:

$(W + X + Y)(W + \bar{X} + \bar{Z})(\bar{W} + \bar{Y})(\bar{R} + V + W + \bar{X} + Y + \bar{Z})(\bar{X} + \bar{Y} + \bar{Z})(W + Y + \bar{Z})$

4.16 Simplify the following functions using the map method.
(a) $f = X\bar{Y} + X\bar{Z} + \bar{X}Y + \bar{X}Z$
(b) $f = (\bar{A} + C + \bar{D})(A + \bar{B} + \bar{C} + D)(\bar{B} + \bar{C} + \bar{D} + E)$
$\times (\bar{A} + B + \bar{C} + \bar{D})(A + C + \bar{D} + \bar{E})(A + \bar{B} + \bar{C} + \bar{E})$

Present the resulting f both as a sum of products and as a product of sums. [Required for part (b) only.]

4.17 Simplify using the theorems.
(a) $f = \bar{A}C\bar{D} + B(DG + A\bar{E}) + E + D(\bar{C} + \bar{A}C)$
(b) The function in Exercise 4.16(a)

4.18 Prove the theorem $(Z + X)(Z + \bar{X} + Y) = (Z + X)(Z + Y)$. Do not employ perfect induction.

4.19 Find the simplified sum-of-products function, F, realized by the circuit in Figure E4.19.

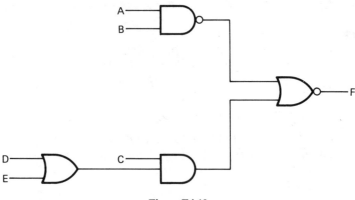

Figure E4.19

4.20 Prove that any EXOR function is true if an odd number of its variables take on true values.

4.21 (a) As concisely as possible, express the function shown in Figure E4.21.
(b) Using only EXOR gates, draw the circuit for the function in part (a).

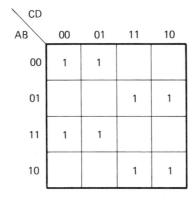

Figure E4.21

4.22 (a) Using EXOR gates, draw the tree-type logic diagram for the function

$$f = A \oplus B \oplus C \oplus D \oplus E \oplus F \oplus G \oplus H$$

(b) Redraw the circuit in part (a) using the fewest possible two-input EXNOR gates.

4.23 If $A \oplus B = C$, find $B \oplus C$ in simplified form.

4.24 Design a logic circuit to convert from BCD to the (2, 4, 2, 1) code.

4.25 How should Figure 4.27 be changed if odd parity is employed?

4.26 It can be shown that a Gray-to-binary conversion algorithm for n-bit numbers is given by $B_i = G_i \oplus B_{i+1}$, where G_i is the ith bit of the Gray code and B_{i+1} is the bit of the binary word that is to the left of the bit B_i. The conversion proceeds from the most significant bit, B_{n-1}, toward the least significant bit, B_0, and it is assumed that $B_n = 0$.

(a) Check the method for the Gray-code elements 0110, 1100, and 0100.

(b) Employing an algebraic method, find the formula for converting from binary to Gray code.

(c) Check the method in part (b) for the binary elements: 0011, 0110, and 1001.

(d) For $n = 4$, draw the circuits for both conversion systems.

Chapter 5

Sequential Circuits

We have been dealing exclusively with *combinational circuits*, where the output depends only on the present inputs. Digital systems, however, contain another broad class called *sequential circuits* to which counters, registers, accumulators, and many other subsystems belong. Here, the output depends on input history (memory) as well as on the present inputs themselves. The present discussion concerns the principles of classical sequential circuits and it serves two major purposes:

1. A brief introduction to the elegant concepts of switching theory, which are fully developed in more advanced texts
2. Insight into the nature of sequential components and circuits, which should lead to a deeper understanding of modern integrated circuits, microprocessors, and digital systems

The two main classes of sequential circuits are *synchronous*, meaning that circuit action is synchronized by means of an input pulse, and *asynchronous*, which contains no synchronizing inputs signals. We place our major emphasis on the clocked subclass of synchronous circuits because they are very important in digital design. (Other useful subclasses, that are beyond the scope of this text, do exist; and the interested reader should refer to the bibliography for these as well as for more information on the asynchronous class.) The elements of our circuit model are shown in Figure 5.1. Note the feedback path from the memory to the combinational network, which is characteristic of all types of sequential circuits. The combinational network not only generates the output signals, but it also sends signals into the memory and changes its content under certain conditions. The clock pulse controls the memory and only allows it to

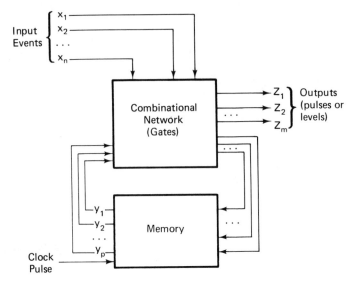

Figure 5.1 Basic Model of a Clocked Synchronous Sequential Circuit.

take new data from the combinational network at specific times. The memory output signals (levels) give information about the *state* of the memory and are labeled y_k, where $k = 1, 2, \ldots, p$. As implied in Figure 5.1, the state of the memory can be visualized simply as a binary number, consisting of p bits, which indicates the history of signals the memory has received. Of course, a synchronous sequential circuit always requires at least one input and at least two memory states.

It is our objective to study some general methods for the design of both the combination network and the memory connections needed in sequential circuits. An introductory computer text, however, can not hope to cover all important aspects of every topic, and this is particularly true of a subject as extensive as sequential design. Thus, it is expected that serious students of computer hardware will later take an advanced course devoted to switching theory. Meanwhile, those desiring more details will want to consult the Bibliography. Included here are analysis of sequential circuits, formal synthesis procedures, informal design techniques, insights concerning the internal construction of flip-flops leading to their implementation with logic gates, consideration of nonideal components, dependency notation, and a brief study of integrated circuits. You should take note that an important alternative to the classic sequential methods of the current chapter is developed in Part II (Modern Structured Logic Design, particularly in the latter part of Chapter 7, where control memories are considered).

5.1 SOME BASIC FLIP-FLOPS

To make our work more specific, from this point on, the more restricted term, *flip-flop*, is used to represent each memory bit. A flip-flop is an electronic circuit whose output state, y, can assume one of two logic levels (0 or 1), and it has one or more

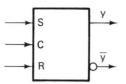

Figure 5.2 Clocked R-S Flip-Flop Symbol.

inputs which can cause the output to change. For tutorial reasons, our discussion begins with the clocked *R-S* flip-flop, whose symbol* is shown in Figure 5.2. Later, however, several other types will be presented. (The arrows in Figure 5.2 merely indicate the direction of signal flow for this diagram; they will not be employed in later figures.)

In Figure 5.2, *S* is the *set* input, *R* is the *reset* input, *C* is the clock input, and the state of the flip-flop, *y*, and its complement, *y*, are available. When $S = 1$, $R = 0$, and the clock pulse occurs, the state of the flip-flop, *y*, is *set* to 1, independently of its previous value. Similarly, the reset input permits changing the flip-flop's state to 0. These operations are summarized by the *timing diagram* given in Figure 5.3. Note how the occurrence of $S = 1$, $R = 0$, and the negative edge of the clock pulse of time t_1 causes the circuit to be set ($y = 1$), and how $S = 0$, $R = 1$, and the clock edge at t_2 causes the circuit to be reset ($y = 0$). The class of flip-flop considered here is negative-edge triggered. This means that the state change occurs at the negative-going edge of the pulse ($1 \rightarrow 0$ transition),† or more accurately, a short delay following the negative edge (usually within a few nanoseconds). This type of flip-flop has the characteristic that the next state is determined only by the inputs *before the negative edge*, and it is

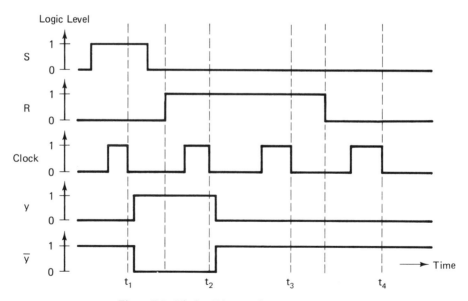

Figure 5.3 Timing Diagram for an *R-S* Flip-Flop.

*Shortly we will augment this symbol by introducing a method called dependency notation so that more detailed characteristics of various types of flip-flops can be concisely represented.

†The process mentioned is very common, but others will be considered later.

Sec. 5.1   Some Basic Flip-Flops

not influenced by inputs from other flip-flops once the edge occurs. Thus, it is not influenced by the memory transient states. In other words, the order in which various flip-flops in the memory change state is unimportant, as long as all flip-flops have reached steady state before the next input occurs. Figure 5.3 (at t_3) demonstrates that, once the flip-flop is reset, another reset condition has no effect on the flip-flop's state.

Let us consider how the R-S flip-flop behaves when it receives an input.* Table 5.1 summarizes all possible cases: S is the set input, R is the reset input, y_n is the state of the flip-flop before the logic zero†-going edge of the clock pulse, and y_{n+1} is the state after clock pulse.

TABLE 5.1 OPERATING CHARACTERISTICS OF THE R-S FLIP-FLOP

| Input | | Next State |
S	R	(y_{n+1})
0	0	y_n
0	1	0
1	0	1
1	1	Not allowed

The type-D flip-flop is introduced now since it is very important in practice and quite convenient in design. The type-D flip-flop has only two inputs, the data or D terminal and the clock terminal, as shown in Figure 5.4(a). The logic level, 1 or 0, existing at the D input is transferred to the flip-flop's output, y, immediately following the logic zero-going edge of the clock pulse. This characteristic is represented in Figure 5.4(b).

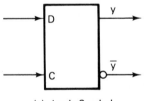

(a) Logic Symbol

| Input | Next State |
D	(y_{n+1})
0	0
1	1

(b) Characteristics

Figure 5.4 Type-D Flip-Flop.

*Discussion of the internal construction of flip-flops will be delayed until Section 5.8, where their realization from gate primitives will be presented.

†Because the most positive voltage is frequently defined as the logic-1 signal, this edge is also called the negative-going edge, or simply the negative edge.

5.2 ANALYSIS

Once one understands the operating characteristics of some flip-flops, the principles of sequential analysis are not difficult to learn. The following timing diagram method of analysis is a common one, but variations may sometimes be preferred. Our method is a step-by-step process controlled by the clock. Each step is based on the flip-flop inputs existing *before* the zero-going edge of the clock pulse generating a set of flip-flop outputs, which appear a short time after the pulse ends. These outputs combined with any gating in their path produce a new set of flip-flop inputs to be processed by the next clock pulse.

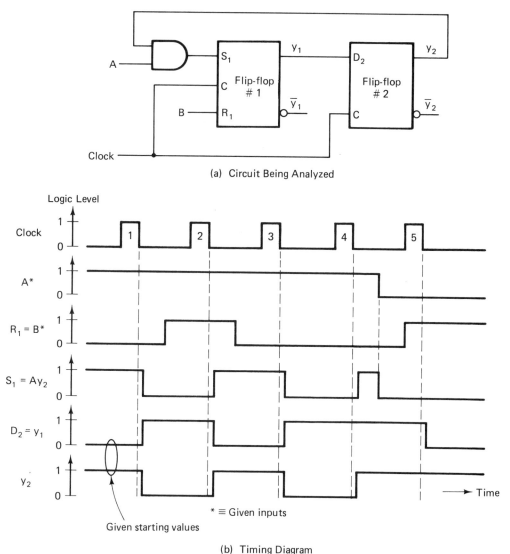

Figure 5.5 Timing Analysis of a Sequential Circuit.

Figure 5.5 shows an elementary circuit analyzed by the above method. Examine the timing diagram in Figure 5.5(b) carefully and note how the given A and B inputs and the previous results affect the new flip-flop values. [You will, of course, need to use the flop-flop characteristics of Table 5.1 and Figure 5.4(b) in making your calculations.] In particular, consider the effect of clock pulse number 1, cpl. Just before the zero-going edge of cpl, $A = 1$ and $y_2 = 1$, resulting in the AND output of 1, which becomes the S_1 flip-flop input. Also, with $R_1 = B = 0$, the end of the clock pulse causes flip-flop number 1 to be set. Note that the same clock edge causes flip-flop number 2 to reset because before the edge, $D_2 = y_1 = 0$. Be sure to understand that it is the flip-flop states, outputs, *before* the zero-going clock edge that determines the next state of each flip-flop and that the *new states have no effect on the flip-flops* until the end of the following clock pulse. In a similar way, study the results of the other four clock pulses and verify the details of each line in the timing diagram.

5.3 DESCRIPTION OF SEQUENTIAL CIRCUITS

Our sequential design method requires, as a starting point, a concise tabular description of the required action. Usually, this table is constructed in a multistep process starting with a *state diagram*, which is a geometric picture of the requirements.

State Diagram

Basically a state diagram consists of several numbered nodes which represent the individual memory states previously mentioned. A transition produced by an input, x_i, is indicated by an arrow, and the presence of an output is represented by the variable Z. Each x_i input represents a specific condition or group of conditions which can be directly represented by pulses on independent input lines or they can be coded by logic levels. Inputs x_1, x_2, x_3, and x_4 could be produced by the 0 and 1 levels* of the variables w_1 and w_2; e.g., $x_1 \rightarrow w_2 w_1 = 00$, $x_2 \rightarrow w_2 w_1 = 01$, etc. Both cases are considered in switching theory books,[2,4] but it is more appropriate for our purpose to *employ only level inputs*, and this convention will be followed unless an exception is specifically mentioned.

Figure 5.6 illustrates the state diagram for a circuit which accepts x_1 and x_2

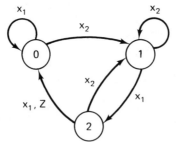

Figure 5.6 State Diagram for the Sequence $x_2 x_1 x_1$.

*The terms pulse and level are relative concepts—typically a pulse has a period of the same order of magnitude as the clock signal while a level is of a much longer time duration; e.g., level duration ≥ 5 times the clock period.

inputs and produces an output whenever the sequence $x_2 x_1 x_1$ occurs. Note the following facts concerning the diagram:

1. All inputs are possible at each memory state. Thus, the complete set of x_i signals are shown leaving each node.
2. When an input does not cause a state change, its arrow forms a loop at the original state, e.g., the x_2 input loops at state 1.
3. The diagram contains sufficient elements so that any arbitrary sequence will terminate at a unique node; but only the specified sequence will produce the output pulse.
4. This circuit is nonterminal, i.e., it will continue to produce an output pulse each time the proper sequence is received.
5. Under certain conditions, many inputs (e.g., a string of x_1's) can be received with no output produced.
6. An output pulse is produced only when a Z appears adjacent to the transition arrow, and at all other times the circuit output is zero.
7. Physical meaning may be attached to each state:
   (a) State 0—a new sequence has not yet started.
   (b) State 1—an x_2 has occurred.
   (c) State 2—an $x_2 x_1$ sequence has occurred.
8. The x_1 and x_2 events must not occur simultaneously. (Figure 5.1 will not allow it.)

Consider the state diagram of a modulo-3 counter which has the ability to be cleared, i.e., it can be reset to zero. (This counter produces the following sequence: 0, 1, 2, 0, 1, etc., and the output, Z, occurs when the counter makes the transition $2 \rightarrow 0$.) In Figure 5.7 note that the events to be counted are labeled C, and the reset is labeled R. As we would expect, the counter may be reset from any state; but in contrast to receiving a C input when in state 2, resetting from state 2 produces no output.

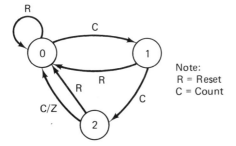

Note:
R = Reset
C = Count

Figure 5.7 Modulo-3 Counter.

Since our emphasis is on level-type inputs to sequential circuits, we now repeat Figure 5.7 within that context. The reader should have no trouble relating the input/output notation of this figure with the symbolic notation of Figure 5.7. The form of Figure 5.7 is probably better for conceptualizing sequential circuits and the form of Figure 5.8 is probably better for detailed design. We will employ both in future work.

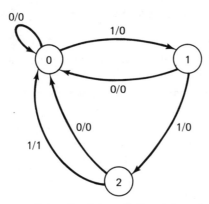

Note: Reset → w = 0, Count → w = 1
Entries: Input/Output = w/Z

Figure 5.8 Counter of Figure 5.7: Binary-Coded Level Inputs.

The State Table and State Assignment

Once the requirements of a problem have been expressed in the state diagram, the information is usually translated into a *state table*. This is the next step in the systematic procedure developed for determining an economical circuit for exciting the system's memory. The state table contains a row for each state in the state diagram and a column for each system input. The entry at the intersection of a particular row and column represents the next state the system will enter upon the occurrence of the input at the top of the column. Table 5.2 is the state table for Figure 5.7.

TABLE 5.2 STATE TABLE FOR THE MODULO-3 COUNTER

Present State	Input, Next State	
	R	C
0	0	1
1	0	2
2	0	0/Z

Now let us determine how much memory is needed to satisfy the example given in Table 5.2. Since we are working in bits, an n-bit memory has 2^n possible states. But our example requires three states; thus, two bits are needed, and there will be one unused state which will lead to the presence of some don't-care map entries. (Unused states $= 2^2 - 3 = 1$.)

The state numbers of Table 5.2 must now be coded in binary. This process is known as the *state assignment problem*. The present assignment is very straightforward because it has previously been specified that the memory follows the binary code: 00, 01, 10, etc. Thus, the state coding is: $0 = 00, 1 = 01, 2 = 10$. Table 5.3 is now constructed using this assignment, where the table entry $y_2 y_1$ represents the memory state variables, and y_1, of course, is the least significant bit. (Since the counter should never enter the $y_2 y_1 = 11$ state, it is shown as a don't-care.)

TABLE 5.3 STATE TABLE: BINARY CODE ASSIGNMENT

Present State (y_2y_1)	Next State* (y_2y_1/Z)	
	$w = 0$	$w = 1$
00	00/0	01/0
01	00/0	10/0
10	00/0	00/1
11	—	—

*Reset $\longrightarrow w = 0$; count $\longrightarrow w = 1$.

In addition to the state assignment, Table 5.3 also contains the binary-coded input and output levels shown in Figure 5.8.

We are fortunate that, in the design of many sequential networks for digital computers, a natural state assignment is frequently obvious. An example occurs when a 3-bit counter following the conventional binary code is required. Here, the natural assignment is state $0 = 000$, state $1 = 001$, etc., and finally, state $7 = 111$. In other cases where a natural assignment is not obvious, an arbitrary assignment can be made and at least a suboptimum solution will be obtained. Some other assignment, however, may lead to a more economical circuit. A general solution for the optimum state assignment is very difficult, but additional information on the subject is found in Exercise 5.7, and References 3 and 5.

Elimination of Redundant States[5,8]

Sometimes a state table for a particular sequential problem will contain redundant states, i.e., more states are employed than the minimum required to describe the circuit. Redundant states should be avoided because they often result in a larger memory than necessary (more bits), and they usually lead to a more complicated design by reducing the number of available don't-care conditions. Consider the situation shown in Table 5.4(a); states 2 and 3 are clearly equivalent since the same inputs lead to the same states and the same outputs. With $2 \equiv 3$, the circuit description may be redrawn as shown in Table 5.4(b). Here, one state is saved and the memory requirement drops from two bits to one bit.

In general, how may equivalent states be recognized? The simplest approach is to assume that two states are equivalent, unless it depends on a proven nonequivalence.

TABLE 5.4 EQUIVALENT STATE DESCRIPTIONS

(a) Original			(b) Reduced Version		
Present State	Next State		Present State	Next State	
	x_1	x_2		x_1	x_2
1	1	3	1	1	2
2	2	1, Z	2	2	1, Z
3	2	1, Z			

A proven nonequivalence can be either a direct difference in outputs or a dependence on two different states which can be proven nonequivalent.

Table 5.5 is an example of *chain dependence*, where finally the proven nonequivalence of states 3 and 4 results in none of the states being equivalent. In contrast, Table 5.6(a) demonstrates chain dependence where there are no proven nonequivalent states. Thus, all states are equivalent and the result [shown in Figure 5.6(b)] is a single state which can be realized entirely by a combinational circuit, without requiring any memory bits.

TABLE 5.5 NONEQUIVALENT STATES

Present State	Next State x_1	x_2
1	2	1, Z
2	3	1, Z
3	4	1, Z
4	2, Z	1, Z

TABLE 5.6 EQUIVALENT STATE DESCRIPTIONS

(a) Original			(b) Reduced Version		
Present State	Next State x_1	x_2	Present State	Next State x_1	x_2
1	2	1, Z	1	1	1, Z
2	3	1, Z			
3	4	1, Z			
4	1	1, Z			

The above discussion shows that considerable simplification can result from the recognition and elimination of equivalent states. Also, the burden on the designer in constructing the original state table from a verbal specification is reduced since it is not necessary to insure that each added state is really unique. In fact, a good rule to follow when trying to decide whether a particular state should be added is: If in doubt, add an extra state. One must be sure, however, to always check each state table for redundancies before making the state assignment.

‡Mealy versus Moore Model

Until now, we have employed a general approach to sequential circuits known as the Mealy Model, which is shown in Figure 5.9(a). Here, the output may depend on the input as well as the current state. On the other hand, a less general approach called the Moore Model [shown in Figure 5.9(b)] is often more convenient for computer circuits such as shift registers and counters. Observe that system outputs come directly from the flip-flop outputs and not from a combinational network as they do in the Mealy Model. Since the output is only a function of the present state in the Moore Model, the arrangement of the state diagram and the state table is slightly different,

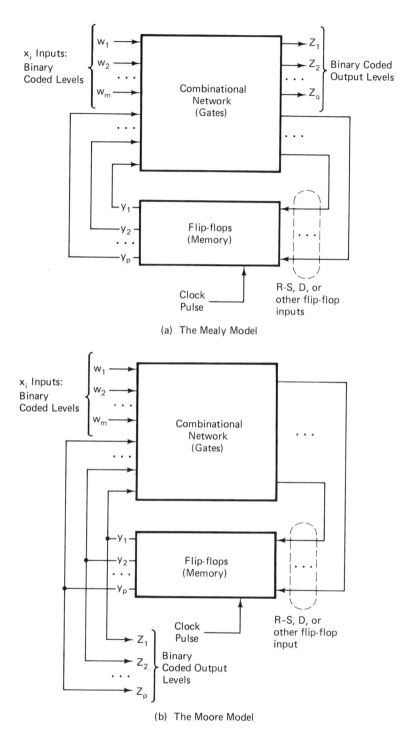

Figure 5.9 Two Models for the Clocked Class of Sequential Circuits. (The levels W_1 through W_m are synchronized with the clock.)

but the basic principles are still the same for both and the methods for eliminating the redundant state are also unchanged.

Design problem. Develop the state diagram and state table for a modulo four up-down counter with reset.

Solution. Since the output depends only on the present state (counter's value), a Moore Model is appropriate. Therefore, assume the following input variables: U representing an up count, D representing a down count, and R representing the rest condition. Figure 5.10(a) shows the resulting state diagram. Observe that the first number inside the node circle continues to be the state number, but the number following the / symbol now represents the output. The nature of the Moore Model makes this slight change from the Mealy notation guide reasonable. In the state diagram, each of the states' numbers is considered to be counter's current total—

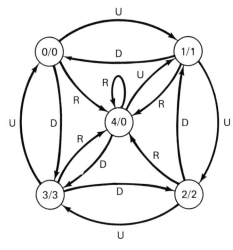

(a) Symbolic State Diagram—
Level Outputs
Where e.g., 4/0 ≡ state/output

Input	Coding $w_2 w_1$
R	$\begin{Bmatrix} 0 & 0 \\ 0 & 1 \end{Bmatrix}$ OR
D	1 0
U	1 1

(b) Coding Table—Levels

	Present State	Inputs $w_2 w_1$	Next State				Output, Z
			01	01	10	11	
	0		4	4	3	1	0
	1		4	4	0	2	1
	2		4	4	1	3	2
	3		4	4	2	0	3
Original nonterminal version	4		4	4	3	1	0
Modified terminal version	4*		4*	4*	4*	4*	0

(c) State Table

Figure 5.10 Modulo-4 Up-Down Counter with Reset.

except that State 4 is assigned to the reset state, which is why its output is zero. In going from the state diagram's symbolic inputs to the state table the input levels must be coded. Our arbitrary choice of these values is shown in Figure 5.10(b). Figure 5.10(c) shows the corresponding state table. Initially, neglect the line in the table labled 4*. (It will be employed later in a variation of the problem.) Again it is seen that a slight, but reasonable, change has been made in going from the Mealy to the Moore representation in that the output, Z, now occupies a separate column.

No redundancies are detected in the table between States 0, 1, 2, and 3 since their outputs are all different. State 4, however, is easily seen to be redundant (i.e., $4 \equiv 0$) because the outputs as well as all the entries on lines 0 and 4 agree. Thus, State 4 should be removed from the tables and all fours replaced by zeros. Although early in the solution you may have been suspicious about including State 4, its use is really not bad practice. Since extra states can be much more easily detected than missing states, a good rule to follow is: If in doubt initially, include questionable states. Our original design problem could have specified that the reset condition must lead to a terminal state (a state that locks out the normal inputs until a master reset is produced by an external means such as an asynchronous clear signal). The modification merely causes the original line 4 to be replaced by the slightly changed 4* line at the bottom of the state table. Under the new condition State 4 would no longer be redundant because the comparison between States 0 and 4* at the $w_2 w_1 = 10$ input would require that $3 \equiv 4*$. This is impossible, however, because the outputs do not agree (one output is 0 and the other is 3). Thus, the modified problem, including a terminal condition, would require that State 4 be present. If it had not been included in the original table, its absence would have been difficult to detect later in the process.

5.4 CIRCUIT SYNTHESIS[2,4]

Special Karnaugh maps, called memory excitation maps, are next constructed so that the Boolean functions required to drive the memory may be determined. We ask: If the flip-flops are in a particular state and a given input occurs, what signal should be received by each flip-flop to bring it to the next desired state? This question can be answered by means of a group of memory excitation maps, Karnaugh maps, whose coordinates (w_i and y_i values) represent the present state and whose entries represent the next state with a map for each flip-flop. Such maps for the modulo-3 counter are shown in Figure 5.11.

In addition to the don't-care symbol, there are other special symbols appearing

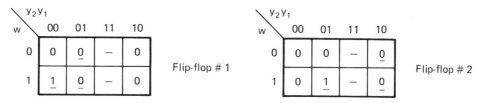

Figure 5.11 Memory Excitation Maps for the Modulo-3 Counter.

TABLE 5.7 MEANING OF MAP SYMBOLS

Required Flip-Flop Transition	Map Symbol
0 → 1	$\underline{1}$
1 → 0	$\underline{0}$
0 → 0	0
1 → 1	1
Optional	—

on the excitation map to label the required transition for each flip-flop. These are summarized in Table 5.7.

The following facts should be observed concerning Figure 5.11; they are representative of all such maps:

1. Each map possesses a square for every possible memory state and every input level w_i; in this case map coordinates wy_2y_1.
2. Each map is for a particular flip-flop.
3. Memory states $y_2y_1 = 11$ possess don't-care entries throughout because the present state 11 does not exist in Table 5.3.

Thus, the map in the second row of the figure describes what happens to flip-flop 2 when the input w occurs. Moreover, if the flip-flops had been in state $y_2y_1 = 10$, Figure 5.11 shows that flip-flop 2 would have been reset by the $w = 0$ input, i.e., a $1 \rightarrow 0$ transition, represented by $\underline{0}$, would have occurred at square $wy_2y_1 = 010$.

We next inquire how the information from Table 5.3 can be translated into Figure 5.11 and, particularly, how the entry discussed in the previous paragraph can be obtained. Consider the present state $y_2y_1 = 10$ in Table 5.3, the $w = 0$ input, and flip-flop 2 (this flip-flop is represented by y_2, the first member of each table entry pair). Here, the y_2 transition is from the present state 1 to the next state 0, which is represented by the symbol $\underline{0}$. Thus, the entry $\underline{0}$ is placed on the Karnaugh map for flip-flop 2 at coordinate $wy_2y_1 = 010$. In a similar way, the translation of the remaining data from Table 5.3 to Figure 5.11 should be verified.

From the properties of the *R-S* flip-flop, given in Table 5.1, a new table can be constructed which allows the Karnaugh map entries in Figure 5.11 to be translated in such a way that the *S* and *R* excitation equations can be obtained. These flip-flop input requirements are listed in Table 5.8 and may be understood through the following reasoning. When a $\underline{1}$-term appears on the map, that flip-flop is to be set. But Table 5.1 tells us that $(S = 1, R = 0)$ will produce the required action in an *R-S* flip-flop; thus, these entries are shown on the second line of Table 5.8. When a 0-term appears on the map, that flip-flop is to stay at zero. But Table 5.1 tells us that either $(S = 0, R = 1)$ or $(S = 0, R = 0)$ will satisfy the requirement; thus, on the third line Table 5.8 shows $S = 0$ and $R = $ —, which is consistent with the don't-care condition of $R = 0$ or $R = 1$. The other entries in Table 5.8 follow from Table 5.1 in a similar way, and the reader should verify them as an exercise.

TABLE 5.8 REQUIRED INPUTS FOR THE *R-S* FLIP-FLOP

Karnaugh Map Entry	Required Inputs	
	S	R
$\underline{0}$	0	1
$\underline{1}$	1	0
0	0	—
1	—	0
—	—	—

Consider the $\underline{1}$-line in Table 5.8. The ($S = 1$, $R = 0$) entries have their usual logical meaning, i.e., in order to satisfy the Karnaugh map, the $\underline{1}$ *must* be included in the map coverage for the set function but it *must not* be in the coverage for the reset function. On the other hand, the 1-line in Table 5.8 tells us that the 1-entry on the Karnaugh is a don't-care for the set function. In any case, it must not be included in the reset function. The other lines of the table follow in a similar way. (It is important to understand that *all* underlined terms *must be covered*, with $\underline{1}$ in S and $\underline{0}$ in R.)

From the information given in Figure 5.11, let us now obtain the required logic equations to drive the flip-flops. First consider the set equations. Once the symbols in Figure 5.11 are properly interpreted as either a 1, 0, or—according to the S column in Table 5.8, the required map coverage can be made by simply following the usual Karnaugh map rules developed in Chapter 4 for combinational circuits. Remember, however, that the individual maps are actually isolated, e.g., the rows of maps are independent because they belong to different flip-flops. Thus, within a row of maps, the coverages must be ORed. The resultant map coverages for the set inputs are shown labeled S_1 and S_2 in Figure 5.12. In the figure, consider the set equation for flip-flop 1, which comes from the first row of maps. There is only a single $\underline{1}$ on this

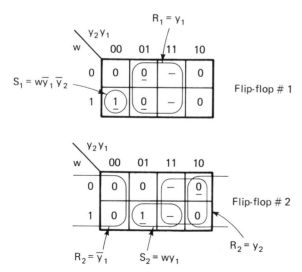

Figure 5.12 Maps of Figure 5.11 Showing the *R-S* Coverage.

map, and the squares contiguous to it are $\underline{0}$ and 0, which are both translated as 0 by Table 5.8. Thus, the symbol $\underline{1}$ is circled as an isolated quantity and the resulting coverage is read as $S_1 = w\bar{y}_2\bar{y}_1$. In the case of flip-flop 2, the don't-care term can be used; this results in $S_2 = wy_1$. The process for the reset equations is similar except that the R column of Table 5.8 is employed to interpret the map symbols, and the resultant map coverage is labeled R_1 and R_2 in Figure 5.12. The set and reset equations, along with the equation for the output, are summarized below. Figure 5.13 shows the corresponding logic diagram.

$$S_1 = w\bar{y}_1\bar{y}_2, \qquad R_1 = y_1$$
$$S_2 = wy_1, \qquad R_2 = \bar{y}_1 \text{ or } y_2$$

For a pulse output: $Z = (\text{clock})w(y_2\bar{y}_1 + y_2 y_1) = (\text{clock})wy_2$. Here $Z = 1$ only during the clock pulse causing the transition: $y_2 y_1 = 10 \to 00$. For a level output: $Z = wy_2$. Here $Z = 1$ during the period before the clock pulse causing the transition: $y_2 y_1 = 10 \to 00$.

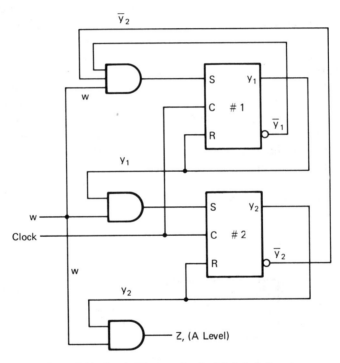

Figure 5.13 Logic Diagram for the Modulo-3 Counter.

Several facts should be observed concerning these equations:

1. Subscripts 1 and 2 on S and R refer to flip-flops 1 and 2.
2. The equation for Z is obtained by inspection of Table 5.3, i.e., $Z = 1$ occurs only when the flip-flops are in the state $y_2 y_1 = 10$ or in the $y_2 y_1 = 11$ don't-

care state and $w = 1$ occurs. More complicated cases could be evaluated simply by use of another Karnaugh map.

3. The w input must be synchronized with the clock so that w does not change within several nanoseconds of being in coincidence with the clock's negative edge. (More details are included in Section 5.8.) In many systems, especially digital computers, this requirement is not difficult to satisfy since the level inputs usually come from other flip-flops served by the same clock.

One of the general and very useful characteristics of the above technique is that one set of maps may be used to determine the excitation equations for a large variety of flip-flop types. All that is needed is a new Table 5.8 for each different type of flip-flop. These tables and further examples will be discussed later.

Summary

The circuit synthesis method, which has just been presented, may be summarized by the following six steps:

1. Draw a state diagram to meet the requirements of the circuit.
2. Translate the diagram into a state table.
3. Make a state assignment. (An arbitrary one is acceptable, but it may not be optimum.)
4. Translate the data from step 3 into the required memory excitation maps.
5. Employing a table of required inputs for the desired flip-flop (such as Table 5.8), select the minimal map covering and read the flip-flop excitation equations.
6. Draw a logic diagram for the circuit.

As long as the sequential circuit requirements fit the broad model of Figure 5.1, the synthesis method is quite general and, in principle, is not limited by such things as the number of flip-flops or the type of state diagram specified.

5.5 SYNTHESIS OF COUNTERS

Counters belong to one of the most important groups of circuits found in digital computers. They are particularly common in the control and arithmetic sections, where they are used to keep track of the sequence of instructions in a program (an instruction counter), to totalize the number of shifts in multiply and divide operations (a step counter), to distribute the sequence of timing signals (often a ring counter), and to accomplish a host of similar applications. Counters are of the proper degree of difficulty (not trivial, but also not overly complex) for presenting synthesis techniques (and later heuristic design methods). Moreover, an in-depth look at the design of one class of circuits is an excellent way to learn the basic engineering principles which apply to all digital logic.

Example: Synthesis of a Three-Bit Up-Counter

The clocked toggle, T, flip-flop is introduced in this example, together with one of the most frequently used counters. After we construct the state diagram (Figure 5.14) and the state table (Table 5.9) and make a conventional binary assignment for a three-bit up-counter, we obtain Table 5.10. The following facts should be observed from the diagram and the table.

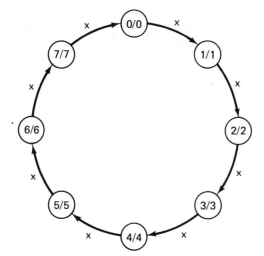

Figure 5.14 State Diagram for a Three-Bit Up-Counter.

TABLE 5.9 STATE TABLE FOR A THREE-BIT UP-COUNTER

Present State	x Input, Next State
0	1
1	2
2	3
3	4
4	5
5	6
6	7
7	0

1. The counter is of the nonterminal type, i.e., with an x input, each clock pulse causes the counter to increment and the next state following $y_3y_2y_1 = 111$ is 000.
2. Since each output level corresponds to the present state and may thus be taken directly from the flip-flops, a Moore Model is assumed; i.e., $Z_3Z_2Z_1 = y_3y_2y_1$.
3. For simplicity, the circuit is not internally cleared.

The excitation maps corresponding to Table 5.10 are shown in Figure 5.15. One interesting fact about the requirements of this problem is that with the counter's

TABLE 5.10 STATE TABLE FOR A THREE-BIT UP-COUNTER

Present State ($y_3y_2y_1$)	x Input, Next State ($y_3y_2y_1$)
000	001
001	010
010	011
011	100
100	101
101	110
110	111
111	000

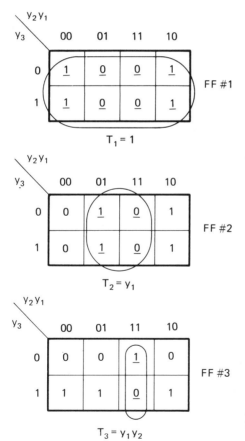

Figure 5.15 Excitation Maps for an Up-Counter.

reset being done externally* the input x is constant at $x = 1$; thus, X does not occur on the map. An alternate way of looking at this is that each map is ANDed with the input but since $x = 1$ it does not appear.

In order to interpret the maps, we must understand the operating characteristics of the T flip-flop. These are presented in Table 5.11, and the resulting map requirements are given in Table 5.12. (As an exercise, the reader should derive the requirements of Table 5.12 from the facts given in Table 5.11.)

TABLE 5.11 OPERATING CHARACTERISTICS OF THE T FLIP-FLOP

Input	Next State (y_{n+1})
0	y_n
1	$\bar{y}_n$

TABLE 5.12 MAP COVERING DEMANDED BY THE T FLIP-FLOP

Karnaugh Map Entry	Required Inputs T
$\underline{0}$	1
$\underline{1}$	1
0	0
1	0
—	—

When reasoning similar to that previously used for the R-S flip-flop is applied to the T flip-flop, and Table 5.12 is employed to read the maps in Figure 5.15, the resulting equations are

$$T_1 = 1, \qquad T_2 = y_1, \qquad T_3 = y_1 y_2$$

(As expected, the equations are independent of the input x.) These lead to the circuit diagram in Figure 5.16.

Inspection of the excitation equations and the figure leads us to the conclusion that this circuit could be extended to as large a counter as desired, with the excitation for the nth bit being given by $T_n = y_1 y_2 \ldots y_{n-1}$. Understand that in a circuit like this, the clock need not be a high-frequency signal as it is in most computers; but it can be any unit which produces a pulse (of arbitrary period) indicating an increment in the count. Thus, a person entering a theater can be counted through a pulse generated by a photoelectric cell, and the pulse can be many seconds long in case of a slow-moving ticket line.

*External reset is easily accomplished by employing the asychronous flip-flops to be studied in Section 5.8. Also, an example to be discussed at the end of the current section consists of an internal reset integrated with the other inputs. After some additional examples we will soon be ready for this somewhat more complex case.

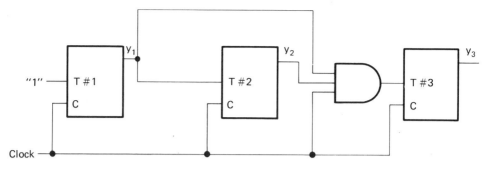

Figure 5.16 Three-Bit Up-Counter.

Example: Synthesis of One Stage of a BCD Counter with Clocked J-K Flip-Flops

Having gained a basic understanding of clocked sequential circuits through design with the elementary *R-S* and *T* flip-flops, we now move to the more practical clocked *J-K* type, which combines some characteristics of both previous types. One stage of a binary-coded decimal, BCD, counter with external clear is to be designed. (This circuit counts from 0 to 9 in binary, and then returns to 0.) Again we assume a Moore Model with output $Z_4 Z_3 Z_2 Z_1 = y_4 y_3 y_2 y_1$.

TABLE 5.13 STATE TABLE FOR A BCD COUNTER

Decimal Number	Present State ($y_4 y_3 y_2 y_1$)	x Input, Next State ($y_4 y_3 y_2 y_1$)
0	0000	0001
1	0001	0010
2	0010	0011
3	0011	0100
4	0100	0101
5	0101	0110
6	0110	0111
7	0111	1000
8	1000	1001
9	1001	0000
10–15	1010 to 1111	Don't-care

The required state conditions are shown in Table 5.13, and the excitation maps corresponding to Table 5.13 are shown in Figure 5.17.

Table 5.14 presents the characteristics of the *J-K* flip-flop, and the resulting map requirements are given in Table 5.15.

Now we are ready to obtain the equations to satisfy Figure 5.17. The use of clocked flip-flops means that the actions indicated in Table 5.14 are not allowed to

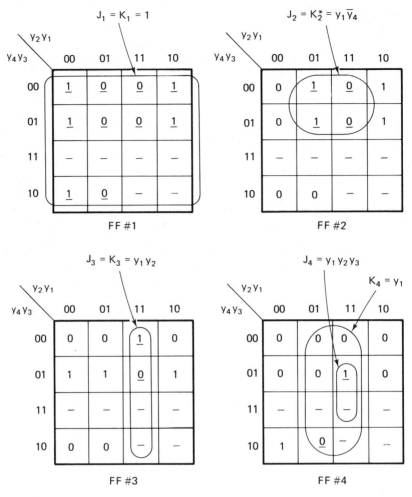

Figure 5.17 Flip-Flop Excitation Maps—BCD Counter.

TABLE 5.14 CHARACTERISTICS OF THE *J-K* FLIP-FLOP

| Inputs | | Next State |
J	K	(y_{n+1})
0	0	y_n
0	1	0
1	0	1
1	1	$\bar{y}_n$

occur until a pulse is supplied to the clock input, *C*. Effectively, this means that the *C* input is ANDed with the *J* and *K* inputs. (In most systems, the *J* and *K* inputs must be stabilized at the values shown in Table 5.14 at least a few nanoseconds *before* the clock pulse occurs. The action shown in the table takes place when the *clock pulse*

TABLE 5.15 MAP COVERING DEMANDED BY THE J-K FLIP-FLOP

Karnaugh Map Entry	Required Inputs J	K
$\underline{0}$	—	1
$\underline{1}$	1	—
0	0	—
1	—	0
—	—	—

makes the $1 \rightarrow 0$ transition, i.e., operation occurs on the negative edge.) Again, with external clear and a constant $x = 1$, the input does not appear explicitly in the equations.

The resulting equations for this example are:

$$J_1 = K_1 = 1, \quad J_2 = K_2 = y_1 \bar{y}_4, \quad J_3 = K_3 = y_1 y_2,$$
$$J_4 = y_1 y_2 y_3, \quad K_4 = y_1$$

These lead to the circuit diagram shown in Figure 5.18.

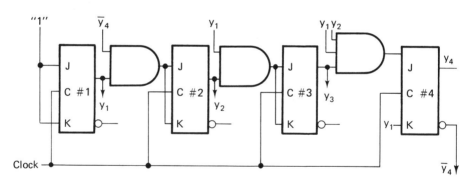

Figure 5.18 One Stage of a BCD Counter.

The equations $J = K = 1$ mean that flip-flop 1 will change state each time the trailing edge of the clock pulse occurs; thus, a clocked J-K flip-flop wired in this way behaves like a T flip-flop with $T = J = K$. As a matter of fact, the T flip-flops appearing in Figure 5.16, and all other occurrences of them, can be easily realized with clocked J-K flip-flops wired as shown in Figure 5.18. In the future, however, we will continue to employ T flip-flops where appropriate, since they uniquely (and concisely) represent a particular mode of flip-flop operation. On the other hand, in actual circuit construction, the more powerful clocked J-K flip-flops would usually be used. Incidentally, clocked flip-flops usually possess a direct reset input which is independent of the clock circuit; therefore, counters and other systems constructed from these devices may be cleared without additional gating. (This, and the possibility of a direct set input, will be considered more fully in Section 5.8.)

(Note: In Figure 5.10 (c), 4 ≡ 0)

Present State (y_2y_1)	Next State (y_2y_1) Input w_2w_1				Output Z_2Z_1
	00	01	10	11	
01	00	00	11	01	00
01	00	00	00	10	01
10	00	00	01	11	10
11	00	00	10	00	11

(a) State Table for the Counter of Figure 5.10

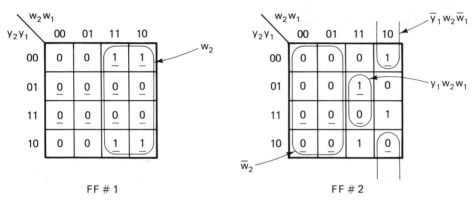

(b) Excitation Maps for the Counter in (a) above

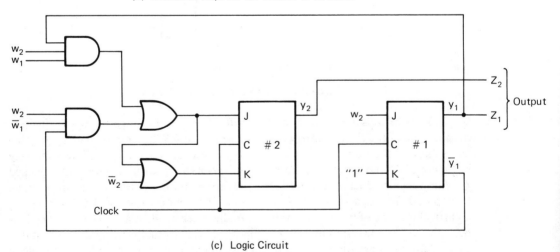

(c) Logic Circuit

Figure 5.19 Design of an Up-Down Counter with Internal Reset.

Example: Synthesis of an Up-Down Counter with Internal Reset

Now consider the up-down counter with internal reset of Figure 5.10. This is a more interesting design because it takes three inputs resulting in a level decoding scheme requiring two variables, w_1 and w_2. When the binary state assignment is made in Figure 5.10(c), the table in Figure 5.19(a) results. Following our usual methods, the data are then arranged into the two Karnaugh maps shown in Figure 5.19(b). Observe that w_i, the input variables, are the coordinates for half of the map; thus, contrary to some previous examples, the inputs will play a major part in the current solution. After covering the maps of Figure 5.19(b), for *J-K* flip-flops, the following equations are determined:

$$J_1 = w_2, \quad K_1 = 1$$
$$J_2 = \bar{y}_1 w_2 \bar{w}_1 + y_1 w_2 w_1, \quad K_2 = \bar{w}_2 + J_2$$

(The outputs come directly from the flip-flop states $y_2 y_1$.) These then result in the circuit of Figure 5.19(c). Note how circuity is saved by employing the J_2 function as part of K_2.

5.6 HEURISTIC DESIGN METHODS

In principle, one could use the synthesis method to obtain any required sequential circuit; however, the method becomes unwieldy when many flip-flops are involved. On the other hand, the design effort required by the method to be introduced here is not nearly so strongly influenced by the number of flip-flops. In practice, solutions are often obtained that can be easily extended to as many flip-flops as desired. Moreover, it is desirable to have quicker techniques for common circuits as well as for circuits that are fairly simple in structure. Rather than presenting another formal technique such as the circuit synthesis method developed in Section 5.4, the object of this section is to give the reader insight into some of the subjective techniques employed by the computer engineer. This should enable the novice to develop his own design skills more rapidly.

The circuits in Section 5.5 suggest a number of intuitive design techniques which, though fairly obvious, are nevertheless important. Such intuitive techniques, which do not guarantee an optimum solution (or for that matter any kind of solution), but which are frequently successful and fast, are called *heuristics*. (An example of a common heuristic in mathematics is the concept of converting all functions to sines and cosines when proving trigonometric identities.) Although the number of flip-flops may be large (10 or 20 or more), the inherent structure of most individual computer circuits is surprisingly simple. For this reason, heuristic methods have been very successful and are widely used in computer design (although they are not always called by that name).

The following list summarizes some of the heuristic techniques:

1. *Iteration.* Once a good circuit has been developed, it can frequently be iterated to fulfill the requirements of a much larger circuit. (An example is the iteration of one stage of a counter to produce a 12-bit unit.)

2. *Module synthesis.* The synthesis method itself may be one of the most important heuristics because it can be used to optimize a portion of a circuit, a module. This module may then be iterated to produce a very large system. (An example is the synthesis of one stage of a BCD counter, which is then iterated to produce a counter with 10 decimal places.)

3. *Information analysis.* Through analysis of the information needed to excite a particular flip-flop, the necessary gating to obtain this information from the other flip-flops is often apparent. (The way a single AND gate couples the information into the nth T flip-flop from all the others in a synchronous up-counter is representative of this concept.)

4. *Evolution (ANDing or ORing).* Often we have had experience with a circuit which almost satisfies the current requirements. The idea here is to complete the design by modifying the original circuit in a fairly straightforward way or by combining two or more circuits. Of course, the obvious way to combine circuits is to use the AND or OR function.

5. *Requirements segmentation.* By segmenting the circuit requirements into small manageable sections, we can often determine the gating and flip-flops necessary to realize the individual sections by inspection or by using one of the other heuristics. Two major segments which are frequently employed are flip-flop excitation and output generation. In the first of these, the designer concentrates on the inputs and current flip-flop states to generate all the next flip-flop states. In the other, emphasis is placed on inputs and the current flip-flop states to generate the outputs.

   Taking a more formal approach, we can divide the segmentation heuristic into the following components:

   (a) Break the requirements into the smallest identifiable entities for individual realization and later recomposition.

   (b) Try to use circuit entities already mechanized as components in those not yet realized.

   (c) Work on the most primitive requirements first, with the hope that these may later be used in satisfying more complex requirements.

   (d) Look for regularity and symmetry in the overall requirements, and take as much advantage of these as possible.

   (e) Where possible, use circuits or parts of circuits which have already proved their worth as a base for designing new circuits.

The above heuristics will now be employed in the design of several practical computer circuits. The emphasis will be on heuristics 1 through 4, because these apply to circuits which essentially have only a single requirement and may be illustrated in a page or two. Heuristic 5 has its major power in the design of large computer blocks where there is a multiplicity of requirements, and it will be mainly applied in later chapters, particularly in our study of the control system (Chapter 6). However, certain aspects of heuristic 5, particularly (b), (d), and (e), also apply at lower levels.

Example: Heuristic Design of a Ring Counter

Let us now attempt to use the heuristic methods to design a ring counter, which passes a single binary 1 from flip-flop to flip-flop until the last stage is reached, at which point the 1 is passed back to the first stage. Consider clocked *J-K* flip-flops and a four-bit system.

From the statement of the problem, and using the information heuristic, it is apparent that, whatever state the *n*th flip-flop is in, the next one should be in that same state after the input pulse has passed. This and the properties of the clocked *J-K* flip-flop lead us to the circuit for one stage in Figure 5.19.

Using the iteration heuristic, we obtain the complete circuit for the ring counter, as shown in Figure 5.20.

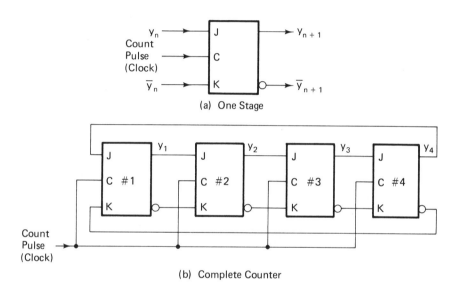

Figure 5.20 Four-Stage Ring Counter.

Example: Heuristic Design of a Modulo-5 Counter

Suppose we desire to design a modulo-5 counter with clocked *T* flip-flops. From experience or from simple heuristic considerations, we can construct a three-bit up-counter (see Figure 5.16). This will be identical to the modulo-5 counter until the state $y_3 y_2 y_1 = 100$ is reached. However, the next state must be 000. The AND heuristic is used to detect the 100 state and the occurrence of the input, x, through the function $T_3 = y_3 \bar{y}_2 \bar{y}_1$. (On further consideration, this may be reduced to $T_3 = y_3$ since the only time $y_3 = 1$ occurs is when the counter is in the state 100.) If the system is in state 100 and an input occurs, two things must happen: FF3 must be reset and FF1 must not be complemented. The shaded gate at the input to FF3 in Figure 5.21 show how the former requirement is satisfied through the OR heuristics. Similarly, the FF1 uses the information analysis heuristic, function $T_1 = \bar{y}_3$, to allow the input pulse to complement FF1 in all states except 100. In principle, all such functions which are con-

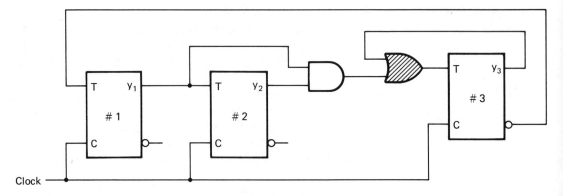

Figure 5.21 Modulo-5 Counter.

structed heuristically should be checked for possible simplifications; no simplification, however, is possible in this example.

Unclocked T Flip-Flops

Although our main emphasis is on clocked flip-flops, unclocked devices are also useful. One of the most common flip-flops of this type is the unclocked T. (Many, including this text, use the term *T flip-flop* to indicate the unclocked device and the term *clocked T flip-flop* for the other possibility.) There really is not much difference between the two variations of T flip-flops, and the unclocked variety can easily be made from one of several clocked devices. (The clocked T can also easily be made from other flip-flops.) For example, a clocked J-K can be converted to a T merely by permanently making the connection $J = K = 1$ and by relabeling the original clock terminal as the new T terminal. (Of course, the old y and $\bar{y}$ outputs continue to serve exactly the same function to the new device.)

Strictly speaking, the use of T flip-flops changes the circuit classification from a clocked-type to a pulse input (or unclocked) synchronous circuit. For many practical applications, however, the distinction is unimportant. This is particularly true in our work, which usually will include counters or other simple circuits requiring only one input, pulse as is done in the next example. The interested reader is referred to the bibliography[5,8] for complete details on pulse-type synchronous circuits, including cases where multiple pulse inputs are required.

The unclocked R-S flip-flop is another type that is occasionally employed. It behaves like the clocked R-S except that the device typically changes state when a negative-going edge occurs at either the set or reset inputs. Thus, the R and S inputs carry both WHAT and WHEN information rather than carrying WHAT information alone. For example, with $y = 0$ and $R = 0$, if S goes to a logic one and then returns to zero, a few nanoseconds later the flip-flop will set (the output will go to the one state).

Example: Heuristic Design of a Gray-Code Counter

We will now attempt a fairly challenging problem—the design of a Graycode counter using T flip-flops. This code follows the sequence shown in Table 5.16.

The equation for T_4 is fairly simple. We see by inspection of Table 5.16 that an

TABLE 5.16 THE GRAY CODE

Decimal	Gray Code ($y_4y_3y_2y_1$)
0	0000
1	0001
2	0011
3	0010
4	0110
5	0111
6	0101
7	0100
8	1100
9	1101
10	1111
11	1110
12	1010
13	1011
14	1001
15	1000

input causes FF4 to change state when $y_4y_3y_2y_1 = 0100$. This results in $T_4 = x\bar{y}_4 y_3 \bar{y}_2 \bar{y}_1$. (*Note:* It is assumed here that the counter is not designed to turn over, i.e., no transition from 1 to 0 is required for FF4.) On the other hand, how can the equation for T_1 be generated? Must the states $y_4y_3y_2y_1 = 0000, 0011, 0110, \ldots, 1001$ be detected, resulting in a function with eight terms? (Incidentally, these are eight isolated terms on a Karnaugh map—no simplification is possible.) Before allowing the circuit to become this complicated, let us use the information heuristic and ask ourselves what is regular about the requirements on y_1. Inspection of Table 5.16 reveals that y_1 changes state on the odd-numbered inputs. Certainly, a very easy way to detect this condition is to run the input through an auxiliary flip-flop. Thus, there is a trade-off between the cost of an extra flip-flop and the cost of many gates. It appears that, under most conditions, the auxiliary flip-flop method would be favored.

After the auxiliary bit is included, the code is as shown in Table 5.17. Is there any additional regularity about these data which may be used to further simplify our design? Careful examination of the table reveals that the y_n bit changes state only if $y_{n-1} = 1$ and the other less significant bits equal zero.

Thus, the equation for the ith flip-flop is $T_i = y_{i-1}\bar{y}_{i-2} \ldots \bar{y}_1 \bar{A} x$, where $1 < i \leq 4$. (For $i = 1$, $T_1 = Ax$.) The resulting circuit is shown in Figure 5.22. (Temporarily omit the dashed line bypassing the shaded gate.)

If we desire to have the counter return to zero on the 16th input pulse, we can use the OR heuristic to find the term which must be added to the T_4 input. From Table 5.17, this term is $x\bar{y}_3\bar{y}_2\bar{y}_1\bar{A}$. From Figure 5.22, the original input to T_4 was $xy_3\bar{y}_2\bar{y}_1\bar{A}$; therefore, $T_4 = x\bar{y}_3\bar{y}_2\bar{y}_1\bar{A} + xy_3\bar{y}_2\bar{y}_1\bar{A} = x\bar{y}_2\bar{y}_1\bar{A}$, which means that the dashed line may be used to replace the shaded gate as the input to FF4. Thus, requiring the counter to reset on the 16th pulse actually allows a gate to be eliminated from the circuit.

TABLE 5.17 THE GRAY CODE INCLUDING THE AUXILIARY BIT, A

Decimal	Code $(y_4y_3y_2y_1A)$
0	00001
1	00010
2	00111
3	00100
4	01101
5	01110
6	01011
7	01000
8	11001
9	11010
10	11111
11	11100
12	10101
13	10110
14	10011
15	10000

Note that Figure 5.22 is still considered to be a synchronons circuit because each flip-flop is actually synchronized by the input pulse, x, even though that pulse was previously gated by one or more ANDs.

Example: Heuristic Design of a Multifunction Register with D-Type Flip Flops

The register to be designed must allow:

1. Parallel-input loading from another register
2. Serial output by shifting the data toward the least significant bit, where it can be extracted
3. A register clear capability

The functions are selected individually by having a logical-1 level appear on one of three lines, and the actual operation is synchronized by means of a clock pulse. Circuits of this type are extensively used in practice, and a typical realization is in the medium scale integration (MSI) package 8274 manufactured by Signetics Corporation.

Having defined the problem and the recalled nature of the D flip-flop, we now seek a heuristic approach to the design. Consider heuristic 5, requirements segmentation. It is apparent that there are three identifiable tasks (load, clear, and shift) that can be realized separately and then combined with the OR heuristic. Moreover, the symmetry in each of these three tasks indicates that module synthesis can be employed for the ith bit, and then iteration can be employed for as many bits as desired.

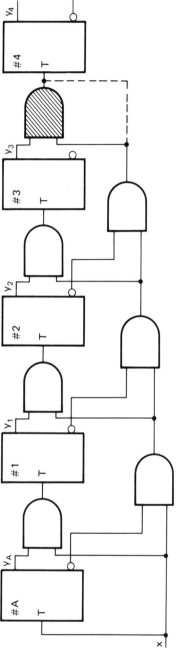

Figure 5.22 Heuristic Design for a Gray Code Counter.

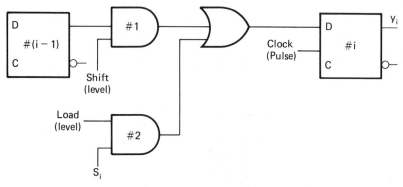

(Level from ith bit of source register.)

Figure 5.23 Multifunction Shift Register—Partially Complete Design.

From information analysis, construction of Figure 5.23 showing the shift and load circuit for the ith bit is straightforward. Assume that it is desired to load data from a source register, S. A logical-1 level will appear at the load input to AND gate 2 a short time before the clock pulse occurs. If the input from bit i of the source register is a 1, it will be passed through the AND gate and the OR gate. Then, when the clock pulse occurs, a 1 will be deposited in the D flip-flop. If the input from bit i S_i, of the source register is a 0, AND gate 2 will have a 0 output, as will the OR gate. Thus, a 0 will be correctly placed in the flip-flop by the following clock pulse. (The shift operation is accomplished in a similar way.)

Now consider the clear function—a 0 is to be placed in the flip-flop when the clear level is 1. Figure 5.23 shows that when the shift and load levels are 0, the output of the OR gate will be a 0. Thus, this condition alone is sufficient to cause the clock pulse to deposit a 0 in the flip-flop, and the design seems to be complete.

When the circuit is tested, however, a flaw will be discovered—if the shift, load, and clear levels are all 0 and a clock pulse occurs, the flip-flop will be cleared. Therefore, the circuit fails to hold the data until one of the three functions is selected. Information analysis of the hold requirement reveals that when neither shift, load, nor clear is true and bit i is in the 1 state, the D input to the flip-flop must receive a 1 so that a 1 will be retained in the flip-flop following the clock pulse. The equation for this task is

$$D_n = y_n \overline{(\text{shift} + \text{load} + \text{clear})}$$
$$= y_n (\overline{\text{shift}})(\overline{\text{load}})(\overline{\text{clear}})$$

Figure 5.24 shows this addition to the original circuit. Testing under all of the specified conditions will verify correct operation of the circuit. Moreover, iteration may be employed to construct a register with as many bits as desired; the only addition that might be required is some special circuitry for shifting in and out at the two ends of the register.

Summary

The following four steps, which should be executed in sequence, serve as a broad summary of the heuristic approach:

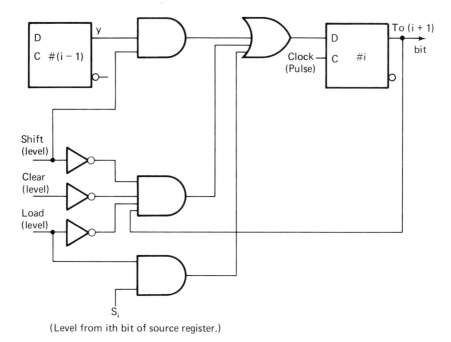

Figure 5.24 Multifunction Shift Register—Complete Design for Module *i*.

1. Concisely describe the desired circuit action. A state diagram or other graphical representation is frequently helpful here.
2. Design the described circuit using the heuristics.
3. Check the combinatorial portions for possible simplifications using the theorems or a Karnaugh map.
4. Test the logic by trying critical input-memory state combinations. (Even here, heuristics can be applied to take advantage of symmetry and the like in avoiding brute-force exhaustive testing of all combinations.)

The above method is open-ended in that professional designers are constantly adding new heuristics to their repertoire and striving to improve old ones. Moreover, once mastered, techniques of this class often are employed almost subconsciously. Thus, this section has provided an introduction to the design art rather than furnishing an algorithmic procedure.

5.7 NONIDEAL COMPONENTS

Earlier we employed ideal digital components for gates and flip-flops. Like ideal analog components, such as pure inductors, they facilitate learning the basic principles and adequately describe performance for some design situations, but fail in many practical systems. As long as relatively static logic levels are being considered, the gate's nonideal nature will probably not cause serious problems; but as soon as high-frequency signals (pulse conditions) exist, the propagation delay of gates can result in serious circuit malfunctions. Propagation delay can cause a noise spike or *glitch* to be produced at the output of a logic circuit even though usual Boolean algebra predicts that a constant logic level is expected.

Figure 5.25 shows the timing diagram of a typical inverter gate. In the voltage waveform of Figure 5.25(a), note that there is a delay, t_{PHL}, between the time the input voltage reaches its 50% level and the time at which the output reaches the corresponding 50% level. This interval is known as propagation delay, with the subscripts having the following meaning: $HL \equiv$ the high-to-low transition (i.e., the output going from the high to the low state), $LH \equiv$ the low-to-high output transition, and $P \equiv$ propagation. Note, that it is not necessary that $t_{PHL} = t_{PLH}$. For typical modern gates, the delay is fairly constant for members within a particular family or subfamily, but as one goes from the fastest to the slowest families in current common applications, one is likely to find the following rather wide range: 0.5 ns $\leq$ (propagation delay) ≤ 200 ns. Figure 5.25(b) shows logic values. Observe that a certain amount of detail is lost in this diagram since logic values can only occur as one of two levels, while the corresponding voltage values are continuous.

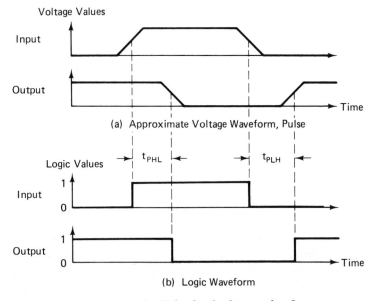

Figure 5.25 Propagation Delay for the Output of an Inverter.

Hazards

When a logic output is developed from signals whose electrical paths include unequal propagation delays, a potentially serious problem called a *hazard* can occur. Figure 5.26 shows how it can happen when one of the inputs, in this case variable A, is changed in a simple logic circuit. Figure 5.26(b) shows that the hazard is caused by the additional delay in the path for the A variable as it passes through the inverter and the AND and OR gates, resulting in a temporary logic error in the circuit output Z.

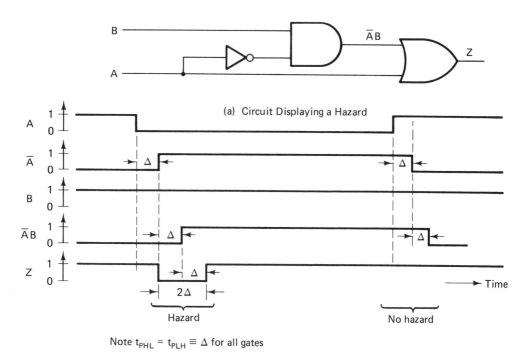

Figure 5.26 Hazard in the Z Output.

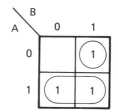

Figure 5.27 Karnaugh Map for Figure 5.26(a).

Fortunately, hazards of this type can be easily detected on a Karnaugh map, as we now see. Figure 5.27 is the Karnaugh map for the circuit being studied. The 1's are seen to consist of two separate coverings, and the hazard occurs when the inputs go from $AB = 11$ to $AB = 01$, which is a case of moving from the lower coverage to the upper coverage. It turns out that whenever there is more than one group of contiguous coverages on the map a hazard will be found. The hazard can be eliminated, however, if the map coverages are rearranged in such a way that each adjacent pair overlaps by at least one square. Thus, as the inputs change and we move among each adjacent pair of minterms of the function, the path must always stay within the boundary of at least one coverage.

In our current example, the modification required is the removal of the $A\bar{B}$ minterm and its replacement with the term B. Then the function becomes $Z = A + B$, which the Karnaugh map shows to be equal to the original function, but the circuit is simplified since the AND as well as the NOT gates may be removed.

The above situation, where a hazard is removed and the circuit simultaneously simplified, is fairly unusual. The more likely situation is where removal of the hazard requires adding a redundant coverage to the map. The function $Z = \bar{A}C + AB$ shown in Figure 5.28 is a simple example of this case. Here, a hazard exists as the input variables change from $ABC = 111$ to $ABC = 011$. The solution, of course, is to add the overlapping redundant coverage BC to the map and the resulting function is $Z = \bar{A}C + AB + BC$. In this case removal of the hazard cost an extra AND gate along with an extra input to the original OR gate. The function is now no longer simplified, but the extra circuitry is the cost that must be paid to eliminate the hazard. Although hazards represent practical problems in their own right, they are even more important as a medium to introduce the whole class of timing problems in digital circuits.

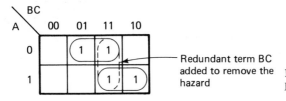

Figure 5.28 Simplified Function Displaying a Hazard.

Nonideal Flips-Flops

Unfortunately, flip-flops, like gates, are not just a group of ideal components guaranteed to yield reliable circuits under all circumstances. On the contrary, they can produce some puzzling failures unless great care is taken in their application. Shift registers erroneously filling with ones; counters oscillating, and registers catching extra ones are typical design problems when some of the hidden characteristics of flip-flops are not anticipated. As might be expected, these problems are often accentuated by the presence of any hazards. On the other hand, all commercial flip-flops do have some reliable applications and the designer need not be unduly apprehensive if he (she) is aware of the component's limitations.

The key to successful designs with nonideal flip-flops is care in employing two parameters: *WHAT* and *WHEN*. WHAT refers to the data inputs accepted (e.g., J-K, D, R-S, T, etc.), and normally it is not the main difficulty. The WHEN parameter, established by the clock input, determines the time interval over which the input data are accepted. This parameter has some subtle aspects about it which can load to unreliable designs. Those accepting data during a large portion of the clock pulse are called, for obvious reasons, pulse sensitive flip-flops and are particularly susceptible to picking up ones (or zeros) existing in the data for only a short time. This phenomena, called *ones* (*zeros*) *catching*, is normally very undesirable since it usually results in erroneous flip-flops states. Another large class of flip-flops is described as *edge sensitive* because they accept data only for a very narrow time interval near either the negative- or positive-going edge of the clock pulse. This property means that the device is insensitive to noise spikes or hazards that frequently occur at other times.

How can the designer determine the WHEN parameter from the flip-flop's common name? Usually this is fairly easy to do, but one must be alert because the industrial terms are not entirely standardized and if in doubt the manufacturer's data sheet should be carefully studied to resolve any uncertainties. For practical purposes, clocked flip-flops can be divided into two classes: *edge triggered* and *pulse triggered*. The device known commercially as an edge-triggered flip-flop may be easily identified as the edge-sensitive type. Typical members of this group are the 74103 package, which contains dual *J-K* negative-edge-triggered flip-flops, and the 7474, which contains dual *D*-type positive-edge-triggered flip-flops. Potentially the most misleading flip-flop is known commercially as the *master-slave*. This term merely means that it contains two separate flip-flop sections: a master, which receives data during the clock pulse, and the slave, which is usually loaded from the master shortly after the negative edge of the clock pulse and holds that output until the following clock pulse. The difficulty here is that the term master-slave says nothing about the WHEN parameter—it may be either edge sensitive or pulse sensitive. A commercial device labeled master-slave with data lockout is positive edge sensitive when the master is filled. On the other hand, the plain term master-slave normally is a pulse-sensitive flip-flop. Typical members of the master-slave with data lockout class are the 74110 and 74111, which are *J-K* devices. Typical master-slave devices which should be classified as pulse sensitive are the 74L71, which is an *R-S* device, and the 74107, which is a *J-K* device. In the remainder of this chapter the nonideal nature of flip-flops will be considered in detail from a logic circuit viewpoint.

5.8 FLIP-FLOP REALIZATION METHODS

Now that you see how valuable flip-flops can be and have gained some experience in employing them in sequential design, you are probably wondering how they are constructed. In the previous sections of this chapter we considered only the terminal characteristics of the various flip-flops—they were treated as black boxes. Moreover, they were assumed to be ideal devices whose output changed state only on the negative-going edge of the clock pulse. This approach allowed us to gain some experience before becoming involved with more complicated issues.

Flip-flops can be produced directly from transistors,[10] but the logic principles upon which various types of flip-flops are based can be best understood if gates are used as building blocks. Moreover, in many practical integrated circuits, gate flip-flops are used. Figure 5.29 demonstrates how an elementary *R-S* circuit can be constructed from two NAND gates and two inverters. The following facts should be observed concerning the logic diagram and the associated truth table.

1. The negated input OR form for each NAND is used to make it easier to visualize circuit action.
2. Pairs of inverters in the diagram have a slash drawn through them to emphasize that they effectively cancel each other.
3. The variables *a*, *b*, *c*, and *d* at the gate inputs represent logic values after inversion.
4. In the truth table, the last column indicates the sequence followed as the logic variables change in response to a particular input.

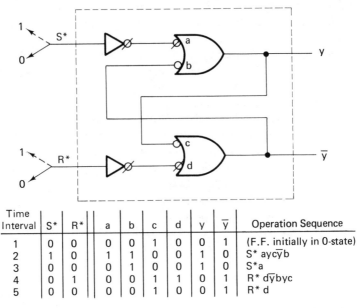

Time Interval	S*	R*	a	b	c	d	y	ȳ	Operation Sequence
1	0	0	0	0	1	0	0	1	(F.F. initially in 0-state)
2	1	0	1	1	0	0	1	0	S* aycȳb
3	0	0	0	1	0	0	1	0	S*a
4	0	1	0	0	1	1	0	1	R* dȳbyc
5	0	0	0	0	1	0	0	1	R* d

Figure 5.29 Elementary *R-S* Flip-Flop Logic Diagram and Truth Table.

5. The flip-flop starts in the 0 state, and during time interval 2 the switching of S^* from 0 to 1 causes the variables ($S^*aycȳb$) to change in a sequence, which eventually terminates with the flip-flop in the 1-state.

6. When S^* returns to 0, only the variable a changes, and the flip-flop remains in the 1-state.

7. Similarly, the switching of R^* from 0 to 1 causes the variables ($R^*dȳbyc$) to change in sequence, ending with the flip-flop in the 0 state.

8. This is called an *elementary flip-flop*† because it changes state when the input changes, as opposed to the circuits of Section 5.3 and 5.4 which changed on the trailing edge. The symbol * is used on the *S* and *R* inputs in Figure 5.29 to denote that an elementary flip-flop is considered.

If the inverters in Figure 5.29 are replaced with NAND gates, it is easy to convert the device into the clocked elementary *R-S* flip-flops shown in Figure 5.30. Because of the inverter canceling, we see that an $S^* = 1$ input immediately sets the device, $y^* = 1$, as long as the clock input, C^*, is high. The one major disadvantage of this configuration, however, is that the flip-flop outputs do not change exclusively on the leading edge of the clock pulse or on the trailing edge. On the contrary, there is no real limit on the number of times it may change state, as long as the clock is high and the S^* and R^* inputs keep changing. This switching action is, of course, due to the latch section embedded in the circuit, and it may result in the pickup of erroneous ones or zeros. It is not, however, a true case of *ones catching* because that effect requires

---

†The particular form of circuit shown in Figure 5.29 is known as a *latch*, but our term *elementary flip-flop* is somewhat more general.

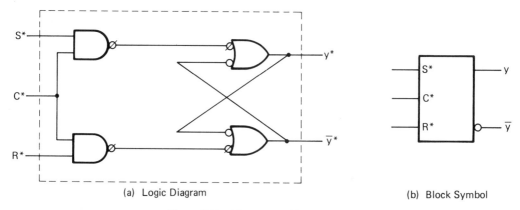

Figure 5.30 Elementary Clocked R-S Flip-Flop.

that the one (or zero) picked up reach the output only after a significant delay, usually after the the clock pulse is gone.

One improvement that can be made to the elementary flip-flop is to provide *asynchronous*† preset, P_r^*, and clear, C_r^*, terminals. In this way, complicated external gating is not required when it is desired to directly load particular bits with ones or zeros. This action can be accomplished by the addition of another output to each side of the flip-flop as shown in Figure 5.31. Because of the inversion at the input of each OR, note that the nonactuating state of these asynchronous inputs is $P_r^* = C_r^* = 1$. The device is immediately cleared when $P_r^* = 1$ and $C_r^* = 0$, and it is immediately preset when $P_r^* = 0$ and $C_r^* = 1$. The symbol for the complete elementary flip-flop is shown in Figure 5.31(b). The inversion symbols have been pulled out of the logic

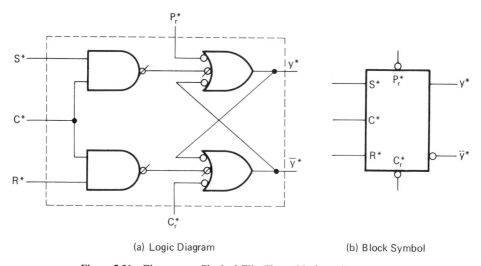

Figure 5.31 Elementary Clocked Flip-Flop with Asynchronous Inputs.

†These are called asynchronous because the flip-flop is made to change state independently of the clock pulse.

Sec. 5.8 Flip-Flop Realization Methods

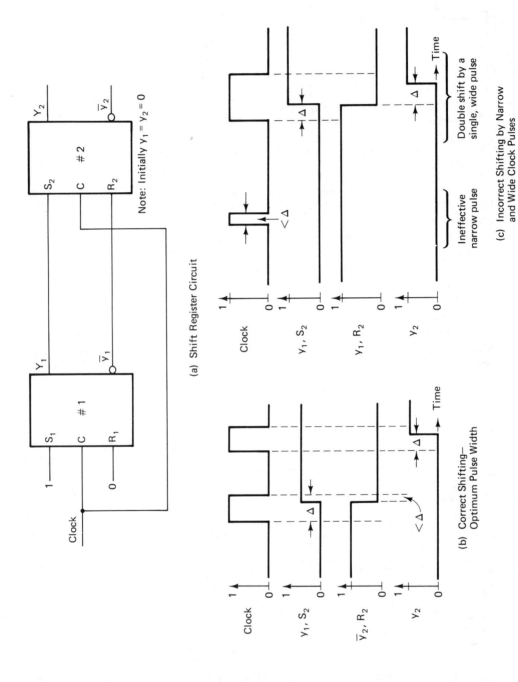

Figure 5.32 Clock Pulse Width Sensitivity of an Elementary *R-S* Flip-Flop.

diagram and shown explicitly in order to emphasize that asynchronous action is initiated by a low (i.e., logical zero) input. (Since they are independent of the clocked features which are now emphasized, simplification is achieved by omitting asynchronous inputs from the flip-flops in most of the remainder of this chapter.)

Pulse Width Sensitivity

Can a reliable shift register be constructed from the elementary R-S flip-flops of Figure 5.30? Consider the situation shown in Figure 5.32. Here the correct operation of the shift register depends on whether the clock pulse width is "tuned" to propagation delay, Δ, of the R-S flip-flop. First, note that for both the correct and incorrect timing diagrams of Figure 5.32, the flip-flops change state during the clock pulse and not slightly after the negative-going edge, which means that there is a considerable difference between these and the ideal flip-flops of Sections 5.1 through 5.6. Since the clock pulse in Figure 5.32(b) is only slightly wider than Δ, correct shifting is produced; and it requires exactly two pulses to shift a logic one from the input of flip-flop 1 to the output of flip-flop 2. On the other hand, Figure 5.32(c) shows that a very narrow pulse produces no shifting at all and a wide pulse produces an erroneous double shift from a single clock pulse. Not only is there a potential reliability problem when trying to construct shift registers from elementary flip-flops, but counters and most other units can also result in improper performance. Even if one were successful in tuning the clock to a specific flip-flop, manufacturing tolerances for the same flip-flop type and design differences for the various types within the same logic family could lead to very unsatisfactory results.

Master-Slave Flip-Flops

One solution to the pulse tuning problem found in the circuit of Figure 5.32 is to employ flip-flops whose inputs respond to signals only while the clock pulse is present, but whose outputs change *only* a short time after the clock pulse has passed. These may be constructed by means of a cascade connection of two elementary clocked R-S's in the so-called *master-slave* arrangement shown in Figure 5.33.

In this circuit, the presence of the clock pulse allows the master to go toward the state dictated by its inputs. The occurrence of the clock pulse also immediately disconnects the slave from the master (through the inverter at the slave's C^*-input)

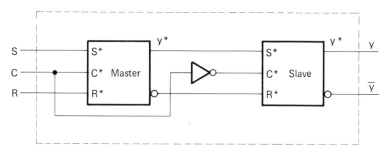

Figure 5.33 Master-Slave R-S Flip-Flop.

so that the flip-flop's output holds the previous state until the clock pulse goes low. At that time, the slave's inputs are opened, the master's are closed, and the master's state is transferred to the slave.

By examining Figure 5.32 [part (c) in particular] in light of the master-slave characteristics one should be able to visualize that the new configuration completely solves the tuning problem for wide pulses. This flip-flop, however, is far from ideal because it displays a classic case of ones catching. If during any clock pulse, the desired circuit inputs of Figure 5.27 are $S = R = 0$, the next state of the flip-flop should be unchanged from the present state. Note that with the original output of $y = 0$, the correct behavior can be completely upset by a brief noise spike on the S input, which occurs anytime during the clock pulse. The noise will set the master, and after the negative-going edge of the clock, the slave output will erroneously be in the $y = 1$ state. This objectionable ones-catching characteristic is nicely eliminated by the next flip-flop to be discussed.

Edge-Triggered Flip-Flops

The master-slave flip-flop corrects the clock tuning problem, and with no hazards or other noise conditions, it can be employed to construct many useful circuits. On the other hand, the *edge-triggered* flip-flop can in addition operate under noisy conditions and is the preferred device for most applications. It solves both the clock tuning and the ones-catching problems. In essence, this method employs the (positive- or negative-going) edge of the clock pulse to transmit the original input state to an elementary flip-flop and to simultaneously lock out any input changes. A simplified* positive-edge-triggered (positive-going) flip-flop is shown in Figure 5.34. Gates 3 and 6 form an elementary flip-flop. As long as the clock is low, gates 2 and 5 have low outputs. With the inputs at $S = 1$, $R = 0$, when the clock goes high, the outputs of gates 1 and 2 will go high, causing the flip-flop to be set. Simultaneously, the output of gate 2 will feed a high level back around the loop to gate 1, locking both of them to a high

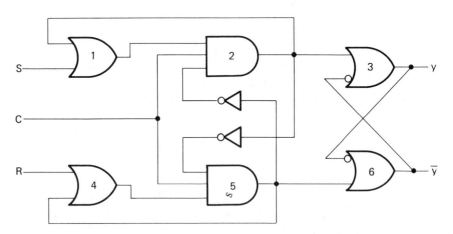

Figure 5.34 Simplified Positive-Edge-Triggered R-S Flip-Flop.

*It is termed a simplified flip-flop because, for tutorial reasons, its lock-out capability omits one case, which will be considered in the exercises.

state independent of the *S* input. Also, the inverted output of gate 2 locks gate 5 into a low state. Thus, further inputs are ineffective and the entire circuit is static until the clock returns to a low state. Upon transition to clock low, the elementary flip-flop continues to hold its state, but the remainder of the circuit unlocks—gates 2 and 5 return to low outputs, and the outputs of gates 1 and 4 are determined only by their external inputs. The flip-flop is now ready to change state as soon as the leading edge of another clock pulse occurs.

Since the WHAT, data, value is sampled at the positive-going section of the clock pulse and the output is locked to this value a few nanoseconds later (after the propagation delay), one can see that neither the clock tuning nor the ones-catching problems exist here. In addition, it is feasible to combine the edge-triggering feature into our previous circuit. Commercially these are known as *master-slave flips-flops with data lockout*. They sample the data on the positive-going edge of the clock, but the new state is delayed from reaching the output terminals until after the negative-going edge. (The construction of a logic diagram for this device will be considered in the exercises.) The sequential circuit designs of all earlier sections can be reliably implemented with these devices and a typical package is the 74110, which is a *J-K* type.

Example: Conversion of an R-S to a J-K Flip-Flop

It is easy to convert the gating within the above *R-S* flip-flop to the more popular *J-K* type. (This will be the subject of some of the exercises at the end of the chapter.) On the other hand, it is not difficult to use external gating to achieve also the conversion of one type of flip-flop to another. Figure 5.35 demonstrates how an *R-S* may be converted to a *J-K* flip-flop. Note that the cross-coupling between the outputs and the input gates allows a high input to pass only if the flip-flop is in the opposite state to the one initiated by the input.

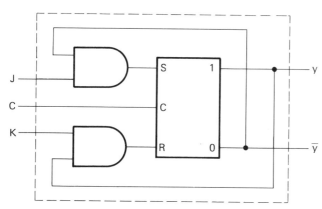

Figure 5.35 Conversion of an *R-S* to a *J-K* Flip-Flop.

Example: Construction of a Master-Slave J-K Flip-Flop from Elementary Flip-Flops

How may elementary flip-flops be used to construct a master-slave *J-K* rather than an *R-S*? Recall that the major difference in performance between the two units is that the *J-K* flip-flop allows $J = K = 1$. This is easily arranged by feeding cross-

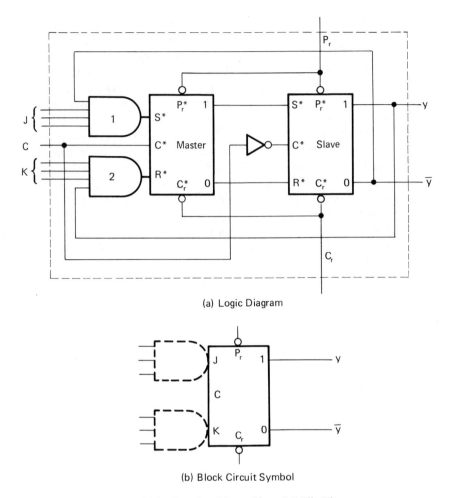

Figure 5.36 Complete Master-Slave J-K Flip-Flop.

coupled signals from the output to each input AND gate as shown in Figure 5.36. The concept here is that an input is allowed to pass on to the master flip-flop only if the opposite side of the slave is currently in the 1 state. Of course, such a constraint is consistent with $J = 1, K = 0$ for setting the device; $J = 0, K = 1$ for resetting; and $J = 1, K = 1$ for complementing. As in the master-slave R-S unit in Figure 5.33, the clock pulse is fed to the slave through an inverter so that the slave does not receive the input from the master until the negative edge of the clock pulse has passed. It should be observed that the cross-coupled feedback required to create the J-K characteristics comes from the slave output, since that output is the one which determines the true state of the flip-flop. Note that extra terminals are provided at both the J and the K inputs. These permit an AND operation at the J and K inputs without an external gate. (Any unused inputs should be tied to used ones so that the input gates will function properly.) Figure 5.36(b) shows the block diagram symbol employed for this unit, including the implicit AND gate terminals. (The AND gates are shown dashed because they are physically inside the flip-flop itself.)

Setup and Hold Time

Many flip-flops require both an interval before and an interval after one of the clock edges during which the data must be stable to ensure reliable operation. The setup time, t_{su}, is the minimum stable data interval before the clock edge, and the hold time, t_h, is the minimum stable interval after the clock edge. The designer must pay careful attention to these parameters to avoid timing problems in sequential circuits. In relatively fast flip-flop families t_{su} and t_h are typically 5 to 10 ns and both parameters are listed on manufacturers' data sheets. Often in pulse-triggered flip-flops $t_{su} = 0$ measured from the positive-going clock pulse edge and $t_h = 0$ measured from the negative-going edge. This means that the data must be stable during the entire clock pulse, which is often more than 50 ns.

‡5.9 THE RIPPLE METHOD

Previous designs in this chapter employed synchronous circuits—all flip-flop action was individually synchronized by means of a pulse, usually a clock pulse. We now introduce an asynchronous method. Here an attempt will be made to eliminate all gates by connecting the T input of one flip-flop directly to the output of the previous flip-flop in a chain-like arrangement so that each flip-flop will cause the next one in line to operate.

Example: Synthesis of a Down-Counter

A three-bit down-counter is to be designed, with T flip-flops, which will decrement by one each time an input pulse occurs. After we construct the state diagram with a decreasing binary assignment, a state table (Table 5.18) results.

The excitation maps corresponding to Table 5.18 are shown in Figure 5.37. The information needed to design a ripple counter is contained in these excitation

TABLE 5.18 STATE TABLE FOR A THREE-BIT RIPPLE COUNTER

Present State ($y_3y_2y_1$)	x Input, Next State ($y_3y_2y_1$)
111	110
110	101
101	100
100	011
011	010
010	001
001	000
000	111

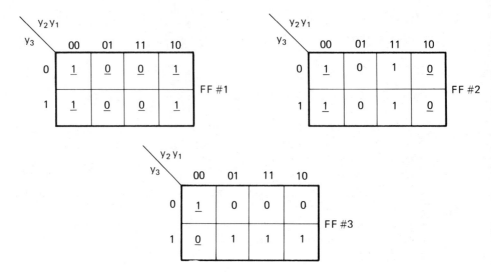

Figure 5.37 Excitation Maps for the Ripple Counter.

maps; however, we must take a somewhat different approach in order to interpret the maps. With our knowledge of the characteristics for the T flip-flop and particularly the fact that it changes state on the negative edge of the input pulse, Table 5.19 may be prepared. Consider the lower half of the table; here $T_j = y_i$. This means that when FFi undergoes the transition $1 \rightarrow 0 = \underline{0}$, FF$j$ will change state; but this is exactly the desired action when the j map contains $\underline{0}$ or $\underline{1}$. Moreover, when FFi undergoes any other transition (i.e., $\underline{1}$, 1, or 0), FFj will remain unchanged; but this is exactly the desired action when the j map contains 0 or 1. Thus, we have verified the lower half of the table. The upper half can be verified in the same way, but the inversion caused by $T_j = \bar{y}_i$ must be taken into account.

The equations for T_2 and T_3 are easily obtained from Table 5.19 and Figure 5.37

TABLE 5.19 MAP REQUIREMENTS FOR THE T FLIP-FLOP-RIPPLE METHOD

Corresponding Map Entries for Flip-Flop Numbers		Required Circuit Connection Between FF i and input T_j
i	j	
$\underline{1}$ / Anything but $\underline{1}$	$\underline{0}$, $\underline{1}$, or $-$ / 0, 1, or $-$	$T_j = \bar{y}_i$
$\underline{0}$ / Anything but $\underline{0}$	$\underline{0}$, $\underline{1}$, or $-$ / 0, 1, or $-$	$T_j = y_i$

when we compare the corresponding squares of the i and $j = i + 1$ maps. We find that, for T_2 and T_3, the upper half of Table 5.19 is satisfied, resulting in the equations $T_2 = \bar{y}_1$ and $T_3 = \bar{y}_2$. Note that we can obtain the equation for T_1 from the conventional map covering demanded by the T flip-flop or by observing that the FF 1 excitation map contains nothing but $\underline{1}$ and $\underline{0}$, which are satisfied by the trailing edge of each x pulse. By either method, the equation is $T_1 = x$. Therefore, the resulting circuit for the above equations is as shown in Figure 5.38.

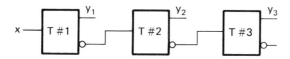

Figure 5.38 Ripple Down-Counter.

Observe how much simpler the three-bit ripple counter is than the ones designed by our previous method. The ripple counter is said to be *asynchronous* in that the individual flip-flops do not operate in synchronization with the x input, but rather (especially when an up-counter changes from 111 to 000) each flip-flop triggers the next in the chain like a row of dominoes falling. The corresponding term for the previously discussed counters is *synchronous* since each flip-flop operates in synchronization with the input signal.

To further explain the ripple method, consider the flip-flop excitation maps shown in Figure 5.39. The requirements for flip-flops 1 and 2 can be satisfied in a straightforward way by the equations $T_1 = x$ and $T_2 = x$; but it is not so easy to satisfy the requirements for flip-flop 3. If an attempt is made to connect the output of FF2 into the T input of FF3, contradictions occur. The second column of the FF3 excitation map requires $T_3 = \bar{y}_2$, but the first column will not allow it. The third column requires $T_3 = y_2$, but the the fourth column will not allow it. On the other hand, if the FF1 output is connected into FF3, the resulting equation is $T_3 = y_1$. This equa-

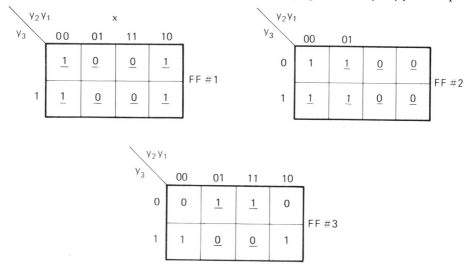

Figure 5.39 Example Demonstrating the Ripple Method.

Sec. 5.9 The Ripple Method

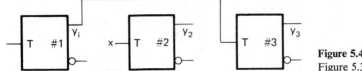

Figure 5.40 Ripple Circuit for Figure 5.39.

tion is consistent, and the resulting ripple circuit is shown in Figure 5.40. Thus, the ripple method is actually more general than it at first appeared, since not only may successive flip-flops be connected together, but also the input of any flip-flop may be connected to the output of any other.

It should be mentioned that the ripple conditions (Table 5.19) are rather specialized, and only if they are satisfied for an entire map can a connection without gates be used for that flip-flop. On the other hand, the fact that important circuits found in computers frequently allow the ripple method makes it an important design technique.

An example of a set of flip-flop excitation maps, which can be only partially satisfied by the ripple method, is shown in Figure 5.41. There, $T_3 = \bar{y}_1$ or $T_3 = \bar{y}_2$, but gates are required to satisfy the needs of the other two flip-flops. These equations are $T_1 = \bar{y}_2 + xy_1$, or $T_1 = x(y_1 + \bar{y}_2)$ and $T_2 = x\bar{y}_1$. (The first term in the first equation for T_1 is produced by the ripple method, but in this case neither gates nor gate inputs are saved.)

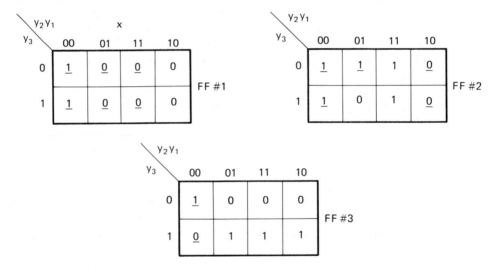

Figure 5.41 Example Allowing Only a Partial Solution by the Ripple Method.

5.10 DEPENDENCY NOTATION

In an effort to produce a set of uniform symbols that concisely convey detailed functional information particularly about flip-flops and higher-level integrated circuits, the International Electrotechnical Commission recently developed the Dependency Notation. A complete discussion of the new notation is beyond the scope of this text, but it is appropriate hereto consider certain aspects of it. (See pp. 373–98 of Reference 11 for a detailed presentation.) In order to provide information concerning the way inputs and outputs affect each other, Dependency Notation employs an *arbitrary number* following the *affecting* symbol and the same number preceding the *affected* symbol. Figure 5.42(a) shows a *J-K* flip-flop with the clock input being the affecting parameter and the *J* and *K* inputs being the affected parameter. One valuable aspect of this feature is that it makes it easy to see directly from the symbol which inputs are clocked and which are asynchronous.

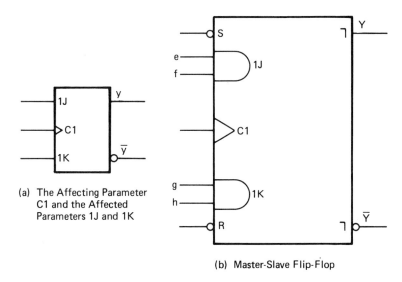

(a) The Affecting Parameter C1 and the Affected Parameters 1J and 1K

(b) Master-Slave Flip-Flop

Figure 5.42 Examples of the Dependency Notation.

Another aspect about the characteristic of the clock input that is not immediately clear on most logic diagrams is whether edge sensitivity or pulse sensitivity is present. The small triangular symbol preceding the "*C*" in Figure 5.42(a) indicates that the positive-going edge initiates any state change (dictated by the *J* and *K* inputs), which is then immediately transmitted to the flip-flop outputs. (If the conventional inversion symbol bubble* precedes the triangle, the normal complementing effect occurs: i.e., the negative-going edge of the clock pulse causes the change.) Of course, the absence of any symbol preceding the "*C*" indicates pulse sensitivity, i.e., the flip-flop responds during the entire pulse.

One final feature which is often not evident in the older notation is the time at which the output changes with respect to the clock input. Here the symbol "⌐" indicates that the output changes on the negative-going edge of the clock. Omission of the symbol indicates that the output is immediately affected by any data clocked into the flip-flop.

Having learned the above notation, you should now be able to correctly interpret the symbol in Figure 5.42(b). You should immediately recognize this as a positive-edge-triggered *J-K* flip-flop whose output occurs on the negative-going edge and that

*With pure Dependency Notation, a wedge is employed to represent inversion instead of the bubble.

it contains an *R-S* asynchronous feature which is affected by a logic zero value. (An alternative way to express this is that a state change is affected by a logic zero.) A master-slave flip-flop has this symbol. The conventional AND gates embedded in the flip-flop symbol should be understood to have their usual meaning. Here "*e*" and "*f*" must be true for the *J* input to be true—the *K* input operates similarly.

The register family is another one to which dependency notation makes a good contribution. The properties of shift registers, counters, etc., often are unclear in the older notation. Of course, all the symbols established with flip-flops apply, but some additions are also needed. Figure 5.43(a) is a three-bit register represented by the older notation and part (b) of the figure shows the improved dependency notation. Registers usually have common reset, load, and other controls which can most conveniently be represented by the common control block shown in Figure 5.43(b). Observe the distinctive indented neck that marks the separation between the common control section and the devices (flip-flops in this case) being controlled (affected).

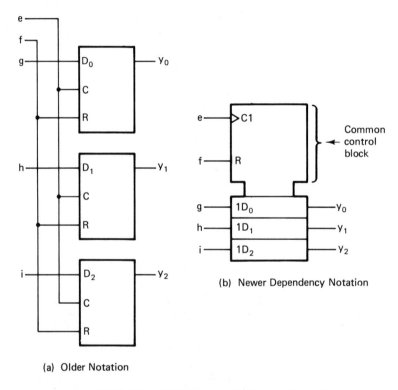

Figure 5.43 Three-Bit Register with Common Controls.

Some additional special notation is employed in representing the details of other MSI/LSI circuits, such as counters, memories, and shift registers; these symbols are presented in Table 5.20. Note that double lines separate the table into three sections: the first shows how a common register can be made into a counter or into a shift register; the second shows how a group of inputs can be employed to select one of several operands; and finally the last shows some miscellaneous dependency symbols.

Observe that the AND operation is achieved in either of three ways: first a comma is shown ANDing two operands that affect a *D* flip-flop input; second, the G symbol forms the AND operation with the parameter it affects; finally, a new logic symbol, which is rectangular instead of shape-sensitive, is introduced to produce an internal AND state. A similar OR gate example employs the symbol "≥ 1," which represents the fact that at least one input must be true for the function to be true. The last two cases show some variations with earlier symbols and also introduce a way to separate independent operations. The memory dependency, *A*, mentioned in the explanation for the Binary Grouping will be discussed in Chapter 6. The reader should take careful note of the location of Table 5.20 since it will be a handy reference for understanding dependency notation figures in later chapters.

TABLE 5.20 DEPENDENCY NOTATION SUMMARY I*

Function	Example	Explanation
Counter	(symbol with +)	Increment (+1) on logical one going edge
	(symbol with −)	Decrement (−1) on logical one going edge
Shift Register	(symbol with →)	Shift right (down) on logical one going edge
	(symbol with ←)	Shift left (up) on logical one going edge
Binary Grouping (k is the MSB)	inputs 0, 1, 2, ..., k with $S \frac{0}{2^{k+1}-1}$	Bits 0 thru k determine a binary number, up to $2^{k+1}-1$, pointing to the operand. S is a symbol representing the operation. If S is: $A \equiv$ Memory Address, $G \equiv$ AND (see below), $M \equiv$ Mode Selection (see Figure 5.44)
Weighted Grouping	inputs 1, 2, 4, ..., 2^m	Only the operand whose number corresponds to the sum of the weights will be actively asserted. Typically the operand is one of $2^{m+1}-1$ output lines.

*Summary II is in Table 6.13.

TABLE 5.20 (CONTINUED)

Operation	Symbol	Description
AND Operation (Symbol ≡ ,)	a —[2, 3D]	Both operands 2 AND 3 must be asserted before the D flip-flop will accept the a input.
AND Dependency (Symbol ≡ G)	a —[G1] b —[G2] d —[1D$_0$] e —[2D$_1$]	The equivalent notation is: a, d → D$_0$; b, e → D$_1$
AND Internal State (Symbol ≡ & ≡ ampersand)	a —o[&]— e b —[ ]	An internal state e is true when: $e = \bar{a}\,b$
OR Internal State (Symbol ≡ ≥ 1)	a —[≥1]— e b —[ ]	An internal state e is true when: $e = a + b$
Compound and Inhibited Internal States	a —[C2] b —[&]— h d —[&] e —[ ]— k g —[$\overline{2}$G3]	Internal states h and k are true when: h = bde, k = de G3 may be asserted if C2 is false (i.e., a = 0)
Separation of Independent Operations (Symbol ≡ /)	a —▷ C1/→ b —[1D$_0$]	Loading the flip-flop, bit 0, and the shift right operation, on all the other bits, are independent of each other.

To see how useful this notation can be, consider the representation, in Figure 5.44, of a four-bit up-down counter, which may be loaded and reset. Figure 5.44(a) contains the complete circuit with the Weighted Grouping Function operating on the Control Function (C1/2, 4+/3, 4−) to select parallel loading, count up, or count

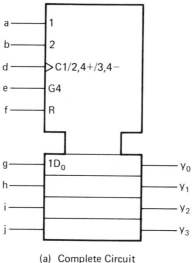

Figure 5.44 Four-Bit Up-Down Counter with Parallel Loading and Reset.

down (weights 1, 2, and 3, respectively). To be specific, 2, 4+ indicates that a weight of 2 (inputs $a = 0$, $b = 1$) and G4 asserted ($e = 1$) selects the increment operation. In addition to these, the table in part (b) of the figure shows how the R and G symbols (inputs e and f) provide for additional reset and no action commands. Finally, part (c) of the figure gives the Mode Dependency, M, as an alternative to the Weighted Grouping method for selecting the operation.

5.11 IMPLEMENTATION OF DIGITAL LOGIC

Modern combinational and sequential circuits, particularly for digital computers, are implemented entirely from integrated circuits. These circuits may be classified in several ways. One method, which is discussed in Chapter 12, is according to the type of electronic circuitry employed: e.g., transistor, transistor logic, TTL. A completely different approach is to place them in one of four groups according to the number of equivalent gates in one package or on a single semiconductor chip. Here, a circuit is termed *small-scale integration*, SSI, if it contains the equivalent of nine gates or less in one package. A circuit is called *medium-scale integration*, MSI, if it contains the equivalent of between 10 and 100 gates; and it is called *large-scale integration*, LSI, if it contains the equivalent of over 100 gates. Finally, if the circuit contains the equivalent of thousands of gates in a single package, it is called *very-large-scale integration*, VLSI. The latter category permits very powerful computers and their memories to be implemented on a single chip. (For details see Bibliography entries 1 and 10.)

SSI, MSI, LSI, and VLSI

A typical SSI configuration is the 74H11 TTL triple three-input AND, which, including the output terminals, requires $3 \times 4 = 12$ pins for the gates plus the ground and power supply, V_{cc}, yielding a total of 14 pins. Thus, it conveniently fits into a standard 14-pin *dual in-line package*, dip, illustrated by the logic diagram in Figure 5.45. (Incidentally, the H in the part number indicates that it is a high-speed circuit—about a factor or 2 faster than standard TTL.)

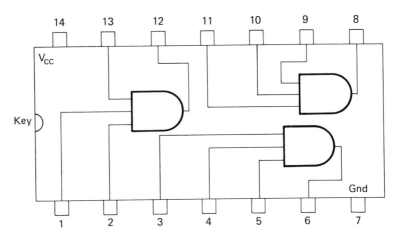

Figure 5.45 Type 74H11, SSI Triple Three-Input AND.

A typical MSI circuit is the 8202 TTL 10-bit buffer register, which is available in the 24-pin configuration shown in Figure 5.46. Figure 5.47 shows a photomicrograph of the actual circuit. It employs *D*-type flip-flops having a direct reset. (When reset $= 0$, all flip-flops are cleared; when reset $= 1$, all flip-flops are enabled.) Since all 10 *D*-inputs as well as all true outputs are available, many different serial and parallel register operations can be performed.

A typical LSI circuit is the 2533 MOS 1024-bit static shift register, which is available in the 8-pin configuration shown in Figure 5.48. This circuit has a large number of possible applications such as a CRT refresh circuit, a delay-line replacement, and a low-cost sequential-access memory. When the clock ϕ_{in}, is at logical 1, the stream-select input allows one of two signals to enter the register; later, when the clock goes low, the register is shifted. As shown by the broken lines in the figure, frequently one of the input channels is employed to provide a recirculate path so that

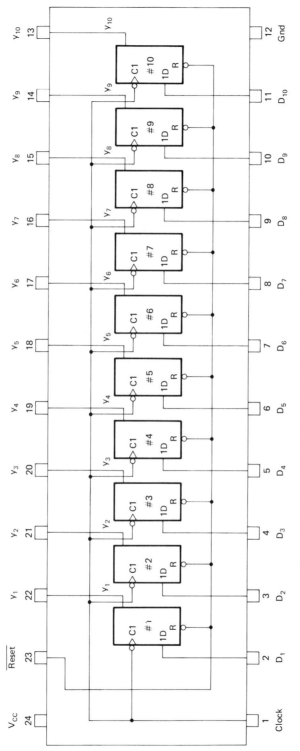

Figure 5.46 Type 8202, MSI 10-Bit Buffer Register. (See Exercise 5.43 for a concise dependency version of this diagram.)

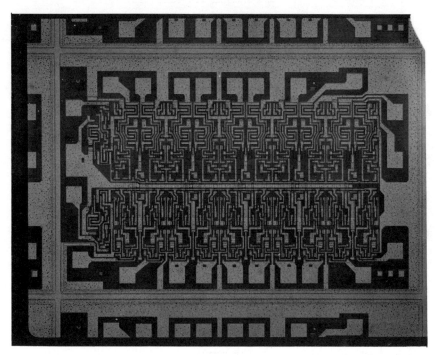

Figure 5.47 Photomicrograph of Integrated Circuit of Figure 5.45. (Courtesy of Signetics Corporation, Sunnyvale, CA.)

data reaching the output of the shift register are reentered. On the other hand, new data can be inserted, where desired, through the other channel by changing the SS input. Comparing Figures 5.46 and 5.48, an interesting point is observed—the number of pins on an IC package is not necessarily an indication of circuit complexity. Here the eight-pin package contains a circuit which is many times more complex than that in the 24-pin package. But the fact that the input and output leads for each bit are available means that a much larger variety of circuits may be constructed with the 24-pin package.

An interesting VLSI chip is the 32-bit microprocessor NS16032, shown in Figure 5.49, which is produced by National Semiconductor Corporation. This processor contains approximately 60,000 transistors on a single chip area of 290 mil^2. The CPU provides a 16-million byte address space and is designed to efficiently compile programs for higher level languages. In the photomicrograph of Figure 5.49, the register file and ALU occupy the left third of the chip with the ALU on the bottom. The queue, loader, preprocessor, and micromachine are in the middle third of the chip. Finally the microcode ROM, microinstruction decoder, and bus interface unit occupy the right third of the chip.

Figure 5.50 shows integrated circuits ranging from 16- to 64-pin versions. Although this certainly is only a subset of those in current use, it does give one a quick impression of the wide variety of circuit packages. Many more details concerning

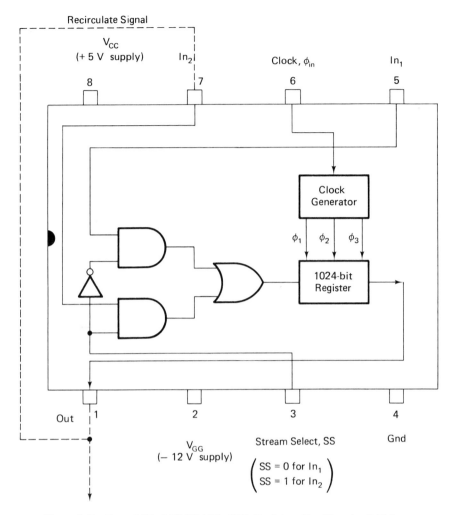

Figure 5.48 Type 2533, LSI 1024-Bit Shift Register. (See Exercise 5.44 for a dependency version of this diagram.)

integrated circuits and their various families will be found in the references for this chapter and in the text of Chapter 12.

Design Criteria

How can one choose from several competing circuit designs for the same application? The obvious answer is to select the optimum design, but this depends on your definition of optimum. The optimum could be based on the fastest system, the one with lowest cost, the most reliable, the one most easily maintained, a weighted combination of several criteria, or many other possibilities. Before integrated circuits became the

Figure 5.49 Photomicrograph of the NS 16032 VLSI 32-Bit Microprocessor. (Courtesy of National Semiconductor Corporation.)

dominant medium for constructing digital computers, the number of gates and flip-flops required by a system was a reasonable universal measure of cost. With the advent of complex logic blocks it is not nearly so easy to formulate a concise universal cost criteria. If the system requirements match the characteristics of an available module or small group of modules, a very complex requirement may be satisfied for only a few dollars. LSI and VLSI circuits that provide data communication, controllers for magnetic disk, and a growing number of other specialized modules are typical. On the other hand, more specialized and lower-volume applications may often be satisfied by sets of modules fabricated with MSI or LSI technology. Chapter 7 goes into considerable detail on this design technique. In addition, a microprocessor system may also be a good solution in many applications—Chapter 8 is concerned with this design approach. Here cost is closely related to module, package, count rather than gate and flip-flop count. On the other hand, special requirements sometimes dictate that circuits be constructed from fairly low-level primitives. In this case, the gate and flip-flop count may again be a valid criteria. For a detailed discussion of cost minimization through the use of higher-level primitives, see the text by Blakeslee.[2]

Figure 5.50 Integrated Circuit Packages Ranging from 16 to 64 Pins. (Courtesy of TRW LSI Products.)

5.12 BIBLIOGRAPHY

1. Barbe, D. F. (ed.), *Very Large Scale Integration (VLSI) Fundamentals and Applications.* Berlin, Germany: Springer-Verlag, 1980.

2. Blakeslee, T., *Digital Design with Standard MSI and LSI* (2nd ed.). New York: John Wiley & Sons, Inc., 1979.

3. Hill, F. J., and G. R. Peterson, *Introduction to Switching Theory and Logical Design* (3rd ed.). New York: John Wiley & Sons, 1981.

4. Huffman, D. A., "The Synthesis of Sequential Circuits." *Journal of the Franklin Institute,* Vol. 257, Nos. 3 and 4, 1954, pp. 161–90 and 275–303.

5. Kohavi, Z., *Switching and Finite Automata Theory* (2nd ed.). New York: McGraw-Hill Book Company, 1978.

6. Lee, S. C., *Digital Circuits and Logic Design*. Englewood Cliffs, N.J.: Prentice-Hall, Inc., 1976.
7. Mano, M. M., *Digital Logic and Computer Design*. Englewood Cliffs, N.J.: Prentice-Hall, Inc., 1979.
8. Marcus, M., *Switching Circuits for Engineers* (3rd ed.). Englewood Cliffs, N.J.: Prentice-Hall, Inc., 1975.
9. Mealy, G. H., "A Method for Synthesizing Sequential Circuits." *Bell System Technical Journal.*, Vol. 34, 1955, pp. 1045–79.
10. Mead, C., and L. Conway, *Introduction to VLSI Systems*. Reading, Mass.: Addison-Wesley Publishing Company, Inc., 1980.
11. Peatman, J. B., *Digital Hardware Design*. New York: McGraw-Hill Book Company, 1980.
12. Rhyne, V. T., *Fundamentals of Digital Systems Design*. Englewood Cliffs, N.J.: Prentice-Hall, Inc., 1973.

5.13 EXERCISES

5.1 Before the first clock pulse, the sequential circuit shown in Figure E5.1 has $y_1 = 1$, $y_2 = 0$. By constructing a timing diagram, determine the circuit performance for six clock pulses.

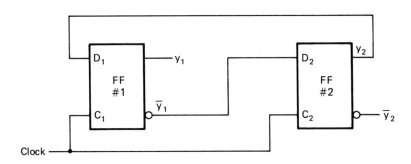

Figure E5.1

Before Clock Pulse Number	1	2	3	4	5	6
Input A	1	1	0	1	1	1

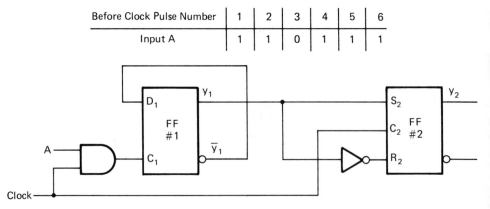

Figure E5.2

5.2 Repeat Exercise 5.1 using Figure E5.2 with $y_1 = y_2 = 0$ before the first clock pulse and with the following values for the input A. (Input A changes approximately halfway between clock pulses.)

5.3 (a) Construct the state diagram and state table, using the binary code assignment, for a modulo-7 counter. Allow the counter to be cleared from any state.
(b) Repeat part (a) for the same counter, but with the maximum count resulting in a terminal state.

5.4 Construct a state diagram and state table, using the binary code assignment, for a modulo-4 up-down counter. This device becomes an up-counter whenever the level $w = 0$ and it continues in that mode until $w = 1$ at which time it becomes a down-counter, etc. Each clock pulse increments or decrements the counter's total, depending on whether it is in the up or down mode.

5.5 Eliminate any redundant states in the following table.

Present State	Next State x_1	Next State x_2
0	3	2
1	1	5
2	0	1, Z
3	4	3
4	3	2
5	0	3

5.6 Repeat Exercise 5.5 for the following table.

Present State	Next State x_1	Next State x_2
0	3	1
1	0	2
2	1	3
3	4	2
4	2, Z	1

5.7 Synthesize a three-bit up-counter using clocked R-S flip-flops.

5.8 Using the synthesis method, clocked T flip-flops, and the binary code assignment, determine the logic circuit for the flow table shown.

Present State	Next State x_1	Next State x_2
0	1	0
1	2	0, Z
2	2	2

5.9 Using the synthesis method and clocked J-K flip-flops, determine the logic circuit for a modulo-6 counter.

5.10 Using the synthesis method and clocked J-K flip-flops, determine the logic circuit for a modulo-7 counter.

5.11 Determine which assignment leads to the simplest sequential circuit for the flow table shown. The three unique assignments appear in the table, and it can be shown that, for two secondary variables, all others are equivalent to these three. Use clocked R-S flip-flops in this exercise.

		Next State	
Present State		x_1	x_2
0		0	1
1		0	2, Z
2		0	2

		Assignments	
State	1	2	3
0	00	00	00
1	01	01	11
2	11	10	01
3	10	11	10

5.12 Carefully examine the flip-flop excitation map shown in Figure E5.12. Then list any map entries (by their $y_3 y_2 y_1$ values) which are not consistent with the requirements of an actual sequential circuit. Explain your reasoning.

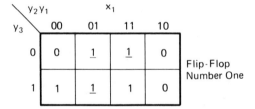

Figure E5.12

5.13 Determine the logic equations for clocked J-K flip-flops, and draw the circuit for a Gray-code counter using the synthesis method. Compare these results with the heuristic design results in the text.

5.14 Synthesize a practical modulo-5 counter. Take into account the fact that flip-flops have a 0.5 probability of coming on in the 1 state. The clocked T variety of flip-flops should be considered.

5.15 For the state diagram shown in Figure E5.15, synthesize the sequential circuit with assignment 1 (see Exercise 5.11). Assume that clocked J-K devices are available.

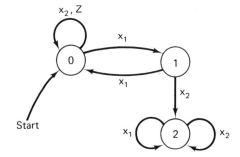

Figure E5.15

5.16 One of the four excitation maps for a special counter having a single input, x, is shown in Figure E5.16. This map is to be satisfied by means of clocked J-K flip-flops.
(a) The map shown is for which flip-flop? (Give its number.)
(b) Using the fewest possible gate inputs, draw the circuit required by this flip-flop.

$y_2 y_1$ \ $y_4 y_3$	00	01	11	10
00	1̲	1̲	0	1̲
01	0	1̲	0	0
11	1	0̲	0̲	1
10	0̲	1	1	0̲

Figure E5.16

5.17 Synthesize a three-bit register (with clocked R-S flip-flops) which can be complemented by a pulse on line x_1 and which can be decremented by 1 by a pulse on line x_2.

5.18 (a) Using the heuristic method and the fewest possible T flip-flops, design a circuit whose outputs follow the following table. (Assume that the circuit always comes on in the start condition.)

Pulse Number	Level Outputs		
	w_3	w_2	w_1
Start	1	0	0
1	0	1	0
2	0	0	1
3	0	1	0
4	1	0	0
5	0	1	0
6	0	0	1
etc.			

(b) Repeat part (a) using the formal synthesis method and the specification that $w_i = y_i$, where y_i is the direct output of the flip-flop i.
(c) Discuss and compare the designs.

5.19 Applying the heuristic method and T flip-flops, draw the circuit for a four-bit ring counter.

5.20 Applying the heuristic method and clocked R-S flip-flops, draw the circuit for a four-bit register which may be complemented by means of a pulse on input line x.

5.21 Using the heuristic method, design a circuit whose flip-flop outputs reproduce the table shown.

Pulse Number	x Input, Next State $(y_3 y_2 y_1)$
Start	100
1	110
2	111
3	011
4	001
5	000
6	100
etc.	

5.22 A computer having a clock frequency of 5.0 MHz and a 20-ns pulse width requires the waveform shown in Figure E5.22 to implement a particular control function. From heuristic considerations, draw a logic circuit which will satisfy these requirements. (Assume that the voltage levels of all parts of the circuit will be compatible, and that the waveform is periodic.)

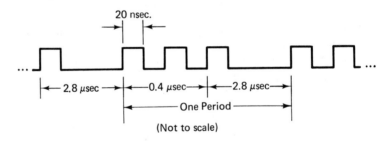

Figure E5.22

5.23 Does the circuit in Figure E5.23 contain a hazard problem? If so, how is the problem eliminated?

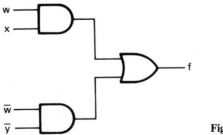

Figure E5.23

5.24 Repeat Exercise 5.23 for the following function:
$$f = \bar{w}xz + xz + wxy + w\bar{y}z + \bar{w}x\bar{y}$$

5.25 Find the hazard in the following function and explain how to eliminate it.
$$f = (w + \bar{y})(\bar{w} + x)$$

5.26 Repeat Exercise 5.23 for the following function:
$$f = xy + \bar{x}\bar{z} + w\bar{x}\bar{y}z$$

5.27 Is it possible to construct a two-state device with only inverters? Explain.

5.28 Draw a gating circuit, similar to Figure 5.34, for a positive-edge clocked J-K flip-flop with asynchronous inputs.

5.29 Redraw Figure 5.34 with NAND gates only.

5.30 Redraw Figure 5.34 for a trailing-edge (negative-going) trigger.

5.31 Draw a gating circuit for a leading-edge (positive-going) clocked type-D flip-flop.

5.32 Draw the gating, similar to Figure 5.35, to convert a clocked J-K block to a clocked R-S block.

5.33 Make a logic diagram showing the additional gating necessary to make the lock-out capability of Figure 5.34 complete.

5.34 Using the fewest possible NAND gates (all in common form), draw the diagram for a positive-edge-triggered D flip-flop.

5.35 Using the minimum possible (AND, OR, and NOT) gate inputs, convert a clocked T flip-flop to one having the characteristics shown in the table. Only uncomplemented versions of the A and B variables are available.

| Flip-Flop Input | | Next State |
A	B	y_{n+1}
0	0	y_n
0	1	$\bar{y}_n$
1	0	1
1	1	$\bar{y}_n$

5.36 Using the ripple synthesis method, determine the logic circuit for the following flow table.

Present State $(y_3y_2y_1)$	x Input, Next State $(y_3y_2y_1)$
000	000
001	001
010	111
011	000
100	100
101	101
110	011
111	100

5.37 Using the heuristic method, design a modulo-5 ripple-down counter.

5.38 Using dependency notation, draw the symbol for a negative edge-triggered D flip-flop which also contains an R-S asynchronous feature which is asserted by a logic one value.

5.39 Repeat Exercise 5.38 for a single flip-flop whose state can be affected either by a J-K section, which is triggered by the positive edge of its clock, or a clocked T section employing a separate clock input, which is pulse sensitive. (The two clock inputs are not allowed to receive pulses simultaneously.) Finally, neither output is available until the negative edge of the active clock input.

5.40 Describe the performance of the flip-flop shown in Figure E5.40.

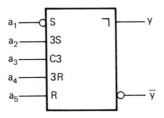

Figure E5.40

5.41 Which of the flip-flops shown in Figure E5.41 exhibit the phenomenon called ones catching? For those that do, illustrate the problem by showing the output, y, the clock pulse, and any other needed signals on a timing diagram.

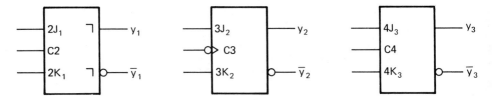

Figure E5.41

5.42 If you were designing an integrated circuit to contain two six-input NOR gates, how many pins would you select for the package?

5.43 Draw a more concise dependency notation version of Figure 5.46.

5.44 Draw a dependency notation version of Figure 5.48.

5.45 Invent a formal method for converting from one type of flip-flop to another and demonstrate your process by converting a clocked J-K to a clocked flip-flop having the characteristics shown in the table. Employ the minimum possible NAND gates. The inputs A and B are available only in uncomplemented form. Draw your complete circuit.

Input		Next State
A	B	y_{n+1}
0	0	y_n
0	1	$\bar{y}_n$
1	0	0
1	1	1

Chapter 6

Computer Architecture: Register Transfer Logic, Organization, and Subsystems

Three major levels of computer design can be identified. The first level, electronics, is briefly considered in Chapter 12; there transistors are the primitives which are assembled into gates, flip-flops, and other useful circuits. The second level is logical design, considered mainly in Chapters 4 and 5; there, the resulting circuits (e.g., gates) of the previous stage become the primitives of the next stage, and the resulting products are registers, counters, etc. Finally, the third level is reached, which is the subject of this chapter. Here, registers, memories, and similar subsystems are primitives, and the resulting products are digital processors. This highest level of design is called *computer architecture*. It has existed informally since the beginning of computers but in the last few years, it has received sufficient systematic development to elevate it to the status of a major topic within computer engineering. By definition, architecture covers a wide range of design topics: from decisions concerning whether a small special-purpose computing system should process data (bits) in series or parallel, to decisions concerning the optimum method for connecting many general-purpose digital computers into a large information-processing network.

We begin the discussion of architecture with the development of additional design tools—tools that allow not only computers but all types of digital circuitry to be described on the register level. These tools are then applied to several practical situations, including a detailed explanation of computer organization. Finally, three of the major subsystems that comprise the architecture of a digital computer are each considered in separate sections.

In order to properly establish the fundamentals, our approach to computer organization is more from the standpoint of a small computer, such as the Radio Shack TRS-80 rather than from the standpoint of a more complicated system, such as the IBM 4300 series. This approach does not sacrifice generality because, once the fundamentals have been firmly established, the reader is then equipped to grasp more complicated structures.

From a single chapter, one cannot hope to become an expert; but sufficient knowledge should be gained here so that the literature and advanced textbooks can be read much more easily. In addition, the reader's design capabilities should be strengthened and his horizons expanded.

6.1 REGISTER TRANSFER LOGIC

When considering problems in computer architecture, we frequently need a concise language to describe the information flow and processing between registers, much as Boolean algebra is available for describing the flow and processing of separate bits. Register Transfer Logic, RTL, is an *informal method* developed to satisfy this requirement. The simplest operation that may be represented in RTL is $A \leftarrow B$. This type of expression is called a *production*, and it indicates the transfer of data from register B to register A. Although this is a very simple production, other more complicated ones will soon be introduced, and sequential combinations of these can be employed to describe complex arithmetic and logic operations in a very general way. More powerful and general RTL methods exist;[1,2,4,6] but the one employed here, which emphasizes familiar engineering notation, is thought to be advantageous—in a first book— for understanding the basic principles. The following list will make our definition of RTL more precise:

1. The *contents* of the registers will be denoted by one or more letters, with the first always being in upper case.
2. Parallel (not serial) transfers will occur between registers.
3. All bits will transfer at the same time, i.e., a synchronous rather than asynchronous case will be represented.
4. The bits of each register are numbered from right to left, with the least significant bit (LSB) labeled zero, e.g., A_0 is the LSB of register A.

Table 6.1 further expands the notation and presents examples for clarification. The variety of ways bits can be transferred should be studied; e.g., attention should be paid to the difference between A_{1-3} and $A_{1,3}$. Also, observe how the subset of bits, E, is labeled in the last example.

Figure 6.1 presents the logic diagram, employing latches (elementary flip-flops), for a typical member of Table 6.1. Note that solid lines are employed for data and broken lines for the transfer signal. The labeling for the bits of register A is standard; but BAC_{E3}, for example, represents a bit from the subset of BAC, which is named E, and this happens to be bit 3 of the BAC register. Table 6.1 specifies a right-justified transfer in which the least significant bit of BAC_E becomes the least significant bit of A. Finally, Figure 6.1 shows a jam transfer between the registers. If a clear and then transfer had been desired, a sequence of two productions would have been needed: $A_{0-2} \leftarrow 0$ followed by $A \leftarrow BAC_E$. Here the first production merely specifies that bits

TABLE 6.1 BASIC REGISTER TRANSFER OPERATIONS

Operation	Meaning	Example (Before transfer; After transfer)
$A \leftarrow B$	The contents of B are transferred to A.	$A = -, B = 110$; $A = 110, B = 110$
$A_2 \leftarrow B_3$	Transfer of bit 3 register B to bit 2 Register A.	$A = 1010, B = 1011$; $A = 1110, B = 1011$
$A_2 \leftarrow A_3$	Transfer bit 3 in A to bit 2 in A.	$A = 1010; A = 1110$
$A_{1-3} \leftarrow B_{1-3}$	Transfer bits 1 through 3 from B to A.	$A = 1 --- 0, B = 01011$; $A = 11010, B = 01011$
$A_{1,3} \leftarrow B_{1,3}$	Transfer bits 1 and 3 from B to A.	$A = 1\text{-}0\text{-}0, B = 01011$; $A = 11010, B = 01011$
$A \leftarrow BAC_E$	Transfer the group of bits called E from BAC to A with right justification.	$A = 10---, BAC = 01\underset{E}{0}11$; $A = 10101, BAC = 01011$

Notes:
1. Whenever the symbol "-" appears in the example column, that bit or group of bits is a don't-care.
2. Unless explicitly stated, the registers on the right side of the production are not changed by the transfer operation.
3. In the example column the information preceding the semicolon represents conditions before the transfer and that following represents conditions after the transfer.

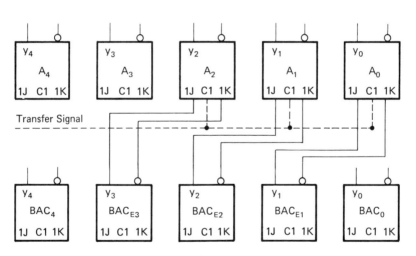

Figure 6.1 Logic Diagram for $A \leftarrow BAC_E$.

0 through 2 are set to zero, i.e., cleared. Corresponding to the logic diagram in Figure 6.1, the very simple register transfer diagram shown in Figure 6.2 can be drawn. It should be recognized that a certain amount of detail is lost in going to Figure 6.2; this may or may not be a disadvantage, depending on the purpose of the diagram.

The clearing of a register is the simplest arithmetic operation that can be performed, but several other possibilities are also shown in Table 6.2. Multiply, divide, and other more complicated processes could also be included in the table, but we

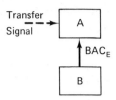

Figure 6.2 Register Transfer Diagram for $A \leftarrow BAC_E$.

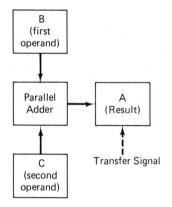

Figure 6.3 Register Transfer Diagram for Parallel Addition: $A \leftarrow B + C$.

TABLE 6.2 ARITHMETIC OPERATIONS

Operation	Meaning	Example
$A \leftarrow 0$	Clear register A.	$A = -$; $A = 00 \ldots 0$
$A \leftarrow B + C$	Transfer the arithmetic* sum of B and C into A.	$A = -$, $B = 011$, $C = 001$; $A = 100$, $B = 011$, $C = 001$
$A \leftarrow B - C$	Subtract C from B and transfer the result to A.	$A = -$, $B = 101$, $C = 010$; $A = 011$, $B = 101$, $C = 010$
$A \leftarrow A + 1$	Increment A by 1.	$A = 0111$; $A = 1000$
$A \leftarrow 1346_8$	Load 1346_8 into A with right justification.	$A = -----$; $A = 01346_8$

*Whether this is 1's or 2's complement arithmetic must be made evident from the context.

will concentrate on defining primitive register transfer operations and will build more complicated ones from combinations of the simpler types. Figure 6.3 represents a typical member of Table 6.2—that of the parallel addition circuit.

Besides the transfer operations already defined, we will find that the logic operations between registers, defined in Table 6.3, are also very useful. A word of explanation is necessary at this point concerning the logic symbols employed: $\vee \equiv$ OR and $\wedge \equiv$ AND. These are the symbols commonly used in mathematical literature; but they will be employed in this text to represent logic operations between registers in order to avoid confusion between the symbol for arithmetic addition, $+$, and the logic OR symbol $\vee$. The realization of a typical member of Table 6.3 is shown in Figure 6.4.

Shift, rotate, and scale manipulations constitute major primitives from which multiply, divide, and some other common operations are formed. These are summarized in Table 6.4.

TABLE 6.3 LOGIC OPERATIONS*

Operation	Meaning	Example
$A \leftarrow B \lor C$	Transfer to register A the OR combination of the corresponding bits of B with those of C.	$A = -, B = 101, C = 100;$ $A = 101, B = 101, C = 100$
$A \leftarrow B \land C$	Transfer to register A the AND combination of the corresponding bits of B with those of C.	$A = -, B = 101, C = 100;$ $A = 100, B = 101, C = 100$
$A \leftarrow B \land C_2$	Transfer to register A the AND combination of bit 2 of C with the individual bits of B.	$A = -, B = 101, C = 100;$ $A = 101, B = 101, C = 100$
$A \leftarrow B \land C_1$	Transfer to register A the AND combination of bit 1 of C with the individual bits of B.	$A = -, B = 101, C = 100;$ $A = 000, B = 101, C = 100$
$A \leftarrow B \lor C_2$	Transfer to register A the OR combination of bit 2 of C with the individual bits of B.	$A = -, B = 101, C = 100;$ $A = 111, B = 101, C = 100$
$A \leftarrow \bar{A}$	Complement the individual bits of A.	$A = 1011; A = 0100$

*The AND and OR operations are undefined unless the two registers contain the same number of bits or one of them specifies a single bit.

TABLE 6.4 SHIFT, ROTATE, AND SCALE OPERATIONS

Operation	Meaning	Example
$A \leftarrow \text{sr}A$	Shift register A right one bit.	$A = 10001; A = 01000$
$A \leftarrow \text{sl}A$	Shift register A left one bit.	$A = 10001; A = 00010$
$A \leftarrow \text{sl3}A$	Shift register A left three bits.	$A = 10001; A = 01000$
$A \leftarrow \text{rr}A$	Rotate register A right one bit.	$A = 10001; A = 11000$
$A \leftarrow \text{rl}A$	Rotate register A left one bit.	$A = 10001; A = 00011$
$A \leftarrow \text{rr2}A$	Rotate register A right two bits.	$A = 10001; A = 01100$
$A \leftarrow \text{sc}A$	Scale register A right one bit.	$A = 10000; A = 11000$
$A,B \leftarrow \text{sl}A,B$	The concatenated register A,B is shifted left one bit.	$A,B = 1001; A,B = 0010$ $\underbrace{\quad}_{A\ B}$

Notes:

1. A group of two lowercase letters represents operations; except for scale, which is only defined for one direction, the first letter denotes the operation type and the second denotes the direction. (If the number of bits for the operation is greater than one, it is indicated by a digit following the two letters.)
   Type: $r \equiv$ rotate, $s \equiv$ shift, $sc \equiv$ scale
   Direction: $r \equiv$ right, $l \equiv$ left
   Example: $rr \equiv$ rotate right
2. In the rotate operation, the least and most significant bits are adjacent to each other as far as bit transfer is concerned.
3. In the shift operation, a zero is moved into the vacated bit.
4. Scale is the same as shift right except that the sign (bit $n - 1$) is left unchanged instead of reading a zero into that position.
5. A concatenated register is the result of joining two (or more) registers in a string so that the LSB of the first-mentioned register is one bit left of the MSB of the second. (The concatenation operator is a comma and the symbol(s) for the next register follows without a space.)

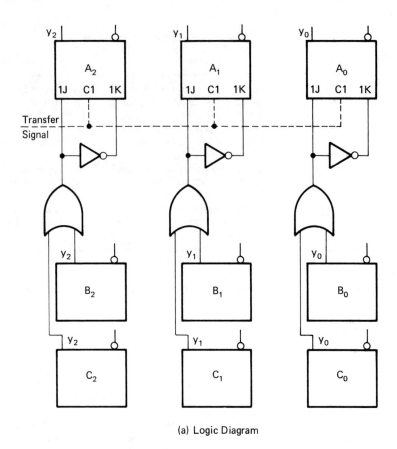

(a) Logic Diagram

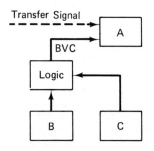

(b) Register Transfer Diagram

Figure 6.4 Realization of $A \leftarrow BVC$.

A typical logic diagram for a member of this group, employing edge-triggered flip-flops, is shown in Figure 6.5. The circuit is very similar to the shift registers that were discussed in Chapter 5; also the circuitry required to mechanize the other productions in Table 6.4 should be obvious.

The final RTL operations to be introduced are shown in Table 6.5. These two groups are very important in that they give the designer the ability to do hardware

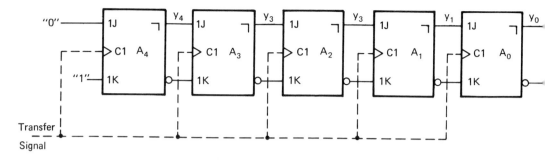

Figure 6.5 Logic Diagram for $A \leftarrow srA$.

TABLE 6.5 MEMORY AND CONDITIONAL OPERATIONS

Operation	Meaning	Example
$A \leftarrow M(203)$	Load A from the contents of memory location 203.	$A = -, M(203) = 312$; $A = 312, M(203) = 312^*$
$M(203) \leftarrow A$	Load memory location 203 from A.	$A = 312; M(203) = -$; $A = 312; M(203) = 312$
IF $(A_3 = 1)\ B \leftarrow 0$	If bit 3 of A is a 1, then clear B.	$A = 01001, B = 001$; $A = 01001, B = 000$
IF $(A \neq 0)\ B \leftarrow 0$ ELSE $B_2 \leftarrow 1$	If A is not 0, then clear B; otherwise set bit 2 of B to 1.	$A = 01001, B = 011$; $A = 01001, B = 000$
IF $(A \geq 1)\ B \leftarrow 0$, $C_1 \leftarrow 1$ See note 1.	If $A \geq 1$, then clear B and set bit 1 of C to 1.	$A = 01, B\text{-}, C = \text{-}0$; $A = 01, B = 00, C = 10$
IF $(A, B_2 = 0, 1)\ C \leftarrow 0$ See note 2.	If A is cleared and if bit 2 of B is a 1 then clear C.	$A = 00, B = 100, C = -$; $A = 00, B = 100, C = 00$
IF $((B_1 = 1 \text{ OR } C = 0)$ AND $D = 0)\ A_1 \leftarrow 1$ See note 3.	If bit 1 of B is a 1 or C is cleared and if D is cleared, then set bit 1 of A to 1.	$A = 01, B = 00, C = 0$, $D = 0; A = 11, B = 0$, $C = 0, D = 0$;

Notes:
1. This production demonstrates that multiple consequences, separated by commas, are allowed with a single IF statement.
2. Registers or parts of registers may be concatenated to form the premise portion of a conditional operation. [The production is a concise form of IF $(A = 0$ AND $B_2 = 1)\ C \leftarrow 0$.]
3. Parentheses may be nested, as required, to clarify complicated conditional relations.

*Some memory types clear the memory location during the read process; this is called destructive readout.

branching, and they provide him with the opportunity to communicate with memory. Instead of working with a fixed memory address, as shown in the table, if it is desired, for example, to load register B from a memory location under program control, the production $B \leftarrow M(A)$ is employed, where the value of A has previously been determined. Figure 6.6 has been constructed corresponding to the second conditional transfer shown in the table. Note how the register transfer diagram employs standard software flow-chart symbols when possible, and how much more concise it is than the logic diagram.

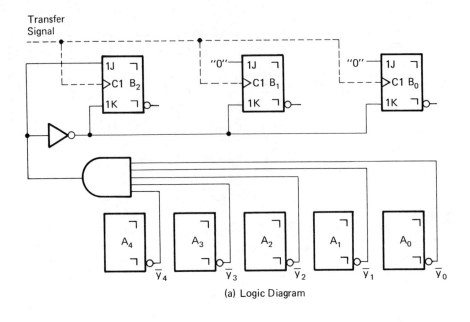

(a) Logic Diagram

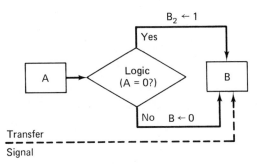

(b) Register Transfer Diagram

Figure 6.6 Realization of IF $(A = 0)$ $B_2 \leftarrow 1$ ELSE $B \leftarrow 0$.

6.2 APPLICATIONS OF RTL

Up to this point, only a single production has been employed to represent an RTL process, e.g., Figure 6.6(b). However, most practical processors need a more complicated representation; they require a sequence of productions, each of which is initiated by a transfer signal. Once this ability has been added to our RTL repertoire, a very powerful method will have been created. In fact, we will have the ability to represent any digital system from small special-purpose digital interface units to full-scale general-purpose computers. Often the transfer signal is a clock pulse or is produced from the clock by means of frequency division; however, with some extensions, RTL can be employed for asynchronous systems.

Example: A Data Storage and Comparison Unit

Frequently in practice (e.g., in the arithmetic section), data are brought in from one unit and processed, and the results are transmitted to the next unit. Figure 6.7 shows an RTL program for the case where one number is read into register A from data source D,* and another number is later read into register B. The numbers are checked for equality, and the process is repeated with new data, as long as the start level, S, remains true. In the figure, the column down the left side indicates the sequencing. To be more specific, each number there denotes the presence of a *level*, one clock period long, which uniquely determines when a particular step in the RTL process is to occur. (Our definition of a level is a signal which remains in a particular logic state, either a 1 or 0, for a length of time at least as long as the period of the clock pulse.)

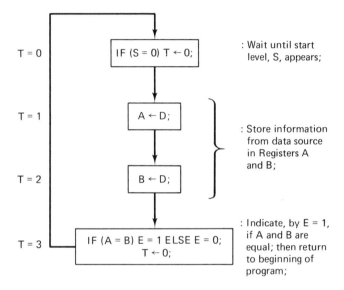

(Note: Except where stated otherwise the production $T \leftarrow (T + 1)$ is implied at each state, T_i.)

Figure 6.7 RTL Program for the Data Storage and Comparison Unit.

For this example, the states† T_0 through T_3 occur in sequence, and then the circuit returns to state T_0. In general, the order can be interrupted and various states repeated or skipped, as is done with the assembly language programs discussed in Chapter 3. As a matter of fact, we could call Figure 6.7 a hardware program since it realizes an algorithm for hardware much as algorithms were previously realized for software. Contrary to most software programs, more than one RTL operation can occur simultaneously (while the processor is in a particular state). The RTL program

*New data reaches D after each clock pulse.
†The abbreviation T_i will frequently be employed to represent the state $T = i$.

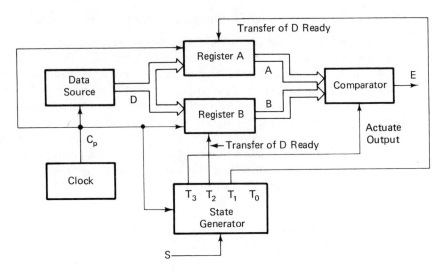

(a) System Block Diagram

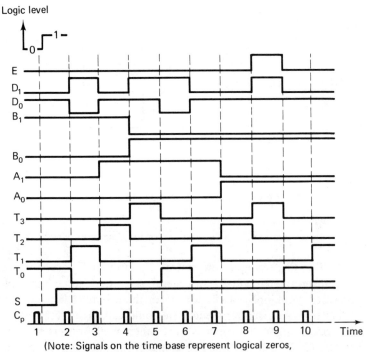

(Note: Signals on the time base represent logical zeros, those above the base represent logical ones.)

(b) Timing Diagram

Figure 6.8 Data Storage and Comparison Unit.

represents this by means of additional statements having the same T_i value but separated by semicolons, as is done at T_3 in Figure 6.7. (observe that a semicolon also terminates the last production of each state, and a colon is placed before each comment.)

Figure 6.8(a) shows the block diagram for the circuit that realizes the program of Figure 6.7. The reader should now carefully study these two figures and visualize the way the data and control signals, specified by the program, are implemented by the hardware. To further clarify the process, a *timing diagram*, employing a simplified case with two bits for each of the A, B, and D registers, is presented in Figure 6.8(b). We have previously employed these diagrams in Chapter 5; but in this application, their full value as a design tool for indicating time sequences and interrelationships between data and control signals will be realized. Since we have been emphasizing clocked J-K flip-flops, these are implied in the timing diagram; and because it is convenient with this type of flip-flop, the information is communicated in a level mode with the actual transfer being completed a few nanoseconds following the trailing edge of the clock pulse. The clock pulse signal, C_p, is the key component in Figure 6.8(b). (The pulses are numbered below the time base for easy reference.) The other signals are synchronized with respect to it; and the state changes, for all signals, occur a few nanoseconds following its negative edge. In agreement with the program of Figure 6.7, the T_i states in the timing diagram occur in sequence, with a reset to T_0 after T_3. Level T_1 furnishes the ready signal to register A, and clock pulse 3 (also later $C_p = 7$) initiates the actual transfer from the data source* (a similar transfer to register B occurs at $C_p = 4$ and $C_p = 8$). Only when the two registers are equal does the output signal, E, go to the logical-1 state (after $C_p = 8$).† It should be noted that the data source is synchronized with the clock, which is evident from Figure 6.8(a) and (b). However, the start level, S, is not synchronized since it occurs between clock pulses 1 and 2.

In order to provide more details concerning the operation of the data storage and comparison unit, the logic circuits for register A, the state generator, and the comparator are presented in Figures 6.9, 6.10, and 6.11, respectively. (The register B circuit is the same as that for register A.) The following facts should be observed concerning these figures:

1. The terminals for the individual logic diagrams are all arranged in the same relative positions as they appear in the block diagram of Figure 6.8(a). (This should be helpful in following the information paths between the various circuits.)

2. For brevity, only two-bit versions of Figure 6.9 and 6.11 are presented; however, extensions of these circuits to as many bits as desired follow exactly the same pattern as shown here.

---

*The fact that the clock simultaneously advances the state generator to T_2 should not be allowed to hide the fact that it is the T_1 level that really opens register A to the incoming data. The register circuit logic details will be discussed shortly to prove this point.

†Since the comparator is a combinational circuit, to be discussed in detail shortly, Figure 6.8(b) correctly shows that, when $A = B$, $E = 1$ during the entire T_3 period.

Sec. 6.2   Applications of RTL

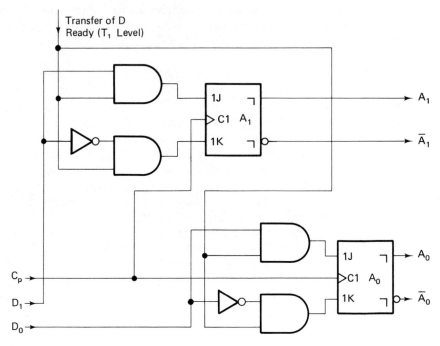

Figure 6.9 Two-Bit Version of Register A.

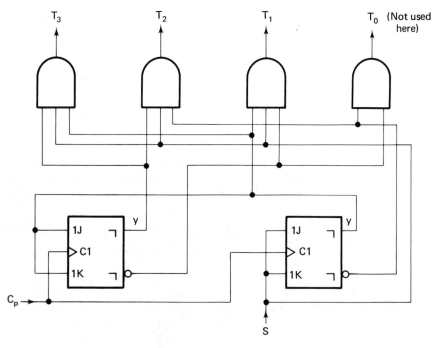

Figure 6.10 State Generator.

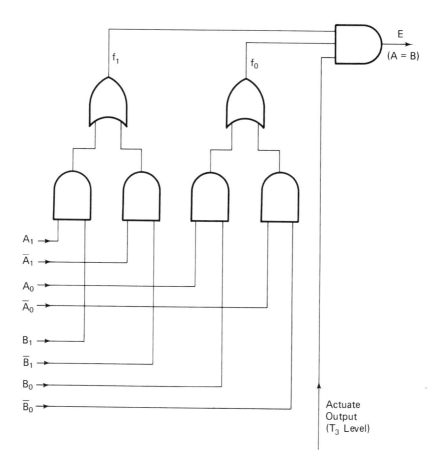

Figure 6.11 Two-Bit Comparator.

3. In Figure 6.9, the AND gates are employed to allow new data to reach the register only when the transfer-ready level ($T_1 = 1$) exists. The inverters are required to generate $\bar{D}_t$ since only the uncomplemented signals are available from the data source. A general version of the data transfer circuit is shown in Figure 6.12. There, any number of data sources, (1 to N) can be individually transferred by means of the AND-OR gating shown.

4. Figure 6.10 shows a standard up-counter which requires the level S for actuation. The AND gates serve as a decoder to transform the counter outputs into the required T_i states. Note that the S level also controls whether or not the decoder's output is allowed to appear.

5. Figure 6.11 shows a comparator; and more specifically, it is an EQUALITY or EXNOR circuit. To determine whether, for example, the least significant bits of registers A and B are equal, the EXNOR $f_0 = A_0 B_0 + \bar{A}_0 \bar{B}_0$ is evaluated. This output is then ANDed with similar circuit outputs for all the other bits to determine whether the registers agree. In this case, the evaluation is valid only at state T_3; thus, the T_3 level is employed to actuate the output AND gate.

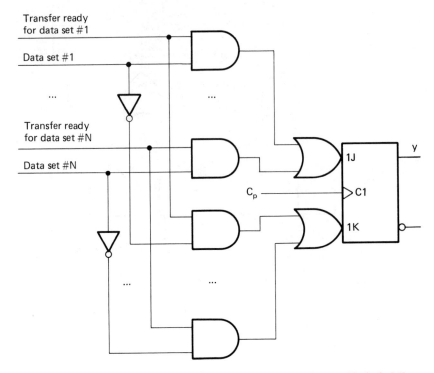

Figure 6.12 General Circuit for the Transfer of Data into a Clocked *J-K* Flip-Flop.

6. The dependency notation in Figures 6.9, 6.10, and 6.12 indicates that master-slave flip-flops are employed.

Example: Character Generator for a CRT Display

Let us now consider another practical example, that of generating alphanumeric characters for a cathode ray tube (CRT) display. There are several ways of accomplishing this, but one of the common methods is to intensify certain positions in a 4-by-6 array, much as is done with small lights in a time-and-temperature sign. For example, the letter A can be represented as:

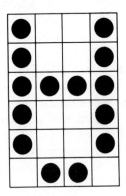

where the filled areas represent intensified points and the remainder represents points not intensified.

The system, shown in Figure 6.13(a), operates essentially as follows. The 24 bits (4 by 6) which form a character are stored in the memory, then transferred into a register in the logic unit, and finally read onto the CRT one bit at a time. The Z signal controls the beam of the CRT, with a $Z = 1$ turning the beam on and $Z = 0$ turning the beam off. The position of the individual bits on the face of the CRT is controlled by the X and Y deflection signals, which in turn are produced by the digital-

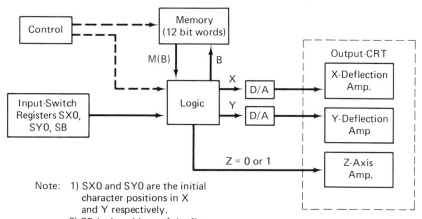

Note: 1) SX0 and SY0 are the initial character positions in X and Y respectively.
2) SB is the address of the first word in memory containing the character pattern.
3) A switch register is a row of miniature toggle switches, which represents a binary number. (The up position represents a logical one.)

(a) System Block Diagram

First word		Second word	
5	11	5	11
4	10	4	10
3	9	3	9
2	8	2	8
1	7	1	7
0	6	0	6

(b) 4 by 6 Character

Figure 6.13 Character Generator and CRT Display.

to-analog (D/A) converters. The D/A converter is a special electronic circuit that develops an analog voltage proportional to the input digital number. For simplicity, we will assume that the CRT in Figure 6.13 has a deflection system such that the entire display is in the first quadrant, with the origin in the lower left corner. To emphasize principles, we have also made a number of other simplifying assumptions, all of which can easily be removed in practice, such as the fact that the appearance of the character will not be objectionable if some points are displayed somewhat longer than others.

Because it is the value employed by several minicomputers and because it is desirable to indicate some of the problems in producing a CRT display, we will assume that the system memory contains 12-bit words. Thus, two words will be required to

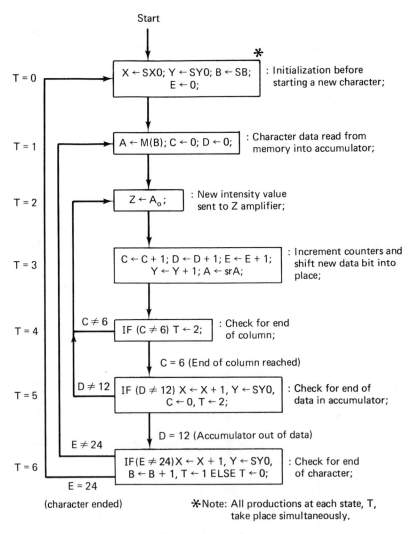

Figure 6.14 RTL for a 4-by-6 Character Generator.

get sufficient bits for a 4-by-6 grid. Figure 6.13(b) shows the assumed structure of each character, including bit numbering.

Now that the basic system has been described, let us examine Figure 6.14, which shows the algorithm used in RTL notation. Again, the comments adjacent to each block should help in understanding the process. Notice that three loops are present in the algorithm and each of these employs a different counter: C, which accumulates the number of points completed in each column of the character; D, which accumulates the number of bits completed in each 12-bit word; and E, which accumulates the number of points completed in the 24-bit character. Examination of Figure 6.13(b) will reveal the physical reasons for these three loops.

Figure 6.14 indicates that once the circuit starts, the character is displayed over and over again, which is necessary to make it visible (unless a storage CRT is used). Also, the fact that there is a return to $T = 0$ after the completion of point 24 means that the input switches can be effective at that time in changing the character's position and even in selecting a different character by changing the SB switch. (Of course, the memory must be preloaded with the proper alphanumeric patterns.)

‡Methods for Resetting the State Generator

In the previous example, there were four branch points where the state generator's normal sequence was broken, and these must be accounted for in the hardware design. There is, however, no unique solution to the state generator reset problem—each particular case has its own requirements and its own optimum solution. On the other hand, Figure 6.15 presents a general approach which will always yield a solution, although for any particular case it may be not the optimum. The figure shows two typical bits, L_i and L_{i-1}, from the generator. A switch register, SW, is employed to store the reset state, but frequently some circuit simplification can be achieved if a fixed source of data is used. The top AND gate associated with each flip-flop forms part of a parallel up-counter, with the accompanying inverter serving to disable the counter function when a new state is transferred in. Finally, the lower pair of AND gates prevents the reset data from affecting normal up-counting operations except when the transfer-ready signal is present. If other reset states are required, they may be constructed from additional pairs of AND gates feeding into the existing OR gates. (Each reset state requires a separate transfer-ready signal.)

Thus, it is technically possible to construct a state generator which can be reset to as many different states as needed by adding pairs of AND gates to all bits for each reset state. Nevertheless, cost and other practical considerations often dictate that alternate approaches be considered. One way to repeat certain states without repeating the complete sequence is to break the process into two parts, either of which can call itself or call the other part. This is the general approach taken in the fetch-execution cycle to be described in the next section. Another possibility is to interrupt the state generator and allow a special down-counter to control the iterative part of the process until it is complete, at which time the regular state generator resumes control. This method is frequently used in multiply operations. Still other approaches will be considered in the exercises (Section 6.10).

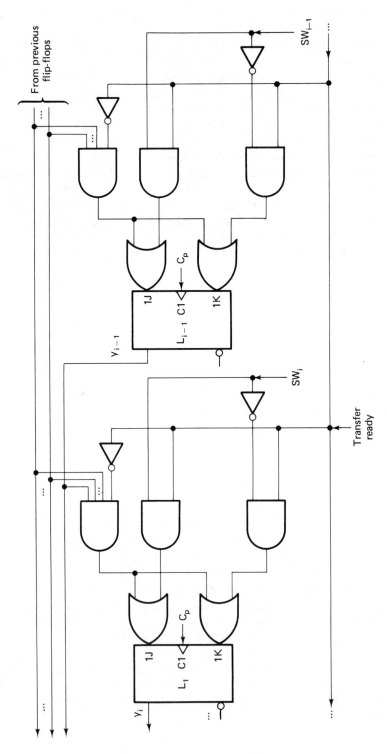

Figure 6.15 General State Generator Configuration.

The principles of the RTL method have been developed, and examples have been given. In the remainder of this chapter and in several later chapters, many applications of the RTL method to the design and analysis of digital systems will be presented. Since RTL is still an informal method that is in the formative stages, following this discussion the reader is encouraged to consult the Bibliography and to examine the current literature for more advanced developments in the field.

6.3 COMPUTER ORGANIZATION

In Chapter 3, in conjunction with Figure 3.1, the block diagram of a general-purpose computer was discussed. Now, in Figure 6.16, we want to expand the control unit and include much more detail, with emphasis on computer organization* from an RTL viewpoint. The nonshaded blocks shown in the figure belong to the control section; all of these are registers except the logic and timing block and the instruction decoder block.

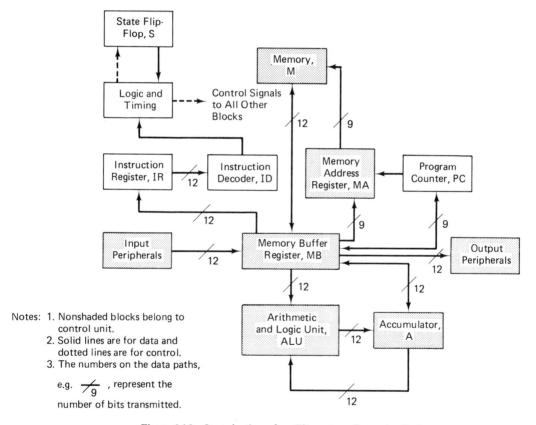

Figure 6.16 Organization of an Elementary Computer System.

*For tutorial reasons, we employ the LINC architecture in this chapter; but in Chapter 8 our primary computer, a member of the 6800 microprocessor family, will be studied.

Before describing the system as a whole, the individual blocks will be explained:

1. *Memory buffer register.* As the name implies, this register serves as an interface between the memory and other units. Thus, all data and instructions entering or leaving the memory must pass through the memory buffer.
2. *Memory address register.* The location in memory being employed as a source or receiver of data is determined by this register.
3. *Instruction register.* As indicated in Chapter 3, a typical machine language instruction is composed of an operator and an address. The instruction register receives those bits of the current instruction which convey operator information.
4. *Instruction decoder.* The contents of the instruction register are decoded here by means of a combinational circuit having as many outputs as there are instructions in the computer's repertoire. The single output possessing a logical 1 indicates which instruction is being executed at a particular time.
5. *Program counter.* This is the register which keeps track of the address in memory where the next instruction is to be found. Normally it is incremented by 1 each time so that the next instruction in sequence will be available for execution; but branch instructions are an exception. For example, in the case of JMP, the program counter has a new location transferred to it from the address section of the memory buffer.
6. *Accumulator.* The accumulator, which is actually part of the arithmetic unit, is the most prominent register in the machine. It always holds one operand for each arithmetic or logic operation, and the result is always placed there.
7. *Logic and timing circuit.* This unit furnishes all of the control signals which allow the computer to execute a program. The master clock, the state generator, and an extensive logic circuit are present in order to initiate transfers between the various registers at the proper times.
8. *State.* In order to keep track of the mode in which the computer is operating, the control unit employs a one-bit register called the *state* flip-flop. (This flip-flop is shown explicitly in the figure to indicate its relationship to the other units; but it is really part of the logic and timing unit and can be thought of as a means to break the state generator's sequence into two sections.)

Now that we have some concept of the various circuits contained in a computer, several typical machine language instructions will be explained in order to give the reader a deeper understanding of this aspect of computer architecture. First, it must be understood that a typical machine language instruction requires several clock pulses to complete because there are many suboperations that the machine must accomplish sequentially* in obtaining an instruction from memory, decoding it, obtaining the data from memory, performing the operation, and preparing to obtain the next instruction.

The process of carrying out an instruction will be divided into two modes called

---

*In Section 6.4 we will consider both the restrictions on performing two or more RTL operations concurrently and those imposed by typical hardware configurations.

the *fetch* (or *instruction*) cycle and the *execution* cycle. The fetch cycle is concerned with obtaining the instruction from memory and preparing for execution, while the purpose of the execution cycle is to obtain the data and carry out the operation. Some other functions of both cycles will become obvious later in this section. As might be expected, the fetch cycle is common to all instructions, but the execution cycle is different.

An RTL diagram for the fetch cycle in a simple computer is shown in Figure 6.17. The reader should study this figure in conjunction with Figure 6.16, completely visualize the process that is taking place, and agree that by the completion of $T = 2$ sufficient information has been transferred so that the execution cycle can be initiated. Note that the convention for the machine state is $S = 1$ for execute and $S = 0$ for fetch; also, the symbols used for memory address register, program counter, etc. are the same ones used in Figure 6.16. At $T = 2$ it should be observed that several transfers are occurring simultaneously, which is perfectly feasible and certainly efficient.

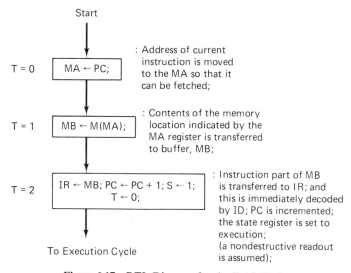

Figure 6.17 RTL Diagram for the Fetch Cycle.

Having discussed the fetch cycle, we are now ready to consider the execution cycle. These will be different for each instruction; but several have been gathered together in Table 6.6 for comparison purposes. First, note the following format differences between this program and previous ones. For simplicity, RTL statements are written in more conventional programming form rather than in flow-chart form. Moreover, the production $T \leftarrow (T + 1)$ is implied at each state unless an explicit branching statement for T is given. Now let us make detailed comments concerning the RTL statements required by the STC instruction in the table:

$T = 0$ The address section of the memory buffer is transferred to the memory address register in order to establish the storage location required by the STC instruction.

TABLE 6.6 RTL PROGRAMS FOR THE EXECUTION CYCLE OF SEVERAL INSTRUCTIONS

	ADD	COM	JMP	STC	ROR
T=0	$MA \leftarrow MB_{0-8}$;	$A \leftarrow \bar{A}$;	$PC \leftarrow MB_{0-8}$;	$MA \leftarrow MB_{0-8}$;	$A \leftarrow rrA$; $MB_{0-5} \leftarrow MB_{0-5} - 1$;
T=1	$MB \leftarrow M(MA)$;	- - -	- - -	$MB \leftarrow A$; $A \leftarrow 0$;	IF $(MB_{0-5} \neq 0)$ $T \leftarrow 0$;
T=2	$A \leftarrow A + MB$; $S \leftarrow 0$; $T \leftarrow 0$;	$S \leftarrow 0$; $T \leftarrow 0$;	$S \leftarrow 0$; $T \leftarrow 0$;	$M(MA) \leftarrow MB$; $S \leftarrow 0$; $T \leftarrow 0$;	$S \leftarrow 0$; $T \leftarrow 0$;

Note: A detailed description of each of the instructions is found in Table 3.2.

$T = 1$ The number to be stored is transferred from the accumulator to the memory buffer register and simultaneously the accumulator is cleared. (As previously mentioned, there is no conflict in having a simultaneous transfer because of the type of flip-flops employed.)

$T = 2$ The actual transfer of the data to memory is accomplished. The state flip-flop and state generator are reset so that another fetch cycle can begin.

As an exercise, carefully study the composition of the entire set of instructions in the table. Again, Figure 6.16 will be of considerable aid in visualizing the details of the data flow. At the end of each execution cycle, note that preparation is made to start the fetch cycle for the next instruction by means of the $S \leftarrow 0$ and $T \leftarrow 0$ transfers. Although certain instructions like COM and JMP require only two clock pulses for completion, the full three clock pulses have been employed for uniformity, and nothing is accomplished during $T = 1$. The philosophy of making most execution cycles of constant length is fairly common in practice since it simplifies the timing circuitry; but of course, there may be cases where the extra cost and complication could be justified by the time saved. On the other hand, the ROR instruction is an interesting one in that the number of clock pulses composing the execution cycle is not fixed at three, as it is with the other four instructions shown; but it is reasonable to make it depend on the number of bits the accumulator is rotated right.

Of course, part of the art of computer architecture is to find more effective ways to compose machine instructions within the host of practical constraints that exist; e.g., certain types of main memory require approximately one microsecond to complete a memory cycle. With these increasing requirements, the number of clock pulses within the instruction cycles may increase; but this will be partially compensated by more sophisticated instruction design.

Since we are interested in the logical design of computers, let us now consider how the operations indicated in Table 6.6 can be realized. Figure 6.18 shows the logic diagram of the STC instruction in the execution cycle. There the accumulator and the other registers have their individual bits arranged in a vertical line. Since the others follow the same pattern as bit 0, only that one is shown for each register. Note that

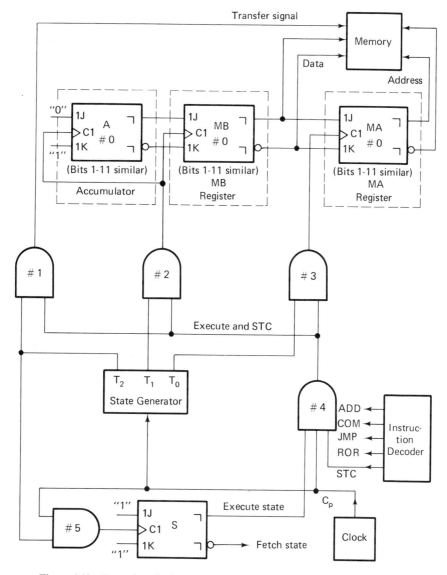

Figure 6.18 Execution Cycle Logic for the STC Instruction in an Elementary Computer.

the instruction decoder indicates which instruction is being executed by providing a logical-1 level output at the proper terminal. A careful study of this figure will reveal several interesting features. For example, gate 4 generates a signal which indicates that the STC instruction is being executed, the machine is in the execute state, and a clock pulse is present. In this simplified case, only three gate inputs are saved over running the STC, $S = 1$, and clock signals directly to gates 1 through 3; but it demonstrates a principle which is frequently employed in practice. In agreement with Figure

6.17 and Table 6.6, note that the T_2 signal from the timing generator goes directly to the S register, causing it to change state each time $T_2 = 1$ regardless of the instruction being executed or the state the machine is in.

In Figures 6.9, 6.10, and 6.12, the clock signal is permanently connected to the C terminal of each flip-flop and the transfer-ready signal from the state generator controls when each flip-flop receives new data. An alternate, but less general, approach is shown in Figure 6.18. Here, it is convenient to employ the state generator to gate the clock signal so that clock pulses reach each flip-flop only when data are to be received.

Several additional concepts concerning computer organization, the composition of machine instructions, and logic will be considered in the exercises and in Chapter 9.

6.4 FURTHER DEVELOPMENT OF RTL

What technique can be employed to make RTL programs more concise? Are all possible RTL statements allowed? It is the purpose of this section to answer practical questions such as these.

Criterion for Grouping RTL Statements

In Chapter 5, we found that state tables sometimes contain redundant states whose elimination can result in significant savings. A similar reduction process can be applied to simplify an RTL program, either by eliminating both a state and its production(s) entirely or by eliminating a state by grouping its production(s) with those of another state. In either situation, the criterion is similar to that in Chapter 5: assume that two RTL states may be grouped (or one eliminated) unless prevented by a *proven incompatibility*. Two major sources of proven incompatibilities are:

1. Any specific limitations on circuit construction dictated by the system specifications, e.g., there may be a requirement concerning which registers must be employed.
2. Constraints established by the desired sequence of circuit actions, e.g., when one register is loaded from another, the first must recieve the data at least one clock pulse before the second register is loaded.

As an example of the criterion's use, consider the RTL program in Figure 6.7. Because the data comparison unit requires that a sequential pair of samples from source D be transferred into registers A and B, an attempt to group the production from T_1 and T_2 into a single state results in a *proven* incompatibility. Thus, it may seem that four is the minimum number of states for the RTL program. The specification, however, did *not* explicitly require that the checking of level S and the transfer $A \leftarrow D$ be done serially, which means that the productions in T_0 and T_1 may be combined in a single state.

What about also eliminating state T_2 by removing register B and making a direct comparison between A and D as the following production indicates?

$$\text{IF } (A = D) \; E = 1 \text{ ELSE } E = 0;$$

This production contains a proven incompatibility, however, because the original specification explicitly stated that the second number is to be read into register B before the comparison. Therefore, the simplified RTL program is the three-state version shown in Figure 6.19. Although register A is loaded from D each time T_0 is executed, only the value contained in A when $S = 1$ is actually compared with the contents of register B. (Because A is loaded at T_0 in the simplified program, compared to being loaded at T_1 in the original program, two different sequential pairs from D are compared in the two programs. Nevertheless, the original specification is sufficiently broad to permit both solutions.)

$$T = 0 \text{ IF } (S = 0) \ T \leftarrow 0; \ A \leftarrow D;$$
$$T = 1 \ B \leftarrow D;$$
$$T = 2 \text{ IF } (A = B) \ E = 1 \text{ ELSE } E = 0; \ T \leftarrow 0;$$

Figure 6.19 Simplified Version of the RTL Program in Figure 6.7.

Limitations on RTL Statements

There are always hardware constraints whenever one constructs electronic systems. Moreover, these constraints are not absolute. Sometimes they are influenced by the environment in which the circuit must operate; they are subject to change as technology improves; and sometimes they may exist because of established conventions or practical considerations such as the desire for a finite inventory of circuit module types. Hardware constraints, as well as certain practical rules which have already been mentioned, cause limitations of RTL production formats. Several typical productions which are forbidden for the purposes of *all RTL programs* in this text are shown in Table 6.7 along with a brief reason for the limitation.

TABLE 6.7 GENERAL RTL STATEMENT LIMITATIONS

Statement Type	Reason for Limitation
1. $M(C) \leftarrow A; \ B \leftarrow M(D);$ 2. $M(B) \leftarrow M(A);$	A conventional, single, unified memory only allows one location to be operated on at a time.
3. $A \leftarrow A + B + C;$	A conventional adder permits only two operands.
4. $A \leftarrow A \times B;$	Multiply is not an RTL primitive.
5. $A \leftarrow A \div B;$	Divide is not an RTL primitive.
6. $B \leftarrow B + M(A);$	A special circuit would be required.
7. $M(A) \leftarrow B; \ B \leftarrow M(A);$	In contrast to the legal simultaneous register exchange $A \leftarrow B; B \leftarrow A;$ we assume memory may not be read and written simultaneously even if the same address is employed.

In addition to the general limitations considered above, there are some specific limitations for digital computers with the classical organization shown in Figure 6.16. These limitations are listed in Table 6.8 and will be assumed to hold for *all digital computers* discussed in this text unless explicitly indicated. (For digital logic not part of a computer, however, the Table 6.8 restrictions do not apply; but those of Table 6.7 are still valid.) Neither of the tables is exhaustive—the reader, however, should be

TABLE 6.8 RTL STATEMENT LIMITATION
(for the System of Figure 6.16)

Statement Type	Reason for Limitation
1. $MA \leftarrow srMA$;	Although the *accumulator and memory buffer* do have circuitry for shifting and other logical operations, the other registers do not.
2. $T = 0$ $MB \leftarrow M(MA)$; $T = 1$ $MC \leftarrow M(MA)$;	Memory can not be read directly into two different buffer registers because only one buffer is assumed to exist. $MC \equiv$ a second buffer register.
3. $T = 0$ $M(MA) \leftarrow MB$; $MA \leftarrow MA + 1$; $T = 1$ $M(MA) \leftarrow MB$;	For *destructive readout only* it is assumed that a particular address must be read before writing can take place.
4. $M(MA) \leftarrow A$; $PC \leftarrow A$;	Data paths do not exist for these separate productions.
5. $MB \leftarrow MA$; $MB \leftarrow IR$;	Although data paths exist between these registers, the arrow on each path indicates that information may not flow in the direction specified by the RTL statement. (Note the production $MB \leftarrow PC$ is allowed because it is needed in Chapter 11, particularly Figure 11.4.)

able to visualize many variations of the examples shown. Moreover, as a new RTL program is being checked, the table should serve as a model for validating any questionable productions. Just like pure hardware constraints, our limitations on productions are not absolute since a different environment or a different technology could add to or subtract from the lists; but for purposes of this text, we will agree to employ the limitations of Tables 6.7 and 6.8.

6.5 THE BUS SUBSYSTEM[13]

The Control Unit, Bus Structure, Arithmetic and Logic Unit, and Memory are the the four major subsystems comprising the general* architecture of a modern digital computer. Control was studied at some length in Section 6.3 using RTL, and additional facets of this topic will be covered in both Chapters 7 and 9. To achieve a balanced discussion, however, it is appropriate now to present separate sections of each of the other major subsystems. Bus structures, considered here, facilitate communication both between computer subsystems and between the computer and the outside world. It is an important factor in achieving a satisfactory architecture for all modern computer systems; and the principles learned will be applied in later chapters.

Instead of following the philosphy shown in Figure 6.16 of running 24 lines† for each transfer between the MB and the many other subsystems, including the input/

---

*The Input and Output Subsystems, shown both in Figure 3.1 and Figure 6.16, are more specialized units that are tailored to the environment.

†The figure 24 comes from transmitting the signal for 12 bits in double-rail form, i.e., both the y and $\bar{y}$ outputs of each flip-flop. Even if single-rail transmission is suitable, and accounting for the fact that in a few cases not all 12 bits need be sent, there are still a tremendous number of lines involved.

output peripherals, with which it communicates, a more elegant scheme called *busing* is frequently employed. The concept of a bus structure is to have a single set of lines, one for each bit, which runs to each subsystem much like a party line employed by the telephone company. Control lines then select which register will be the transmitter of information and which will be the receiver.

It is instructive to consider Figure 6.20, which shows the basic organization of a computer employing a single-bus structure. This figure should be compared with Figure 6.16, which does not make use of a bus. Again, the reader will find it valuable to review Figure 6.17 and Table 6.6 in light of the structure of Figure 6.20. Note that most subsystems both transmit and receive information, as indicated by the two arrows at each box, but that certain units, e.g., the input peripherals, only handle data in one direction. The way the ALU accesses data during, say, an add instruction is interesting in that one operand must enter from the memory buffer, the other must enter from the accumulator, and the result must return to the accumulator. Consequently, three simultaneous data paths are required, but the single bus is capable of providing only one path. Thus, the transmit-receive lines directly between the ALU and A are required to furnish the extra data paths.

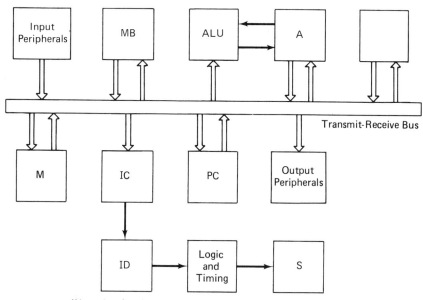

(Note that for simplification the control signals have been omitted)

Figure 6.20 Basic Computer Organization Employing a Single Bus Structure.

We are now aware of one of the advantages of the bus—the fact that there is a considerable savings in equipment possible. But what is the disadvantage? A major drawback is associated with the above-mentioned fact that only one data path is available at a time; yet frequently, particularly in large machines, it is desirable to carry out certain operations in parallel. For example, one might want to bring data to the memory from an input peripheral at the same time an arithmetic instruction is being completed by the ALU. One solution to this problem is to employ a multiple-bus system with certain devices on each bus and with the possibility of communicating between the buses. Another alternative is to provide a dual set of buses, with each device having access to both, and with the proper controls. Of course, the decision concerning the best bus structure depends on the particular requirements; but it certainly does open up a lot of interesting possibilities to the computer architect.

Two major conditions must be satisfied for a successful bus structure:

1. There must be a convenient method for coupling sources to the bus.
2. There must be a good way to ensure that no more than one source tries to "talk" to the bus at one time.

(Of course, any number of receivers may "listen" on the bus simultaneously, the only limitation being that their loading effect must not exceed the driving capability of the source.) In Figure 6.21, an OR gate serves as an elementary method for coupling several sources to a bus. One serious deficiency of this approach, however, is that no

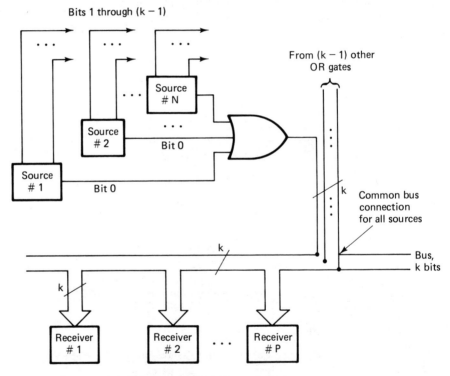

Notes:
1. Each source feeds k bits to the bus.
2. The number of receivers, P, need not equal the sources, N.
3. Only the k OR gates for the sources need to be connected to the bus at a common point.
4. The coupling of Bit #0 from each source to the bus is shown; the other (k − 1) bits are similarly connected.

Figure 6.21 Elementary Coupling of Sources to the Bus.

method is shown for preventing the sources from interfering with each other. What is needed is a way to ensure that all other sources remain disconnected while one member places information on (talks to) the bus. Another disadvantage of Figure 6.21 is that the number of sources may not be expanded indefinitely without modifying the configuration. (OR gates have a limited number of input terminals.) A final aspect of this problem is the fact that the sources are coupled to the bus at a common point; thus, long cables will be required to bring remote sources to the common point on a long bus.

One way to solve the one-talker problem is to use the method presented by Figure 6.22. Here AND gates, controlled by a one of N decoder,* prevent more than one source from using the OR gate at a time. Moreover, the decoder/AND circuit can be employed to ensure a single talker even when other methods are used to couple the sources to the bus.

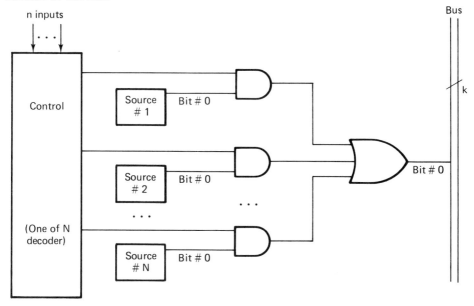

Notes:
1. Each source feeds k bits to the bus.
2. The coupling of Bit #0 from each source to the bus is shown; the other (k − 1) bits are similarly connected.

Figure 6.22 Decoder/AND Circuit Selecting a Single Bus Source.

Open Collector Gates

It is sometimes useful to directly connect the outputs of two or more gates together in wire logic configuration as shown in Figure 6.23. Wired OR is produced in this circuit, but wired AND as well as other functions are also possible. One of the most popular methods for implementing wired logic is by using open collector gates. Here each gate

*A decoder is a circuit for selecting exactly one from N possible outputs. Usually, the decoding is from n inputs to N outputs; thus, $N = 2^n$. (See the beginning of Chapter 7 for more details.)

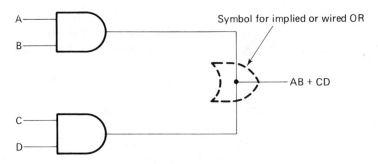

Figure 6.23 Wired Logic.

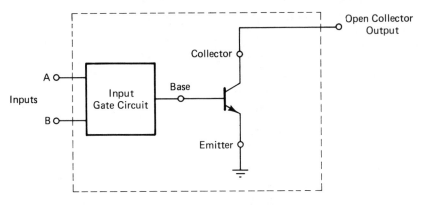

Figure 6.24 Open Collector Gate.

is constructed with the collector of its final transistor, forming a direct output as shown in Figure 6.24. Several gates may then be arranged to form the common (wired logic) output, represented in Figure 6.25(b), by connecting their collectors to a single external resistor–power supply combination. The operation of the composite logic circuit is as follows: If all gates supply no more than zero volts to each transistor's input resistor, each transistor will be off, resulting in the common output being "pulled-up" to the power supply voltage of $+5$ V. On the other hand, any input resistor receiving a voltage somewhat above zero volts (e.g., $+5$ V) will be turned on, causing it to act like a closed switch between the collector and emitter. Thus, one or more on transistors forces the common output to be effectively grounded.

Considering the X and Y voltage inputs to the transistors in Figure 6.25(b) the truth table in Figure 6.25(a) results. By using positive logic on the X and Y inputs and negative logic (i.e., ground $\equiv$ a logical one and $+5$ V $\equiv$ a logical zero) on the common output, F, the right half of truth table reveals that $F = X + Y$, which is the wired OR function. From inspection of Figure 6.25(a), $X = AB$ and $Y = CD$; therefore, the complete circuit yields $F = AB + CD$. (Many more details on the analysis of transistor circuits are given in Chapter 12.)

If an open collector, wired-OR connection were employed in Figure 6.22 instead of a physical gate, each source could be placed on the bus at any convenient

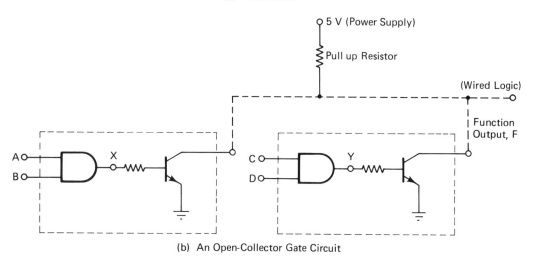

Figure 6.25 Wire Logic Function, $F = AB + CD$.

location rather than at a central point. In addition, more sources could be easily added without having to exchange the original OR gate for one permitting more inputs.

Tristate Logic

While open collection gates have been successfully employed in bus structures for a long time, they are now being replaced in many situations because of their slow speed and the fact that the maximum number of sources permitted on the bus, typically 20, often is too small. Currently, the best solution to the bus coupling problem is the use of tristate logic (also call three-state). It is considerably faster and typically permits more than 100 sources on the bus before practical factors cause any limitations. As the name implies, tristate gates have three possible outputs—the normal logical one and zero, plus an open circuit condition. In the latter state, the gate's output may be modeled by a very large resistance connected to ground. A typical tristate gate is shown in Figure 6.26(a) and its characteristics are summarized in Table 6.9. Note that when the control is in the logical zero state the output is effectively disconnected from the bus so that the input has no effect on the bus. When the control is in the one-state,

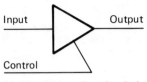

(a) The Tristate Gate Symbol

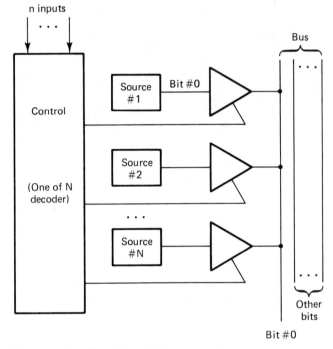

(b) Tristate Coupling of Several Sources to a Bus
(All other bits from each of the N sources are similarly connected.)

Figure 6.26 Application of Tristate Gates.

TABLE 6.9 LOGIC CHARACTERISTICS OF A TRISTATE GATE

Input	Control	Output
—	0	(Open circuit)
0	1	0
1	1	1

however, the output and hence the bus follow the gate input faithfully. Figure 6.26(b) indicates how a number of sources may be properly connected to the bus.

Figure 6.27 demonstrates tristate gates employed to produce a practical bidirectional bus structure with the capacity for a large number of transmit/receive units. The three-position switch forms an elementary circuit indicating how control can be

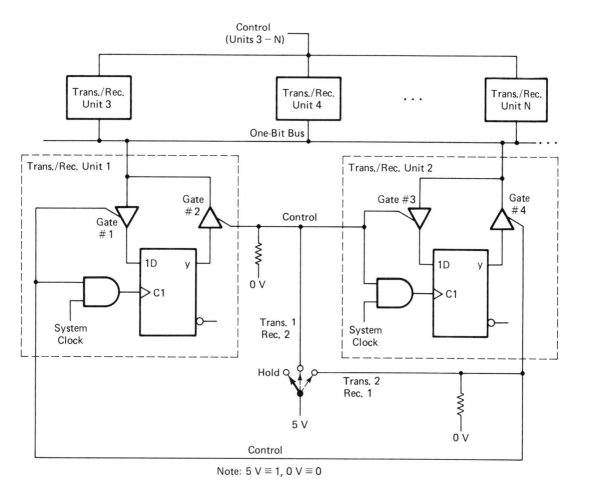

Figure 6.27 Small Bidirectional Bus Structure Employing Tristate Gates.

arranged between two flip-flops so that data are either received, transmitted, or held as the system clock runs continuously. It may appear that the receive ports, gates 1 and 3, could be replaced by a shorting jumper. In practice, particularly when the bus serves several units, the presence of these receive tristate gates is important because they essentially remove the loading of all receive circuits except any which are currently selected. Finally, Figure 6.28 shows the bus structure for four two-bit registers using D flip-flops. The following points should be observed concerning Figure 6.28:

1. To transmit data from register W to register Y, for example, a control pulse would be placed into the Transmit W terminal and into the Receive Y terminal.
2. Wired-OR logic is employed at the point where each AND gate joins the bus, but tristate gates could also be used.
3. By allowing more than one receive terminal to become 1 simultaneously, the same information can be received by more than one register. In contrast, no more than one register can simultaneously transmit information to the bus.

Sec. 6.5   The Bus Subsystem

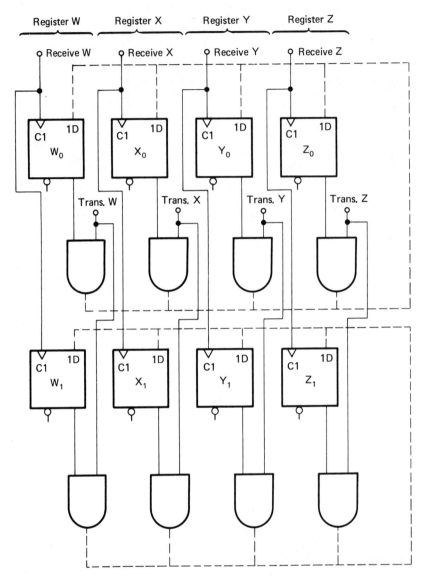

Figure 6.28 Bus Structure for Four Two-Bit Registers (Bus ≡ ------------;
Logic and Control ≡ ―――).

4. Because of the nature of D flip-flops, the signal at the D terminal must be settled a few nanoseconds before the occurrence of the signal at the C terminal. Consequently, the timing must be such that the transmit signal is generated several nanoseconds before the receive signal(s).
5. Only one data word may be transmitted at a time. Thus, there is a trade-off between the simple bus structure and the restriction on the rate at which words may be transferred between registers.

6.6 THE ARITHMETIC AND LOGIC SUBSYSTEM[10]

The Arithmetic and Logic Unit, ALU, is one of the most important parts of a digital computer because this is where the actual "number crunching" is accomplished. There is a wide range in size and complexity of the ALUs employed in computers. Some only include hardware for the add, complement, and shift operations, with all remaining work being accomplished with software subroutines. On the other hand, there are machines that not only have multiply, divide, and floating-point hardware, but also other special wired features such as square root and fast Fourier transform. Later, in Chapter 10, we will study it in greater depth and more extensive functions will be considered. At this point, however, basic hardware for addition and subtraction will be presented to give you a better understanding of how the ALU design effects computer organization and architecture.

Adder Circuits

The adder is the fundamental ALU circuit; it is often employed as a building block for the construction of multiply, divide, and other units. Our goal is to find the combinational circuit, outlined in Figure 6.29, which will realize the $(n + 1)$ sum bits resulting from the addition of two n-bit numbers.

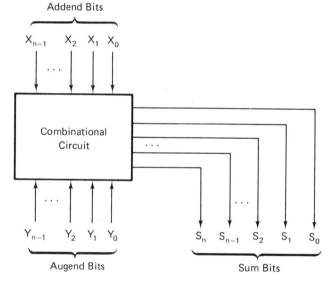

Figure 6.29 n-Bit Adder.

In the Digital Equipment Corporation's PDP-11 family of computers, n is 16; in the Cray-1, it is 64; and in some other machines, it is even larger. Because of the number of variables involved, we cannot hope to design the adder combinational circuit, even for the PDP-11, by employing the same techniques used in Chapter 4. Thus, a heuristic method would seem to be appropriate—perhaps one starting with an elementary *module* which can then be iterated for as many bits as the adder requires. If we try to *segment* this problem into an individual circuit or module, for each bit of the sum, we will soon conclude that the manual addition process—employing one addend bit and the corresponding augend bit, resulting in a sum and carry output for each stage—can easily be implemented. (The resulting circuit is called a *full adder* in computer terminology.)

The addition process can be represented as

$$\begin{array}{c} X_i \\ Y_i \\ \underline{C_{i-1}} \\ C_i S_i \end{array}$$

where C_{i-1} represents the carry-in, C_i represents the carry-out, and the other symbols are defined in Figure 6.29. Table 6.10 presents the required logic relationship between the variables, and Figure 6.30 shows the corresponding Karnaugh maps and the resulting simplified functions.

Note that the most concise form for the sum output is the EXOR function. When one of the full-adder circuits is employed for each bit of the adder of Figure 6.29, the resulting circuit is as shown in Figure 6.31. Since there is no carry into the

TABLE 6.10 TRUTH TABLE FOR THE FULL ADDER

Input			Output	
X_i	Y_i	C_{i-1}	S_i	C_i
0	0	0	0	0
0	0	1	1	0
0	1	0	1	0
0	1	1	0	1
1	0	0	1	0
1	0	1	0	1
1	1	0	0	1
1	1	1	1	1

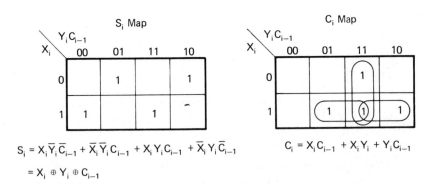

$S_i = X_i \overline{Y_i} \overline{C_{i-1}} + \overline{X_i} \overline{Y_i} C_{i-1} + X_i Y_i C_{i-1} + \overline{X_i} Y_i \overline{C_{i-1}}$

$= X_i \oplus Y_i \oplus C_{i-1}$

$C_i = X_i C_{i-1} + X_i Y_i + Y_i C_{i-1}$

Figure 6.30 Karnaugh Maps and Simplified Equations for the Full Adder.

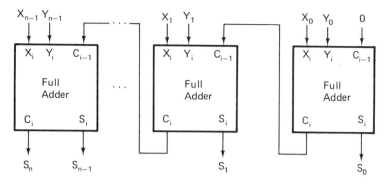

Figure 6.31 Basic Parallel Adder.

least significant bit, the C_{i-1} input to that circuit is constant at logical 0. Also, the C_i output of the leftmost full adder becomes the S_n bit.

Serial addition. Is it possible to do n-bit addition without employing n full adders? Again from heuristic considerations, using an analogy with manual addition, we see that by processing the bits serially only one full adder is needed; but an extra flip-flop will be required to store the previous carry. This scheme is shown in Figure 6.32 where the following process occurs: The current LSB of registers X and Y are combined, in the full adder, with the previous carry, yielding a new sum, S_i, and a new carry, C_i; S_i is shifted into the MSB end of register S, and C_i is temporarily stored in a flip-flop; registers X and Y are shifted right by one bit, placing new values in the LSB position; finally, the process is repeated until the complete sum is stored in register S. The time history of the circuit, when adding two four-bit numbers, is presented in Table 6.11. (Assuming the line marked Start is given, the reader should verify the contents of the table by employing the logic of Figure 6.32.)

As with the parallel adder, $n + 1$ sum bits are generated by two n-bit inputs, and this is the reason that $n + 1 = 5$ clock pulses were employed in Table 6.11 to get S shifted to the proper position in its register. Further consideration will be given to

TABLE 6.11 SERIAL ADDITION EXAMPLES (5 + 7 = 12) USING THE CIRCUIT OF FIGURE 6.32

After Clock Pulse Number	X	Y	FF1	C_{i-1}	X_i	Y_i	S_i	C_i	S
Start	0101	0111	0	0	1	1	0	1	00000
1	0010	0011	1	1	0	1	0	1	00000
2	0001	0001	1	1	1	1	1	1	00000
3	0000	0000	1	1	0	0	1	0	10000
4	0000	0000	0	0	0	0	0	0	11000
5	0000	0000	0	0	0	0	0	0	01100

Result = 12_{10}, as expected.

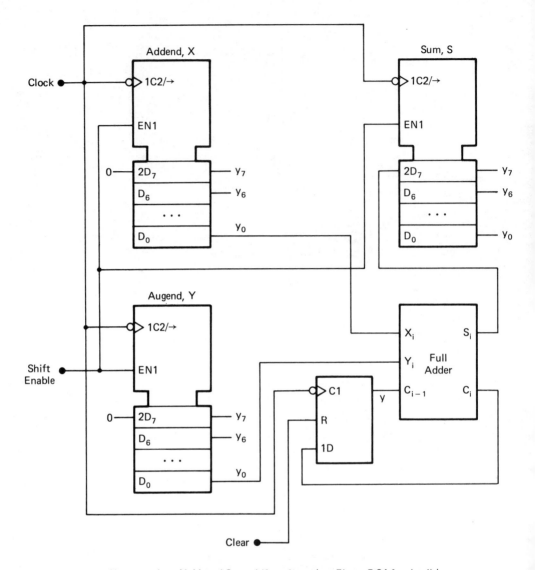

Notes: 1. X, Y, and S are shift registers (see Figure 5.24 for details).
2. The means for loading the registers are omitted to avoid complicating the diagram.

Figure 6.32 Serial Adder.

adders, e.g., gate implementation and the half-adder circuit, in the exercises at the end of this chapter.

Subtraction

The two main ways to perform subtraction are by the complement method, discussed in Chapter 2, and by mechanizing a special subtractor circuit analogous to the adder circuit in Figure 6.31.

The 2's complement method is particularly simple since the carry-out of bit n is discarded, both in producing the complement and in the addition step that follows. By contrast, an end-around carry must be provided in the 1's complement system. This may be accomplished with the circuit of Figure 6.31 by merely connecting the carry out of the nth bit (signal S_n) into the C_{i-1} terminal of bit 0. (The modification needed in the serial adder, Figure 6.32, to allow for 1's complement numbers is left as an exercise.)

Now consider the alternate approach—a circuit which performs subtraction directly by means of the process

$$\begin{array}{r} X_i \\ -Y_i \\ -B_{i-1} \\ \hline B_i D_i \end{array}$$

where X is the minuend, Y is the subtrahend, B is the borrow, and D is the difference. From heuristic reasoning, similar to what was done in the design of the parallel adder, a *full subtractor* circuit is first designed. The corresponding truth table is shown in Table 6.12, and the resulting Karnaugh maps with their simplified equations are shown in Figure 6.33. A complete parallel subtractor circuit looks exactly the same as the

TABLE 6.12 TRUTH TABLE FOR THE FULL SUBTRACTOR

Input			Output	
X_i	Y_i	B_{i-1}	D_i	B_i
0	0	0	0	0
0	0	1	1	1
0	1	0	1	1
0	1	1	0	1
1	0	0	1	0
1	0	1	0	0
1	1	0	0	0
1	1	1	1	1

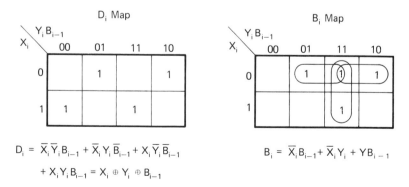

Figure 6.33 Karnaugh Maps and Simplified Equations for the Full Subtractor.

parallel adder in Figure 6.31, except that the sum variables, *S*, are changed to difference variables, *D*, and the carry variables, *C*, are changed to borrow variables, *B*.

6.7 THE MEMORY SUBSYSTEM[7,11]

The memory is another of the major subsystems which is critical to the architecture of digital systems and there has been a truism in the computer field that as memory developments go, so goes computer technology. This is still valid today in that a significant part of a general-purpose computer's cost is memory. Also, the processing speed of most present computers is strongly limited by memory speed.

What portions of the computer are classified as memory? Certainly the *main store*, which holds the current instructions and data, is part of it, with typical members being magnetic core and semiconductor types. The *mass memory* (or *bulk storage*), which contains additional routines and data that must be available at short notice, is also part of it, with disk and drum being representative types. Finally, *archival storage*, for vast quantities of data files and other information, typically on magnetic tape or photographic film, is another form of memory. Moreover, the categories are not as clearly defined as is implied above because, for example, many minicomputer users employ magnetic tape as a form of mass memory while many large-scale computer users employ slow, relatively inexpensive core for the same purpose.

Many different physical phenomena have been successfully employed as a memory medium. These range from simple schemes of punching holes in paper or cards to sophisticated cryogenics. Moreover, there is reason to believe that many engineers and scientists will continue to be employed for a long time in research and development on new types of memory systems.

Even though the phenomenon used may be very different, there are a few basic characteristics by which all memories can be compared. One of these is *volatility*. To say that a memory is *nonvolatile* means that the storage principle employed does not require electrical power to sustain it. Magnetic cores have this property since, once the core is magnetized in a particular direction, it permanently retains that state until changed by an external agent. The catch here is that, in turning off the power to a computer, one must take precautions that electrical transients do not become that external agent. Such precautions, however, usually are not hard to take.

Another basic memory characteristic is whether *nondestructive* readout is possible. It happens that some memory types, which are very good in other respects, destroy the memory contents of a word as it is read. This is not as serious as it at first seems since it is possible to immediately rewrite the word in question from the memory buffer register. The main disadvantage of the rewrite process is that it costs a certain amount of time and money.

The various storage methods are often classified on the basis of *memory access time*, which is the interval between the occurrence of a request for a certain data word and the time that it is available to the CPU. This should not be confused with *memory cycle time*, which is the total time it takes to read and, if necessary, to restore a memory having a destructive readout. The memory access time, in turn, depends on the storage access method. *Random-access memory*, abbreviated RAM, is usually the fastest; and the term, random, itself implies that each word requires the same time for retrieval. Core and semiconductor memories fall into the random-access category, while disk and drum storage are *cyclic access*, since the word of interest comes under the read head in a cyclic manner and the access time depends on where the system is on the cycle when the request occurs. If the word is just about to come under the read head, the access time will be very short, but on the average it is one-half the cycle time. Because information on tape is stored in a sequential fashion, it and similar methods are called *serial access*. Another parameter to consider here is the memory write capability. A memory section for storing permanent data, e.g., a trigonometric sine table, does not require a write capability once it is loaded; thus, cost and time may be saved by omitting the write feature. Such a storage device is called a *read-only memory*, ROM. Moreover, for data only occasionally changed, there are even cases between full write capability and the ROM. One such type is called a *programmable read-only memory*, PROM. Because of the frequency with which computer engineers are involved in the area and because of the opportunity to present many basic principles, this section will be devoted to the design of the main store and related topics.

6.8 SEMICONDUCTOR MEMORIES[7,11]

Semiconductor storage is now the most popular type for main memory applications. As a matter of fact, it has significant advantages, particularly when it comes to speed and to the cost of small memories. It must be understood, however, that semiconductor memories are not homogeneous; they are quite varied in their objectives, construction, and performance. They range in purpose from the random-access memory, RAM, to the read-only memory, ROM. (The latter type will be discussed in Section 7.3.) Bipolar memories are available which are more than an order of magnitude faster than some less-expensive MOS memories. RAMs require a memory cell to store each bit and have two main ways for addressing individual bits: either a linear technique or a coincident technique.

A photomicrograph of a typical 256-word bipolar RAM (the 74S201) is shown in Figure 6.34. The regular pattern in the figure is produced mainly by the transistors and their interconnections forming the memory cells; but a portion of the pattern is due to decoders and other supporting circuitry.

First, consider how the idealized *linear selection* technique shown in Figure 6.35(a) is employed to develop the output data. Note that, by means of the decoder, one of the 16-word lines (w_0 through w_{15}) selects the outputs of its four flip-flops to be placed on the four common data lines, and that a wired OR connection allows the AND gates servicing each line to be tied together. Here all $16 \times 4 = 64$ bits of the memory are represented on a single plane.

The other addressing scheme, *coincident selection*, is shown in Figure 6.35(b). By contrast with linear selection, here four planes are required to represent the same 64-bit memory. Moreover, the reason for the name *coincidence selection* is obvious from the diagram since an AND gate can place information on the data output line only when a logical 1 exists on both its X and Y inputs from the decoder. Since each 1-of-4 decoder has a single logical-1 output, only one memory cell can be selected on

Figure 6.34 Photomicrograph of a 256-Word Bipolar RAM. (Photo Courtesy of Texas Instruments, Inc.)

each plane, and the result at the data output terminal is the state of the selected memory cell in that plane.

Thus far, only the read process has been discussed; we now consider the complete system. The semiconductor memory industry has invested a great amount of effort in developing efficient memory cells since the savings achieved is multiplied by the number of cells in the memory, and this can be a large number indeed—several thousand or more. To establish the basic principles, the idealized coincident cell and its plane, shown in Figure 6.36, are considered. The cell in part (a) of the figure accomplishes data read with an AND gate exactly as was done in Figure 6.35. Moreover, information is stored with a similar coincidence technique in that the x_i and y_i decoder signals combined with the write pulse to control whether or not the clock input of the flip-flop is actuated. Finally, the input information is present on a data input line; but of course it is not stored unless a pulse reaches the C1 terminal. Part (b) of the figure shows how cell wiring is organized to produce a memory plane. (In order to understand memory operation thoroughly, the reader should carefully trace all wiring paths.) Each plane has separate data input and output lines, which thread through each cell in a way analogous to the method employed by the sense lines in a core memory. The x, y, and write lines, however, are common to all planes. Note that, as a means of simplifying the memory plane circuits, a "feed-through" wiring convention is employed. This means that all lines supplying a particular cell pass directly through the

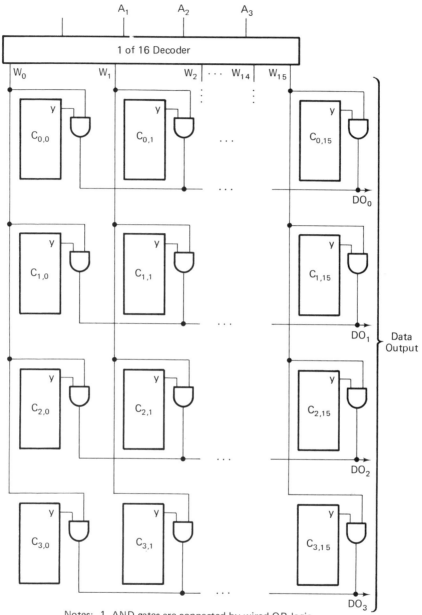

Notes: 1. AND gates are connected by wired OR logic.
2. A_0 through A_3 are from a four bit address register.

(a) Linear Selection System (all output bits) for 16 Word × 4 Bit Memory

Figure 6.35 Memory Selection Techniques.

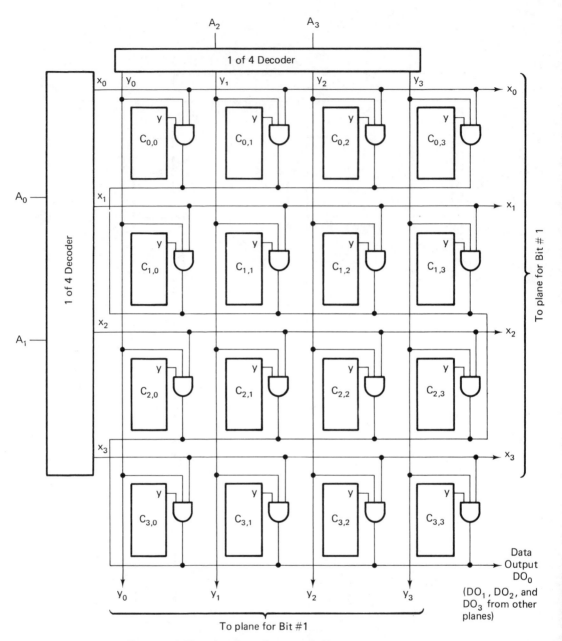

Notes: 1. The other three planes are similar.
2. AND gates are connected by wired OR logic.
3. A_0 through A_3 are from four bit address register.

(b) Coincident Selection System
for Bit Zero of a 16 Word x 4 Bit Memory

Figure 6.35 (Continued)

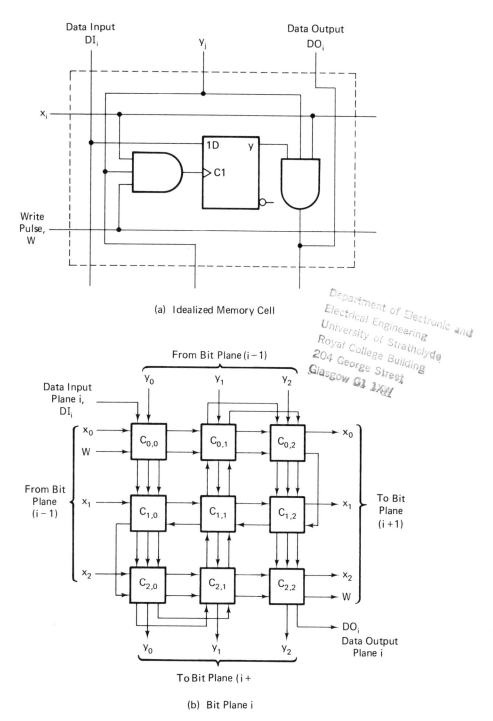

(a) Idealized Memory Cell

(b) Bit Plane i

Figure 6.36 Nine-Word Semiconductor Memory Employing Coincident Addressing.

TABLE 6.13 DEPENDENCY NOTATION SUMMARY II*

Function	Example	Explanation
Three-State Output	▽	Conventional three-state output, which has a very high output impedance when disabled.
Open Collector Output	◇	Output can furnish a relatively low impedance when enabled. An external pull-up resistor to the positive supply is required.
Enable (Symbol ≡ EN) Basic Case	a — EN — y_1, y_2, y_2	All outputs y_i are disabled unless EN is asserted, i.e., $a = 1$ enables all y_i; $a = 0$ disables all y_i.
More Complex Cases (Note that the internal state [e.g., of a flip-flop] is not affected when an output is disabled.)	a — EN, 4 — y_1, y_2, y_3	If input EN is not asserted, all outputs will be disabled, but other inputs are not affected. If EN is asserted, all outputs depend on whatever else affects them. For example: If y_1 and y_2 are from flip-flops, EN = 1 causes their states to be displayed, but y_3 will remain disabled unless 4 is asserted.
	a — EN1 2 — y_3 b — 1EN2 1 — y_4	Selected inputs and outputs are enabled: y_4 is enabled by $a = 1$; y_3 is enabled by $a = b = 1$. When $a = 0$, the input at b is completely disabled.
Memory Addressing	0, 2 } $A\frac{0}{7}$	Bits, 0, 1, and 2 determine an address, A, between 0 and 7, which is the location of the operand.

*Summary I is in Table 5.20.

cell and emerge in the same position on the opposite side. [Refer to part (a) of the figure for details.] As a consequence of this technique, there is no preferred direction for signal flow; it is left to right or right to left, depending only on convenience (up and down are similar). The only time a line does not emerge from a cell is when there is no further requirement for it. The termination of the data input signal within $C_{2,2}$ is typical. (A similar situation occurs with the origination of the data output signal in $C_{0,0}$.)

Consider the situation where a 1 is to be stored in $C_{2,1}$. Here the decoder outputs would be $x_2 = y_1 = 1$ and $x_i = x_j = 0$ for all other i and j. $DI_i = 1$ and the write pulse would be present. Since only $C_{2,1}$ is selected, it would be the only cell which would store the data. The output from $C_{2,1}$ would appear at DO_i first, indicating the original bit contents and then, after the write pulse, the new bit contents.

Additional Dependency Notation

Table 6.13 presents some additional Dependency Notation, which is particularly applicable to the memory systems of the next section. The first members of the table—the three-state and open collector symbols—are to be used with the corresponding output circuits described in Section 6.5, and the next three examples show how the Enable symbol may be employed to control these output circuits. The memory addressing notation in the table was first introduced as a general case in the Dependency Notation summary of Table 5.20, but a specific case is included here because of its application to the elementary memory systems to be presented in the next section.

An Elementary Memory System

Semiconductor memory systems are commonly constructed from groups of moderate-size, elementary memory chips, frequently dual-in-line packages. By employing a number of these packages, the designer has great flexibility in constructing memories of various word and bit configurations. Manufacturers' literature should be consulted for the many practical devices currently available. For convenience, we will employ a hypothetical eight-bit memory, the TUtorial 1 (TU1), shown in Figure 6.37, which

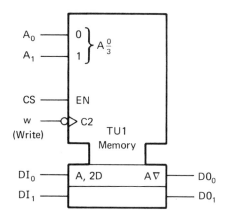

Figure 6.37 Logic Diagram of the TU1 Memory Package.

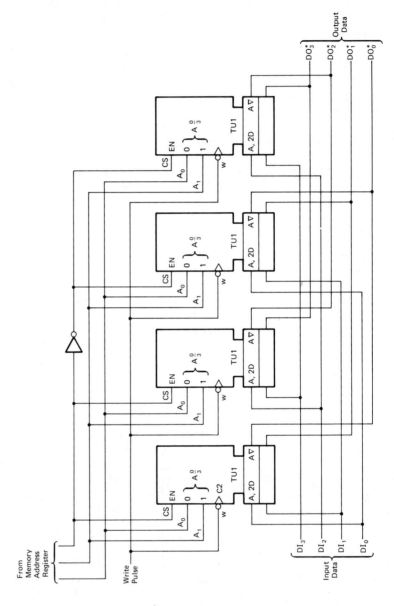

Figure 6.38 8-Word by 4-Bit Memory System Constructed from 4-Word by 2-Bit Packages.

is organized into an array consisting of four words of two bits each. (It is similar in structure to several small commercial bipolar memories, e.g., the Signetics 8225, which is a 16-word 4-bit array). Note how the Dependency Notation allows one to immediately understand the various characteristics of the memory. In particular, the A_3^0 symbol at the top of the diagram, along with the $A, 2D$ and $A\nabla$ symbols at the bottom, reveal how information on the address bus, A_0 and A_1, helps control the storage of input data and retrieval of data to be placed on the output bus.

Since the memory handles two bits in parallel form, at both the input and output, for each addressed word, these are available as DI_0, DI_1, DO_0, and DO_1, respectively. A chip select, CS, and a write, W, control are also provided. The condition $CS = 1$ effectively actuates the chip and allows its data lines to receive and transmit information. (With $CS = 0$, the EN notation reveals that the three-state output, symbol ∇, is in the open circuit state.) Even though $CS = 1$, a pulse must exist on the W line, with a $1 \rightarrow 0$ transition, before the input data are written into the specified memory address. Finally, the chip has an OR* capability, which allows the outputs to be tied to corresponding terminals on other chips so that quite complicated memory configurations can be produced.

As an example of the way memory systems can be constructed from elementary chips, consider Figure 6.38, where an 8-word × 4-bit memory has been constructed from four TU1 chips. Here two TU1's are employed to increase the number of bits per word to four by paralleling both the input and the output lines. In addition, the number of words is increased to eight by having a third address line, A_2, control another pair of parallel chips through an inverter, while the first pair of chips receive the direct A_2 line. (Note that corresponding bits between the pairs of TU1's are OR-tied.) Further consideration will be given to memory systems in the exercises.

6.9 BIBLIOGRAPHY

1. Barbacci, M. R., "A Comparison of Register Transfer Languages for Describing computers and Digital Systems." *IEEE Transactions on Computers*, Vol. C-24, 1975, pp. 133–50.

2. Bell, C. G., J. L. Eggert, J. Grason, and P. Williams, "The Description and Use of Register Transfer Modules (RTM's)." *IEEE Transactions on Computers*, Vol. C-21, 1972, pp. 495–500.

3. Chu, Y. (ed.), "Hardware Description Languages." *Computer* (*IEEE Computer Society*), Vol. 7, No. 12, 1974, pp. 18–66.

4. *Computer*, Special Issue on Hardware Description Language Applications, Vol. 10, No. 16, June 1977.

5. Fletcher, W. I., *An Engineering Approach to Digital Design*. Englewood Cliffs, N.J.: Prentice-Hall, Inc., 1980.

6. Hill, F. J., and G. R. Peterson, *Digital Systems: Hardware Organization and Design* (2nd ed.). New York: John Wiley & Sons, Inc., 1978.

*This is not the conventional wired-OR configuration since all other outputs tied to the active one must be disabled, i.e., in the open-circuit state.

7. Hnatek, E. R., *A User's Handbook of Semiconductor Memories*. New York: John Wiley & Sons, Inc., 1977.
8. Mano, M. M., *Digital Logic and Computer Design*. Englewood Cliffs, N.J.: Prentice-Hall, Inc., 1979.
9. Mano, M. M., *Computer System Architecture*. Englewood Cliffs, N.J.: Prentice-Hall, Inc., 1976.
10. Oberman, R. M. M., *Digital Circuits for Binary Arithmetic*. New York: John Wiley & Sons, Inc., 1979.
11. *The Semiconductor Memory Book*. New York: John Wiley & Sons, Inc., 1978.
12. Stone, H. S. (ed.), *Introduction to Computer Architecture*. Chicago: Science Research Associates, Inc., 1975.
13. Winkel, D., and F. Prosser, *The Art of Digital Design—An Introduction to Top-Down Design*. Englewood Cliffs, N.J.: Prentice-Hall, Inc., 1980.

6.10 EXERCISES

6.1 The initial contents of a group of registers are: $A = 10$, $B = 00$, and $C = 01$. List their contents after execution of the following operation:

$$\text{IF } ((A_1 = 1 \text{ AND } B_0 = 0) \text{ OR } C_1 = 1) \; A_0 \leftarrow 1$$

6.2 The initial contents of a group of registers are: $A = 101$, $B = 0001$, $C = 0010$, and $D = 101$. List their contents after execution of the following operation:

$$\text{IF } (A_1, B = 0, 1) \; C \leftarrow rl2\,D$$

6.3 By employing three-bit, *J-K* registers, draw the logic diagram, similar to Figure 6.1, for the production

$$\bar{B} \leftarrow \bar{A} \wedge B$$

6.4 After the RTL program below halts, determine the following.
  (a) The contents of registers A, B, and C
  (b) The number of times the production at $T = 2$ was executed
  (c) Where the program halts

  $T = 0$ $A \leftarrow 0011$; $B \leftarrow 1011$; $C \leftarrow 0010$;: Only positive numbers allowed;
  $T = 1$ IF $(A_1, B_2, = 1, 0) \; C_2 \leftarrow 1$ ELSE $C \leftarrow \bar{C}$, $T \leftarrow 6$;
  $T = 2$ $C \leftarrow C + 1$;
  $T = 3$ IF $(C < B) \; T \leftarrow 2$;
  $T = 4$ $B \leftarrow 0$;
  $T = 5$ HALT;
  $T = 6$ HALT;

6.5 By employing an algorithm based on repeated subtraction of the divisor from the dividend, write an RTL division program which places the quotient in register Q and the remainder in register R. Assume that:
  (a) Initially the divisor is in register DR and the dividend is in register DD.
  (b) The given values of DR and DD are always positive.
  (c) The fewest possible states, T, are to appear in the program.

6.6 It is desired to add a new instruction called Ṣkip on Ṗositive Ṃemory to Table 3.2. The mnemonic is SPM XXX and the code is 7XXX. The purpose of the instruction is to test whether the content of memory location XXX is positive. If it is, the next instruction is skipped; otherwise, the next instruction in sequence is performed. Write an RTL program for this instruction similar to what is done in Table 6.6.

6.7 The LINC employs an instruction whose mnemonic is LDA, code 1000, with meaning L̰oaḎ A̰ccumulator with the number from memory location Y. The *address* of Y is in $p + 1$. Take the next instruction from $p + 2$. (p is the location of the LDA instruction.) Write an RTL program for this instruction and comment on the number of clock pulses required.

6.8 Write the RTL programs for two new pseudo-LINC instructions: jump to subroutine, JMS, and return from subroutine, RFS. JMS is an instruction located at address p; location $(p + 1)$ contains the address where the start of the subroutine may be found; and after the subroutine is completed the next instruction is taken from $(p + 2)$. Employ the smallest possible number of clock cycles and, within this limit, the smallest number of RTL statements. Any extra registers needed may be labeled RA, RB, etc. (Assume that a subroutine may not be called from within another subroutine.)

6.9 (a) Using clocked J-K flip-flops, draw the logic circuit for the state generator and the IF statement part of the RTL program shown below.

$T = 1$ *;
$T = 2$ *;
$T = 3$ IF $(A = 0)$ $T \leftarrow 1$;
$T = 4$ IF $(A = 1)$ $T \leftarrow 2$ ELSE $T \leftarrow 1$;

Assume that A is a level supplied as an input; the T_i states are in a binary-ordered sequence starting with $T_1 = 00$; and the symbol * represents an arbitrary production which implies $T \leftarrow T + 1$ but has no other effect on the state generator design or on the input A.

(b) Draw the timing diagram for the circuit of part (a); show T_1 through T_4, C_p, and A. Starting the circuit in state T_1 before the first clock pulse occurs, present 12 clock cycles for the case where $A = 0$ through the first four clock cycles and $A = 1$ for the remaining cycles.

6.10 Repeat Exercise 6.9(a) for the program shown below, but employ a ripple counter as a basis for the state generator.

$T = 1$ *;
$T = 2$ IF $(A = 0)$ $T \leftarrow 2$;
$T = 3$ *;
$T = 4$ IF $(A = 0)$ $T \leftarrow 4$;
$T = 5$ through $T = 7$ *;
$T = 8$ $T \leftarrow 1$;

6.11 Write an RTL program to process the following problem. A data source register D contains a new number following each clock pulse. If the $S = 1$ level is present, arrange the next three numbers in three registers, R_1, R_2, and R_3, such that $R_1 \geq R_2 \geq R_3$. When the results are ready, generate a level F, which lasts one clock period. Then return and look for another $S = 1$ level. Minimize the number of states, T_i employed in the programs; but there is no requirement to recover any data lost during the processing of a group of three numbers.

6.12 Write the RTL program for a unit which produces a variable time delay.
   (a) The delay starts whenever the level S goes to 1 after having been at 0.
   (b) The delay is proportional to the contents of register D, but only the outputs of D, i.e., its Y_i and $\bar{Y}_i$ values, are available.
   (c) When the delay expires, the level F is to be set to 1 and stays there for the next 10 clock pulses. After this, the system should be available to start another delay.
   (d) Minimize the number of states, T_i, employed.

6.13 Using the array of Figure 6.13(b), list the pairs of 12-bit words that represent the following characters: A, C, E, 3, and 4.

6.14 Using digital logic not part of a computer, write an RTL program, with fewest possible T states, for a system which adds the contents of memory locations 10 through 15 and places the sum in location 25. Before the program starts, you may assume that the data are in locations 10–15 and that a register F with the condition $F = 0$ is available. Other needed registers should be labeled A, B, C, etc. The memory employs a nondestructive readout.

6.15 How would the RTL programs of Figure 6.17 and Table 6.6 change if a destructive readout memory were assumed? Be specific concerning what RTL statements would be added or deleted at what state, T_i.

6.16 Using the fewest possible states, T_i, and being consistent with the constraints placed on the pseudo-LINC by Table 3.2, Figure 6.16, and other pertinent sections of the text, write an RTL program for a new instruction: Clear on Even location Contents (CEC XXX). If the content of XXX is even, clear that memory location. In either case, the accumulator is unchanged.

6.17 Draw a detailed circuit showing how the individual bit lines are connected from two independent sources onto the bus through tristate buffer gates. The sources and the bus each carry three-bit data and a 1-of-2 decoder provides the control. (A related bus problem is Exercise 7.5.)

6.18 Consult an integrated circuits data manual by a manufacturer such as National Semiconductor, Signetics, or Texas Instruments; report the part number and logic function performed by three open collector and two tristate integrated circuits.

6.19 Find the logic function produced at the output for the following circuits when open collector gates, such as Figure 6.24, are connected to their pull-up resistor: (Assume that positive logic, i.e., ground $\equiv$ logical zero and $+5$ V $\equiv$ logical one.)
   (a) A single open collector gate with an AND input section
   (b) The circuit of Figure 6.25
   (c) The input section is a single wire, which connects directly to the base of the transistor and four such open collector gates are wired together

6.20 A single open collector gate like that of Figure 6.24 is defined to be an open collector NOR if it produces the NOR function when connected to its pull-up resistor.
   (a) If three such two-input open collector NOR gates are wired together, what positive logic function is produced?
   (b) Each gate in Exercise 6.19(c) is an open collector _____. (Fill in the blank.)

6.21 Assuming that complemented inputs are not available, draw the circuit for a basic full adder with AND, OR, and NOT gates; then repeat the process with NOR gates.

6.22 An arithmetic module called a *half adder* has sum and carry outputs but only addend and augend inputs; carry-in is not present.
  (a) Employing NORs, design a logic circuit for the module with a minimum number of gate inputs.
  (b) Employing half-adder modules and any needed gates, draw the minimum block diagram for a three-bit parallel adder.

6.23 What must be done to Figure 6.32 so that 1's complement numbers may be processed?

6.24 Write an RTL program which employs the serial algorithm to convert a number in register X to its 2's complement in register Y.

6.25 Design a two-level combinational circuit which performs the addition of two-bit numbers. Discuss the advantages and disadvantages of the circuit along with its possible extension to more bits.

6.26 In a 2048-word, 32-bit memory employing a coincident selection system:
  (a) How many memory planes are required?
  (b) How many flip-flops are needed per plane?
  (c) How many decoders and what size (1 of ?) are needed?

6.27 Repeat Exercise 6.26 for a memory employing linear selection.

6.28 If we construct a 1024-word, 32-bit memory, how many flip-flops will be required for the memory address register?

6.29 Consider the possibility of constructing a semiconductor memory from shift registers. Present your ideas in terms of a block diagram and a word description. Can this memory truly be called a RAM? Explain. Evaluate your system, e.g., give the advantages and disadvantages of it.

6.30 Using *R-S* flip-flops, draw the basic memory cell and the array for a 4-word × 2-bit coincident-addressing semiconductor memory.

6.31 How can a chip-select signal be implemented for the memory in Figure 6.35?

6.32 Draw the logic symbol similar to our TU1 symbol but for a 16-word × 3-bit memory.

6.33 Using TU1 chips, draw the circuit for a 16-word × 2-bit memory.

6.34 Consider a memory system constructed from a rectangular group of TU1 RAMs consisting of five rows with eight chips per row. Address bits A_0 and A_1 are connected to each chip in the usual way, but bits A_2, A_3, and A_4 go to a 1-of-8 decoder. The decoder outputs, W_i, are connected to the *CS* inputs in column fashion; i.e., W_0 connects to all chips in the first column, W_1 to all chips in the second column, etc. The corresponding chip outputs in each row are joined forming wired OR connections, and the various row outputs form the set of system outputs without further connections.
  (a) How many total bits does the memory system contain?
  (b) How many bits are in each word of the memory system?

Part II

MODERN STRUCTURED LOGIC DESIGN

Chapter 7

Logic Design with MSI/LSI Components

As technology has progressed with larger-scale integrated circuits, two new major methods of performing logic design have become available—MSI/LSI components implementation (identified later more concisely simply as components) and microprocessor implementation. The former technique is the subject of the current chapter and the latter is the subject of Chapters 8 and 9. Actually, the two techniques can complement each other in certain systems; e.g., MSI/LSI components can provide special interface functions for microprocessors. Because of their low cost, great capabilities, and generality, microprocessors are currently the most popular implementation method and should always be given serious consideration for satisfying any system requirement. On the other hand, the components approach also has some significant advantages, such as speed, economy for small tasks, and convenience for certain larger tasks where specialized ICs have been developed, such as for serial data communication.

MSI/LSI digital components themselves can be classified into two major groups, programmable and nonprogrammable or fixed. Programmable components are basically memory arrays whose function (either combinational or sequential) may be arbitrarily configured: by one of the last steps in the IC manufacturing process or by a special programming unit in the field. Nonprogrammable components are ones that implement a specific fixed function which is configured by the design of the semiconductor chip itself; an MSI chip for an up-down counter is a typical example. In this chapter we first describe the fundamentals of some very useful fixed components; then consider the nature of programmable components. Application of both classes of components is presented within the context of Structured Design, whose fundamentals are introduced in Section 7.2.

7.1 FIXED MSI/LSI COMPONENTS: HIGHER-LEVEL PRIMITIVES

The logical operations discussed in Sections 4.5 and 4.6, AND, OR, NOT, NAND, and NOR, are low-level primitives in that many are required to generate a complex function. With the development of integrated circuits, however, a pool of higher-level primitives is now available, making it very easy to accomplish many complex logical design tasks. The EXOR and EXNOR gates, discussed in Sections 4.7 and 4.8, represent a beginning of the higher-level primitives; but they are still relatively low level compared to those to be presented in this section. We will now briefly describe some of the more popular higher-level primitives. The reader is encouraged to consult catalogs from integrated circuit manufacturers for other primitives as well as additional applications information.

Decoders

A *decoder* is a circuit which permits one of N possible outputs to be selected, where $N = 2^n$, and n is the number of bits in a binary word specifying the selected output. The BCD-to-decimal decoder shown in Figure 7.1 is a typical decoder for selecting one output line from ten available lines. It is frequently employed to select digits in a numerical display such as found in digital voltmeters. The dashed input labeled S is the select line. When it is high, the output developed will be the decimal equivalent of the binary number $DCBA$ at the input. For example, when $SDCBA = 10010$, gate 2 will be the only gate with a logical 1 output, and the decimal number 2 will be decoded. When $SDCBA = 00010$, none of the gates will be high because the decoder select bit is off. Finally, when $SDCBA = 11010$, none of the gates will be high because $1010_2 = 10_{10}$ is not a legal BCD number. Decoders are extensively used in memory systems in order to select the word of interest from the many available.

Multiplexers and Demultiplexers

A decoder may be converted to a *demultiplexer* if the select terminal is used as a signal input. When the circuit in Figure 7.1 is employed in this way, the S input stands for signal and one of the ten gates (0 through 9) is selected by means of the input address terminals A, B, C, and D. The signal will pass through only the selected gate; all others will remain at zero. Thus the demultiplexer is a circuit for selecting one of N receivers for the input signal. For example, if line 1 is to become a logical 1, the demultiplexer should receive $SDBCA = 10001$.

The purpose of a *multiplexer* is just the opposite of a demultiplexer, i.e., the multiplexer is a circuit for selecting one of N input data sources to be transmitted on a single output channel. Figure 7.2(a) shows the detailed logic for a 4-to-1 multiplexer, and Figure 7.2(b) shows the corresponding dependency circuit symbol. In Figure 7.2(a), note the function of the enable,* the select, and the data input sections of the

*For the remainder of the text, the symbol E is employed for the *external* chip enable (similar to chip select) to distinguish it from the standard internal enable, EN.

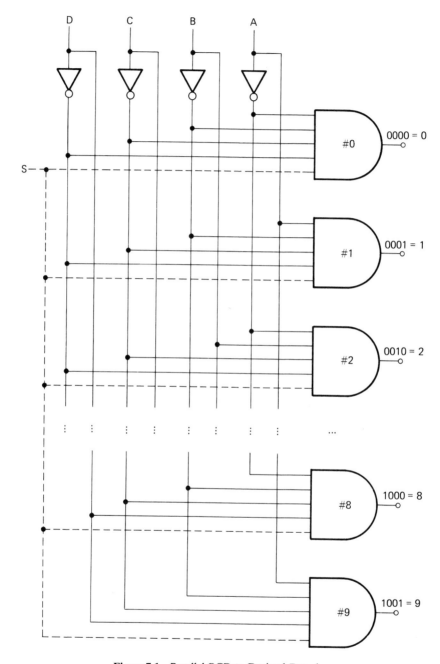

Figure 7.1 Parallel BCD-to-Decimal Decoder.

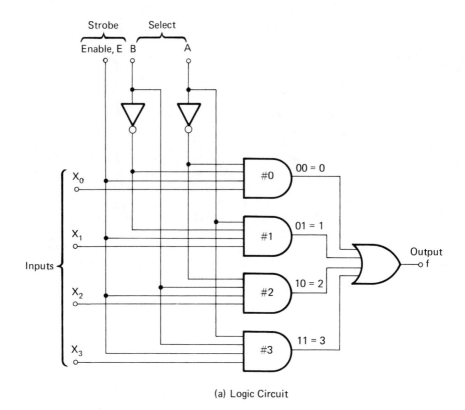

(a) Logic Circuit

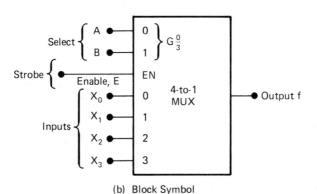

(b) Block Symbol

Figure 7.2 4-to-1 Multiplexer.

circuit. For example, to pass the information on input X_2 through to the output, it is necessary to select output gate 10, which means the control signals become $EBA = 110$. It is important to understand that when the enable signal is low ($E = 0$), the output will remain at zero, independent of the data or select inputs. (Note that the dependency notation requires the subscript i for the X_i input only to be shown on the block symbol.)

Expansion circuits. Since the multiplexer is a primitive, it is not surprising to find that it can be combined with itself and with other primitives to perform a large number of different operations. One of the simplest operations is *expansion*, which combines two 4-to-1 multiplexers to form the 8-to-1 multiplexer shown in Figure 7.3. Note how the enable input is employed to control which multiplexer is active. In this case, in order to pass the data on y_6 through to the output, the control signals must be $CBA = 110$. (Further attention will be given to expansion circuits in the exercises at the end of the chapter.)

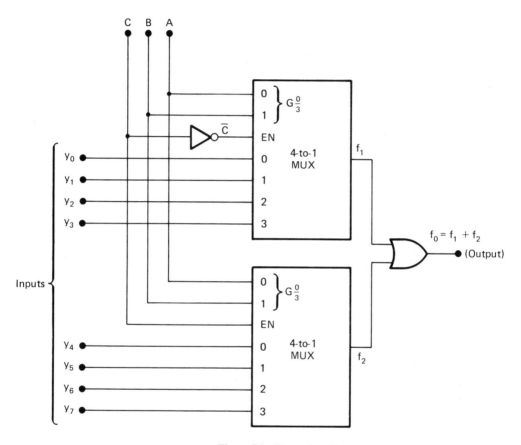

Figure 7.3 Expansion Circuit.

Realization of Boolean Functions

Multiplexers provide a very efficient way to realize Boolean functions. Consider Figure 7.2(a). When $E = 1$, the function realized by that circuit is

$$f(B, A) = X_0\bar{B}\bar{A} + X_1\bar{B}A + X_2B\bar{A} + X_3BA$$

With X_0 through X_3 set to the proper combinations of zeros and ones, $f(B, A)$ generates the minterm expansion of any two-variable function. For example, to generate the EXOR function, it is only necessary to make the following selections:

$$X_0 = X_3 = 0 \text{ and } X_1 = X_2 = 1$$

The corresponding circuit is shown in Figure 7.4. Thus, we see that there is no difficulty in generating any of the 16 functions of two variables with a 4-to-1 multiplexer. Of course, an 8-to-1 multiplexer would be required for the 256 functions of 3 variables, etc.; but in principle, any function can be generated with a multiplexer. Moreover, instead of hard-wiring the X inputs as shown in Figure 7.4, we could connect them to a set of switches or to another circuit so that the function generated could be changed quite rapidly.

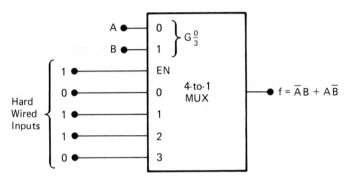

Figure 7.4 Multiplexer Realization of the EXOR Function.

Minimization of multiplexer cost. For each additional "select-variable" required by a multiplexer, the number of data input pins doubles, increasing cost; thus, minimization techniques can be important. Sometimes a Karnaugh map will show that one or more variables can be completely eliminated from a function; this reduction should, of course, be done first. Then, if a least one variable in a function being implemented is available in both complemented and uncomplemented form, the following method guarantees that the design can be accomplished with a multiplexer one size smaller than permitted by our original method. With this modified procedure, first assign the variable in both forms to the data input section; then assign the remaining $m - 1$ variables to the select inputs.

Figure 7.5 shows the way any four-variable function can be produced by means of a single 8-to-1 multiplexer. Here a combination of D, $\bar{D}$, 1, and 0 terms are employed as data inputs, resulting in the function

$$f = CB\bar{A}\bar{D} + CB\bar{A}D + \bar{C}BAD + C\bar{B}A\bar{D}$$

You should now be asking: "How can the required data inputs be systematically determined?" The answer is that, as expected, a Karnaugh map is the key to our solution. In Figure 7.6 first notice the regions labeled with the small numbers 0 through 7. These are controlled by the corresponding terms X_0 through X_7; e.g., region 5 is $CBA = 101$, which is for the X_5 input. Again for region 5, observe that only the right square contains a logical one. Thus, the data input is $X_5 = \bar{D}$; and its configuration satisfies the minterm $C\bar{B}A\bar{D}$, which is the last term in the required

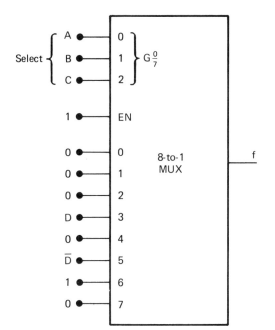

Figure 7.5 Economical Realization of a Four-Variable Function.

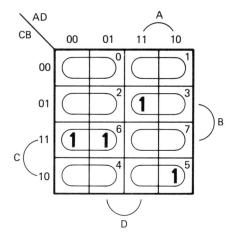

Figure 7.6 Karnaugh Map for the Figure 7.5 Function.

function. On the other hand, both parts of region 6 contain ones; thus $X_6 = \bar{D} + D = 1$, which generates the first two minterms in the function. Our improved method for implementing logic functions should now be clear:

1. Plot the function on a Karnaugh of proper size.
2. After labeling the multiplexer's X_i regions on the map, determine whether each data input is to be a 0, 1, V, or $\bar{V}$, where V is the variable available in both uncomplemented and complemented form.
3. Finally, draw the multiplexer showing the data inputs developed in step 2, and test to see that the desired function is produced.

A general way to look at the above Karnaugh map process is that the select inputs control an X_i region on the map, which then intersects with the region determined by the data input variable(s) to produce a sum-of-products term. In Figure 7.6, the X_5 region intersects with the $\bar{D}$ region to produce the term $C\bar{B}A\bar{D}$. Here with three select variables in a four variable function, there is only one variable (either D or $\bar{D}$) for the intersecting region and every possible four variable function can be realized as long as both forms of the variable are available. On the other hand, if only two select variables were available (i.e., a 4-to-1 multiplexer situation) in a four-variable function, each of the other two variables would be available to intersect each X_i region of the map. In certain functions, the required data inputs will again be at most single variables, resulting in a complicated four-variable function being

Sec. 7.1   Fixed MSI/LSI Components: Higher-Level Primitives realized by a single 4-to-1 multiplexer. Less general but useful designs of this nature are considered further in the exercises.

Multiplexers employed as read-only memories. When a simple *read-only memory* is required, it is frequently very convenient to do the design by employing multiplexers as primitives. Read-only memories will be discussed in Section 7.3, but it is sufficient here to say that they are a type of memory which contains permanent (or semipermanent) information such as a table of trigonometric functions. As an example, assume that the function $R = t^2 + 5$ (where $0 \leq t \leq 3$) is to be used very frequently in a special-purpose digital system. Instead of performing the calculation in the arithmetic unit each time a new value of R is required, a read-only memory could be constructed as specified in Table 7.1. The possible input numbers are shown first in decimal, t_{10}, and then in binary, t_2, in columns one and two. A similar tabulation of the results is given in the next two columns. In the case of the binary results, R_2, note that four bits, a through d, are required; each of these bits represents a different function generated by a separate multiplexer. Thus, in response to $t_2 = BA = $ 00, 01, 10, 11, the multiplexer for bit d generates the bit sequence 0, 0, 1, 1, which corresponds to the function $f(B, A) = B\bar{A} + BA$. The complete multiplexer circuit may be obtained by inspection of Table 7.1, and it is left as an exercise at the end of the chapter.

TABLE 7.1 SPECIFICATIONS FOR A MULTIPLEXER READ-ONLY MEMORY

Number		Results $R = t^2 + 5$					
t_{10}	t_2	R_{10}	R_2				X_i
	BA		d	c	b	a	(Ref.)
0	00	5	0	1	0	1	X_0
1	01	6	0	1	1	0	X_1
2	10	9	1	0	0	1	X_2
3	11	14	1	1	1	0	X_3

In addition to the applications just discussed, there is a wide variety of other important uses of multiplexers. Whenever one needs to generate a series of Boolean functions or to perform a complex switching task, there is a good chance that a multiplexer circuit will be applicable. Again, the reader is referred to manufacturers' catalogs on integrated circuits for many additional practical applications.

7.2 STRUCTURED DESIGN WITH FIXED COMPONENTS

Now that we have a basic understanding of fixed components, it is appropriate to consider how they may be employed in practice. Design is one of the most interesting topics we will consider because it is a challenge to the engineer's imagination and ingenuity.[1,5,7,13,16] Although our emphasis continues to be quantitative, there is no longer a unique solution to our problems. Also, there are many different factors, not considered in analysis that enter into a successful design—some of these are reliability, cost, and maintainability. A quantitative evaluation of these factors is beyond the scope of this text, but where possible they will be mentioned in a general way. All of the designs here as well as in the next two chapters are based on a Structured Design philosophy; thus, it is appropriate to first consider the fundamentals of this concept.

Structured Design

Major features of Structured Design are that the process itself is very systematic and the resulting logic circuits have a high degree of regularity in both their topology and in their hardware algorithms. Moreover, a hierarchy is clearly visible in the breakdown of the system into subsystems, then into modules, and finally into components (individual integrated circuits). The major goals of Structured Design are:

1. *Reliability.* The system contains a minimum of "bugs" and the probability of failure is quite low. It does not require exhaustive tests to ensure proper operation.
2. *Maintainability.* It is not only simple to repair, should a component fail, but also new features and improvements are conveniently added.
3. *Well-documented.* People other than the original designer can quickly visualize the overall circuit plan so that repairs and additions can be completed at low cost.
4. *Designability.* A complete system may be rapidly and economically designed. It may be easily partitioned into two or more sections so that different people can work on the detailed circuit implementation in parallel rather than in series.

Structured Design exists on many levels in the development of modern digital systems. It has recently become very popular in producing reliable software, yet the importance in digital hardware can be traced back at least as far as M. V. Wilkes,[15] who in 1951 introduced the concept of a controlled memory with the attendent microinstructions. (This concept is described in Section 7.4.)

Although the detailed steps in Structured Logic Design depend somewhat on the nature of the components employed, the general features are the same. (See the end of Section 7.2 for the specific design steps with components, and Section 8.5 for those with microprocessors.) The following are the seven main properties of a well-structured system:

1. A high degree of coherence and regularity exists not only in the component arrangement but also in the corresponding systematic pattern of the logic functions and algorithms.
2. The highest-level logic primitives feasible are employed. In particular, a single high-level MSI/LSI integrated circuit is preferred instead of two or more lower-level chips even if only a fraction of the higher-level chips capabilities are actually used.
3. The system is divided into reasonable subsystems and these divided into lower-

level subdivisions to a depth dictated by complexity, with each subdivision being carefully arranged to constitute an operationally complete entity.

4. The hardware algorithms (and software too where employed) are logically straightforward and simple to understand. (This is in contrast to special tricks whose secret even the inventor needs help in remembering after a few months.)
5. Fairly low priority is given to minimization, particularly at the gate level—clarity and a systematic arrangement of elements is much more important. Even a "brute-force"* circuit may be preferred to a logically complex, minimum one if the former's operation is much clearer and if it meets all other requirements of the specification. The exchange of hundreds of lines of programming or a few components for a more reliable and maintainable system is usually a bargain. This is especially true in this age of low-cost memories and other devices.
6. The Top-Down Design philosophy is employed. This is a very systematic approach that starts with a careful study of the requirements and possible trade-offs (including hardware versus software). The system is then partitioned into reasonable subsystems with repeated cycles of evaluation, simplification, and improvement *before* chip selection, detailed circuit layout, and program coding actually begin.
7. The circuit notation emphasizes the structural features and contributes to the reader's understanding of the system's operation. (The dependency notation appears to be particularly good in this regard, especially for large systems.)

Multiple Primitives from a Single Component

In producing a structured MSI/LSI design for a certain problem, it is frequently helpful to think of the available logic components in terms of the several primitive operations that each may perform instead of concentrating only on the function suggested by the chip's common name. For example, a shift register can be thought of in at least seven different roles, each of which emphasizes a different useful capability. To explain the concept of multiple primitives from a single component, we now consider a circuit in which three out of four of its chips perform a function other than that most frequently associated with the device.

One efficient method for sending binary data over long distances is by a phase-shift-keyed, PSK modulator. The principle of this modulator[2] is to transmit a square-wave signal that has been phase shifted (between 0 and 360°) by the input data with respect to a reference carrier of the same frequency. For example: to transmit a three-bit number requires $2^3 = 8$ phase steps of $360°/8 = 45°$. Thus, a phase shift of 225° corresponds to the decimal number $225°/45° = 5$. It is theoretically possible, however, to transmit simultaneously 6 bits, or even larger numbers, over a single pair of wires by this technique. An inexpensive PSK modulator that transmits four-bit numbers is shown in Figure 7.7. The principle of operation is as follows:

*This is a circuit which *appears* to have been designed in such a very direct unimaginative way that some factors leading to circuit economy are neglected. A simple example is a system of three functions which are each implemented independent of common terms available from the other two.

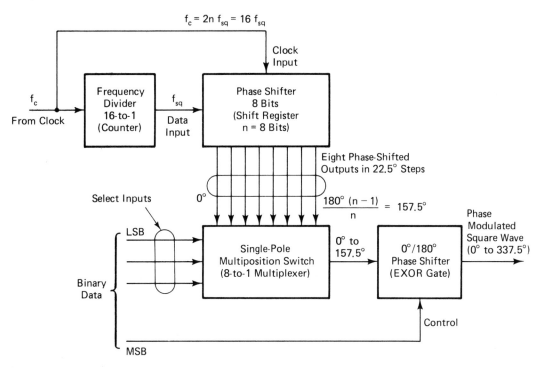

Figure 7.7 Detailed Design of a Phase-Shift-Keyed Modulator. (Note alternative functions produced by common digital components.)

1. The most significant bit of the frequency divider, counter, produces a square-wave carrier whose frequency, f_{sq}, is one-sixteenth of the clock frequency.
2. As the carrier input moves through the phase shifter, shift register, there is a delay, phase shift, of 22.5° per bit with the least significant bit, the input, having a 0° phase shift with respect to f_{sq} and bit four having a 22.5° × 3 = 67.5° phase shift.
3. The single-pole switch, multiplexer, allows the data to select any phase angle step between 0 and 157.5° to serve as the input to the 0°/180° phase shifter.
4. The 0°/180° phase shifter, EXOR gate, is controlled by the most significant bit of the input data. Thus, any phase shift between 0 and 337.5°, in 22.5° steps, may be achieved by the complete system, and the phase shift is an increasing function of the magnitude of the binary data.

The counter acting as a frequency divider, the shift register acting as a phase shifter, and the EXOR gate acting as a 0°/180° phase shifter are not the most common applications of these components. This illustrates our point that a good designer must be alert to alternative interpretations of the logical performance of standard digital components. As an aid to thinking about these alternative interpretations, Table 7.2 has been prepared for several common digital components. You should be aware, however, that this table is far from exhaustive—it is only representative of the multiple

TABLE 7.2 CHARACTERISTICS OF TYPICAL MSI/LSI COMPONENTS

Component (Chip) Name	Functional Operation	Detailed Description
Shift register n = bits in the register f_c = clock frequency T = clock period	1. $180°/n$-step phase shifter 2. Time delay, τ $[\tau = nT \text{ (sec)}]$ 3. Serial-to-parallel data converter 4. Parallel-to-serial data converter 5. Multiply by 2 6. Divide by 2 7. Circulating (ring) memory	1. A square wave of frequency f_{sq} is applied to the serial input and the clock frequency, f_c, is the shift signal; $f_c = 2nf_{sq}$. After the first n clock pulses, $180°/n$ phase steps will be present between successive register bits. 2. Starting with a cleared register and a logical one at the serial input, a delay of τ seconds is required before the MSB goes to one. 3. A serial data input is shifted by n clock pulses. Then the n register output bits are the parallel equivalent of the original serial data train. 4. This is the inverse of the operation in 3. The register is loaded in parallel and n clock pulses produces an equivalent serial data train at the output. 5. After the register is loaded, a single shift left produces the required multiply at the n output terminals. 6. After the register is loaded, a single shift right produces the required division at the n output terminals. 7. After the register is loaded and the output connected to the input, each clock pulse rotates, shifts, the data by one bit. The results are available at the n output terminals.
Counter n = bits in counter T = clock period	1. Frequency divider 2. Sequencer 3. Time delay $\tau = \text{(count)} T$ 4. Up-counter 5. Down-counter	1. Output frequency at the nth bit is the input frequency divided by 2^n. 2. 2^n input clock pulses sequence the n flip-flop outputs through all possible combinations. 3. After reset, clock pulses are summed and the interval until a present count is detected provides a delay of τ seconds 4. Pulses are counted (summed) in binary (BCD, etc.) since the last reset. 5. Each input pulse causes the register to decrement by one.
Multiplexer n = select lines	1. Universal function generator 2. Single-pole output, multiposition switch 3. Parallel-to-serial conversion	1. A logic one or a zero at each of the 2^n data inputs (data inputs are minterms determined by the select variables) either includes or excludes that term. 2. A single output is selected from one of 2^n inputs. (n select lines are required.) 3. Bits of the parallel word are simultaneously entered at the 2^n data inputs. As the select inputs are sequenced a serial version of the parallel data word is produced at the data output.

TABLE 7.2 (CONTINUED)

Component (Chip) Name	Functional Operation	Detailed Description
Demultiplexer	1. Decoder	1. n line-to-2^n-line decoder; e.g., a common 4 line-to-10 line unit is a BCD-to-decimal decoder.
n-select lines	2. Single-pole input, multiposition switch	2. A single input is unidirectionally connected to one of 2^n possible outputs by n select signals.
T = clock period	3. Serial, ordered distributor (serial-to-parallel conversion)	3. As the select inputs are sequenced, producing a scan pattern, serial information at the data input pin is sequentially distributed to the 2^n data output pins, producing a bit-ordered word. (Since the 2^n bits do not occur simultaneously, a register is required to hold the results for serial-to-parallel data conversion.)
EXOR (two-input gate)	1. 0°/180° phase shifter	1. Input A is the signal, C is the control.
	2. Inequality detector	2. If the two inputs are unequal (i.e., 0 and 1 or 1 and 0) the output is true.

primitives possible from some of the currently popular MSI/LSI chips. The reader may want to keep a notebook in a form similar to Table 7.2 and make new entries as good primitives are identified in practice and in various published sources.

Contrary to the well-developed tools, Karnaugh maps, etc., available for SSI design, in the case of MSI/LSI there are no formal methods. One can visualize, however, a four-step, *Top-Down Design Process* which could be employed in obtaining circuits like that in Figure 7.7. We call the four steps Problem Definition, Preliminary Design, Detailed Design, and Component Finalization. Problem Definition is the initial stage, in which the circuit specifications are studied, the input and output signals identified, and the basic system operation defined by means of a very small, general block diagram. For Preliminary Design the basic solution is conceived and expressed by means of a small somewhat simplified block diagram. It is here where one mentally runs through lists of primitives like Table 7.2 searching for the "breakthrough" which will lead to a good solution. In Detailed Design, the careful study takes place; the range of inputs and required outputs are considered; and a complete block diagram of all aspects of the total system is produced. During Preliminary Design as well as during Detailed Design, the engineer is cognizant of currently available chips that are likely to satisfy the block diagrams. Moreover, the question is continually being asked, "Can this operation be done in a better way?" Component Finalization is the last stage and involves identifying the exact integrated circuit (its part number) which satisfies the requirements of Detailed Design.

Figure 7.8 summarizes our four-step design process and indicates by means of decision blocks where a "dead end" may be reached, forcing the engineer back to a previous stage. A problem of this type occurs, for example, when Preliminary Design requires one or more functions that seem either very difficult or impossible to implement in the Detailed Design stage. It is frequently possible to resolve the dilemma

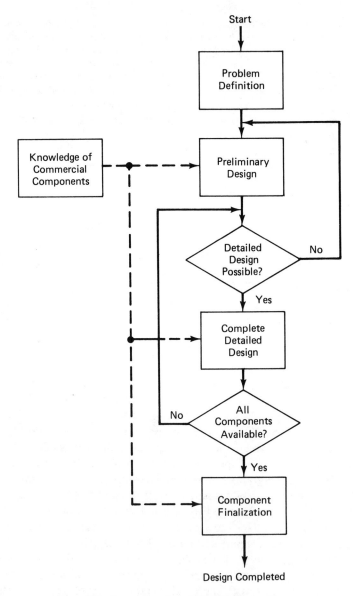

Figure 7.8 Summary of the MSI/LSI Design Process.

merely by modifying the Preliminary Design slightly and then continuing the remainder of the process. Notice that in a particularly bad situation, a "No" at the Chip Available decision could lead back to testing for Detailed Design Possible which could fail, finally leading to the requirement for a completely new Preliminary Design or at least a major modification.

At each step in the design sequence you should be asking yourself, "Is there a simpler way to do this?" The question is equally appropriate no matter which of the four design stages is being considered. With MSI/LSI implementation, to us "simpler" will always mean a system resulting in fewer chips, but in practice the exact criterion depends on many diverse factors. Furthermore, once the first circuit version is complete you should repeat the process starting with a significantly different Preliminary Design. Moreover, in the case of a high-volume production system, a complex one, or a system with very stringent cost requirements, it may be necessary to do several independent designs to be sure that a satisfactory result has been achieved. As the steps of Figure 7.8 are accomplished refer back to the beginning of this section, where seven properties of a well-structured system are listed, and be sure that these required properties are fully assimilated into the design. Note that Figure 7.8 itself is a prime example of the Top-Down process.

Figure 7.9 shows the block diagrams for three out of four of the design steps for our PSK modulator. (The third step, Detailed Design, has already been presented in Figure 7.7.) In Problem Definition, Figure 7.9(a), the PSK modulator is described by means of a single block; nevertheless, the input, output, and the system function should all be clear. Progressing to Preliminary Design, Figure 7.9(b), it is seen that the concept of a shift register to generate the various phase shifts is at the heart of the design. Once that concept was established the addition of a multiplexer to "switch-in" the required phase, as determined by the binary data, was a fairly easy step. If an eco-

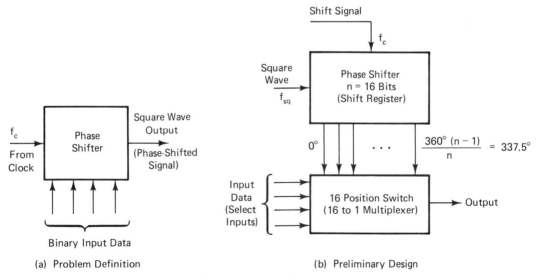

Figure 7.9 Design Steps for the PSK Modulator.

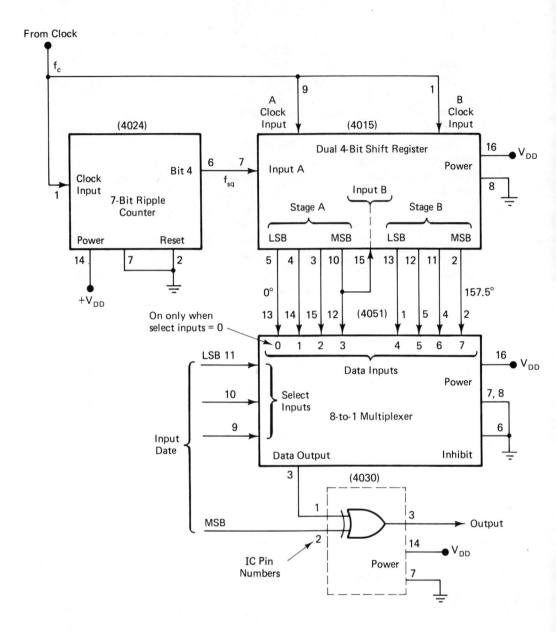

(c) Component Finalization

Figure 7.9 (Continued)

nomical parallel-in serial-out 16-bit shift register were available, the preliminary design of Figure 7.9(b) could also form the Detailed Design. It would, of course, be possible to construct a 16-bit shift register from two eight-bit chips; but realizing that a simple EXOR gate provides a 0°/180° phase shift, the decision was made to follow that course and the result is shown in Figure 7.7. Evaluating trade-offs like that of employing an eight-bit shift register, an eight-bit multiplexer, and an EXOR gate instead of a 16-bit multiplexer and shift register chips is one of the main tasks in Detailed Design.

In both approaches the same number of components are required; but if the two shift registers in the 16-bit approach are balanced against the multiplexer and shift register in the eight-bit approach, we are left with comparing a small SSI EXOR chip with a larger, more expensive 16-bit multiplexer. Thus, the eight-bit approach is slightly more economical both because a smaller component area usually leads to cheaper printed circuit boards and because of the less expensive chip itself. (It could be argued, however, that the eight-bit approach has the disadvantage of being slightly more complicated conceptually.) Using the Detailed Design of Figure 7.7 it is easy to select components from a manufacturer's data book in one of the logic families, such as TTL or CMOS.* Figure 7.9(c) shows the component Finalization step implemented in CMOS. Notice that the pin numbers employed on each chip have been identified from an IC data book. The diagram shows that the shift register is a dual four-bit chip which is then series wired (by connecting the MSB of the A stage, pin 10, to the B-stage input, pin 15) to form a single eight-bit register. By taking the counter output at bit 4, the frequency divide factor becomes $2^4 = 16$, which is the desired value.

The above discussion helps to demonstrate that a system specification can lead to several different final designs, all of which are good results and fully in compliance with the original requirements. Although for tutorial reasons we have picked a fairly simple example to demonstrate the design process, it is certainly also applicable to large problems.

7.3 PROGRAMMABLE MSI/LSI COMPONENTS

There are many situations where circuits constructed from fixed MSI components can provide an excellent design from cost, logic delay, conceptual, and other standpoints. On the other hand, for large circuits and specialized functions, programmable components frequently furnish a better structured, more powerful, lower cost, and general approach. Manufacturers have conceived a number of programmable components, but the two classes to be discussed here, the read-only memory (ROM) and the programmable logic array (PLA), are representative of the available types. Although the nature of programmable components automatically satisfies some of the properties of a well-structured system, the designer still has a good deal of freedom in selecting component parameters. Moreover, if the designer is not careful, a poorly structured system can be constructed with programmable components almost as easily as it can with fixed SSI components.

Figure 7.10 shows an engineer programming a set of PROMs, user-programmable ROMs. Commercial programmers such as this are sold by several manufacturers. These are furnished with a large variety of convenient features, such as a key pad, a display, and a set of editor instructions for reading and writing data; a routine which copies a master data set into a second memory; a process which verifies that

*See Chapter 12 along with References 4 and 14 for information on these families.

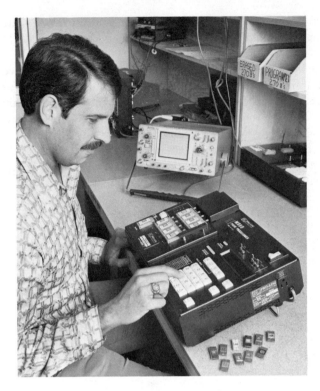

Figure 7.10 A Commercial PROM Programmer. (Courtesy of Pro-Log Corporation.)

a second memory is a faithful copy of a master; the ability to operate under the control of a minicomputer or other supervisory processor; and many other useful operations. Most commercial programmers are built to accept a special unit called a personality module, which configures the programmer to develop the proper signals for burning (programming) a few or in some cases a fairly large subset of PROMs, PLAs, or other programmable devices. Thus, one mainframe and several personality modules can be purchased to yield an economical but broad programming capability.

Read-Only Memories

Its low cost, versatility, and high operating speed have elevated the ROM to the status of a basic building block for digital circuit design. Typical applications include: control sequence generator for a large central processor through a microprogram; storage of frequently used subroutines; character generators; table lookup schemes; large combinational circuits; and translators from one code to another, e.g.. from Gray code to BCD.

A ROM is a memory with many of the same properties of a RAM, except that the data are either permanently or semipermanently stored; therefore, it is not subject to destructive readout, volatility, and similar problems. It can be constructed from diodes, bipolar transistors, MOSFETs, and many other types of devices; also, both linear as well as coincident word selection are employed. ROMs constructed for high-volume applications having permanently stored information are made from photographic masks in large production runs much like ordinary ICs. Moreover, versions with a capacity of thousands of bits are commonly produced.

In order to gain physical understanding of how a ROM may be constructed, consider the elementary diode* memory shown in Figure 7.11. Observe that the ROM transmits a logical 1 to the output if there is a diode connecting the active word select line with the output line; otherwise, it transmits a 0. (The disconnected diodes are shown with one of their leads cut to represent the omission of part of conductor such as would be done with integrated-circuit fabrication.) The reader should have no trouble verifying that word W_1 in the figure does indeed produce voltages at the DO_i output terminals corresponding to 1101. It should also be observed that the diodes connected to the unselected word have no effect on the output since these diodes do not carry current.

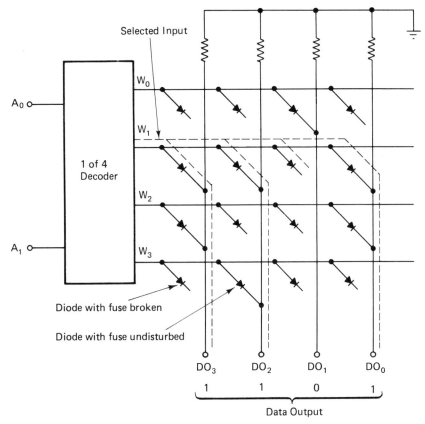

Note: For both word select and data output lines, W_i and DO_i:
"0" = 0 volts and "1" = 5 volts. (Neglect diode drop)

Figure 7.11 16-Bit Diode ROM Employing the Linear Selection Technique.

*Recall that a diode permits conventional current to flow only in the direction of its arrow symbol. For more details, see Section 12.1.

Programmable read-only memories. For small production runs and for developmental systems, it is necessary to place information into a read-only memory in the field quickly and conveniently. For these reasons the *programmable read-only memory* or PROM was developed.

There are two main classes of PROMs: the one-time programmable type and the reprogrammable type. The former involves cutting a special link in the memory array for each 0 created. Typically, this is accomplished by manufacturing an array with all available positions filled with diodes or other components. Then the word of interest is selected by the usual addressing method, and a large-amplitude current pulse is placed on each bit terminal where a 0 is to be created. The pulse causes the fuse link to be broken at the location of each desired 0; but the other diodes are undisturbed. After programming, the memory can be operated in the same way as a factory-fabricated version, and it will produce a 1 output where a diode is intact and a 0 output where the fuse has been broken.

Typically, the reprogrammable PROM employs MOSFETs as memory elements, in place of diodes. The gate of each device, however, is electrically insulated from all other parts of the transistor. Programming is accomplished in a similar way to the fuse type—the desired word is selected as usual, but this time instead of breaking a link to create a 0, an avalanche effect is produced in the selected FET causing a charge to be deposited on the gate, which stores a 1. (Since the gate is electrically well-insulated, the charge will remain for years.) The gates that did not receive a pulse remain uncharged and thus store 0's. After programming, the PROM can be read in the usual way by selecting one word at a time. The FETs storing a 1 have a low-impedance path on the transmission gate between the select line and the bit line, while those storing a 0 have an open circuit.

The charge on the FET's gate requires an energy of about 5 electronvolts to escape; but sunlight and thermal excitation associated with normal operating temperatures provide for less than the required amount. On the other hand, ultraviolet light has sufficient energy to liberate the charge from the gate, thereby clearing the memory. This process is facilitated by means of a small window in the top of the PROM. With commonly available ultraviolet lamps, it requires about 10 minutes to complete the process. A typical PROM, the 16,384-bit 2716, is shown in Figure 7.12. Note how the picture has been composed to simulate the ultraviolet erase process.

Applications of read-only memories. In applying ROMs, there are two alternate paths that can be taken: either a *word approach* or a *Boolean function approach*. Both methods can be thought of as generating a truth table with the input word forming the left half of the table. In the word approach, the designer takes a horizontal view of the truth table and specifies an output word in correspondence to each input word. This view is illustrated in Table 7.3(a), where the ROM design information for converting from BCD to excess-3 code is given. The boxed line shows that 0110 in BCD converts to 1001 in excess-3. On the other hand, the function approach leads the designer to take a vertical view of the truth table, where the truth values of a single output bit for all combinations of the input variables are specified. This view is illustrated in Table 7.3(b), where the ROM information for generating several functions is specified. The boxed column presents the EXCLUSIVE-OR

Figure 7.12 Simulated Ultraviolet Erasing of a 2716 PROM. (Courtesy of Intel Corporation.)

TABLE 7.3 COMPARISON OF ROM DESIGN METHODS

(a) Word Approach: BCD to Excess-3 Converter			(b) Function Approach: Multifunction Generator		

Input Word (BCD) $A_3A_2A_1A_0$	Decoded Word	Output (Excess-3 Word) $DO_3DO_2DO_1DO_0$	Word $A_2A_1A_0$	Decoded Word	Function $F_0\ F_1\ F_2\ F_3$
0 0 0 0	W_0	0 0 1 1	0 0 0	W_0	0 0 0 0
0 0 0 1	W_1	0 1 0 0	0 0 1	W_1	0 1 1 1
0 0 1 0	W_2	0 1 0 1	0 1 0	W_2	0 1 1 1
0 0 1 1	W_3	0 1 1 0	0 1 1	W_3	0 0 1 0
0 1 0 0	W_4	0 1 1 1	1 0 0	W_4	0 1 1 1
0 1 0 1	W_5	1 0 0 0	1 0 1	W_5	0 0 1 1
0 1 1 0	W_6	1 0 0 1	1 1 0	W_6	0 0 1 0
0 1 1 1	W_7	1 0 1 0	1 1 1	W_7	1 1 1 0
1 0 0 0	W_8	1 0 1 1			
1 0 0 1	W_9	1 1 0 0			
Other cases not allowed	W_{10} through W_{15}	0 0 0 0			

259

function

$$F_1 = \bar{A}_2\bar{A}_1A_0 + \bar{A}_2A_1\bar{A}_0 + A_2\bar{A}_1\bar{A}_0 + A_2A_1A_0$$

which is the sum output equation for a full adder. It only requires eight ROM bits ($2^3 \times 1 = 8$); but if mechanized by conventional AND-OR logic, the function requires five gates. In this example, an extremely favorable ratio of $\frac{8}{5}$ bits per gate is obtained. The function F_0, however, has a ratio of $\frac{8}{1}$; and on the average a ratio of about 12 bits per gate is typical. Moreover, in unfavorable cases the ratio can be more than 50 bits per gate. Nevertheless, it is often advantageous to mechanize combinatorial circuits by means of ROMs.

By the proper combination of ROM packages, it is possible to extend both the number of functions available and the number of input variables. Such an expanded circuit composed of three ROMs is shown in Figure 7.13.

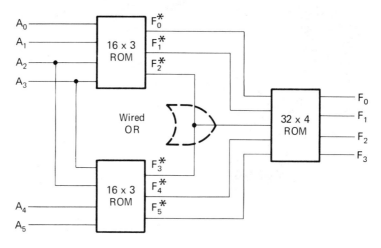

Figure 7.13 ROM Combinatorial Circuit with Extension of Both Variables and Functions.

How are truth tables converted to actual ROM circuitry? The process is really quite straightforward, as shown in Figure 7.14, which is the mechanization of the BCD-to-excess-3 converter from Table 7.3(a). Basically, the realization process consists of placing a circuit element (in this case a diode) between the word select line and the output bit line wherever a 1 appears in the truth table and omitting a component in a position corresponding to a truth table 0. The resulting hardware is essentially a direct mapping of the original truth table onto a ROM circuit similar to that in Figure 7.11, except that a more concise notation is employed—connected diodes are represented as dots whereas open diodes merely produce line crossings. We will *always* assume that the decoder output line is selected from a straight binary representation of the input bits independent of the actual nature of the input code. For example, W_5 is always selected by $A_3A_2A_1A_0 = 0101$, and the input bits happen to represent a 2 in excess-3, but they represent a 5 in BCD. The reader should verify that Figure 7.14 is an exact reproduction of Table 7.3(a) by using the decoded word column in the table as a guide. Observe that the input numbers 1010 through 1111, which are not

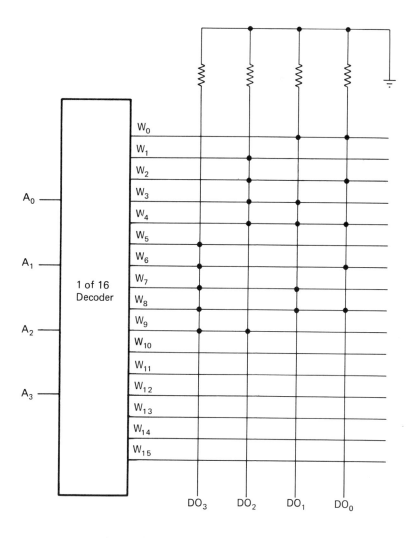

Note: Each dot represents a diode connected in the manner of Figure 7.10

Figure 7.14 Diode ROM Converter for BCD to Excess-3.

allowed BCD codes, correctly yield 0000 outputs since word select lines W_{10} through W_{15} are not connected to the data lines.

‡**Implementation of arithmetic with ROMs.** It is frequently very economical to implement arithmetic operations with ROMs, and IC manufacturers currently have many modules for this purpose.

Let us consider a module for multiplying two 4-bit numbers, which results in an 8-bit product. The resulting truth table will be large ($2^{4+4} = 256$ decoded words), but its implementation will be straightforward. The resulting ROM will be 256 words by 8 bits and yields $256 \times 8 = 2048$ bits, which is well within the capabilities of current

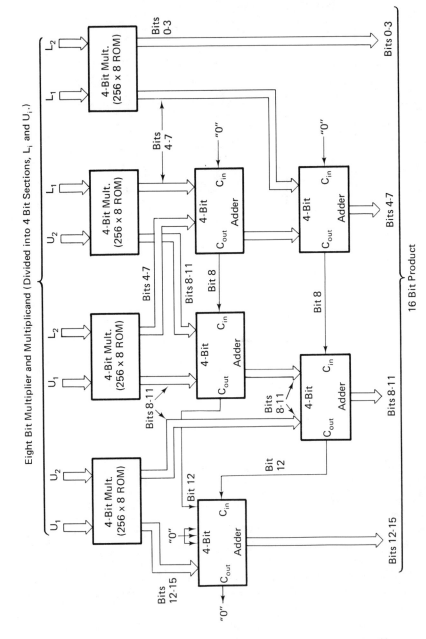

Figure 7.15 Eight-Bit Multiplier Constructed from Four-Bit Multiplier ROMs.

technology. Single chips such as this are currently available from Fairchild Camera, Texas Instruments, National, and others. (The National MM 523 is a typical example.) On the other hand, since the number of memory bits increases with n as $n2^{2n+1}$, single chips for large values of n are difficult to obtain, e.g., $n = 8$ requires 1,048,576 bits, which is considerably beyond current capability for a single chip.

The above result does not prevent ROMs from being employed for multiplication of large numbers; it just means that more sophisticated techniques must be used. Consider the multiplication of two numbers, N_1 and N_2, each consisting of a lower four bits L_i and an upper four bits U_i. (For example, N_1 can be expressed as $N_1 = 2^4(U_1) + L_1$, where U_1 is shifted right four places before summing because it is the upper four bits.) Then the resulting product is:

$$N_1 \times N_2 = [2^4(U_1) + L_1] \times [2^4(U_2) + L_2]$$
$$= 2^8(U_1 \times U_2) + 2^4(U_1 \times L_2) + 2^4(U_2 \times L_1) + L_1 \times L_2 \quad (7.1)$$

This equation reveals that the desired $N_1 \times N_2$ product can be achieved with four multipliers (256×8-bit ROMs) each accepting two 4-bit numbers and developing an 8-bit result. In addition, as Figure 7.15 shows, five 4-bit adders are also required. Now, each multiplier yields an 8-bit result, but the terms in Equation 7.1 are separated by shifts ranging from zero to eight bits. Thus, care must be taken in summing the correct groups of bits in the adders. (The bit sequence labeling at the output of each multiplier and other places within the diagram should be helpful in checking the circuit.) Note that bits 4–7 emerge from the multipliers in three sections, which is why they require two adders. (A similar situation occurs with bits 8–11.) Finally, observe how the various carry outputs (C_{out}) are properly passed to adders summing the next higher sequence of bits; but since two 8-bit numbers never give a result with more than 16 places, the leftmost adder always has $C_{\text{out}} = 0$.

The Programmable Logic Array

In a ROM or PROM all minterms are generated in a fixed decoder and the required terms are combined in a programmable OR structure (array) to produce the desired output function. For many logic design applications, however, only a small fraction of these minterms generated by the decoder are actually used, resulting in a somewhat inefficient circuit. On the other hand, a close relative of the PROM called the Programmable Logic Array, PLA, contains both a programmable decoder (AND logic array) and a programmable OR array, resulting in a circuit which can implement functions from higher-level terms rather than just minterms. Thus, for many applications a PLA is a powerful yet more efficient logic element. The above basic structures for the PROM and the PLA are contrasted in Figure 7.16.

Consider the implementation of the following simple set of three functions by means of a PLA:

$$f_0 = \overset{p_0}{\overline{A}BCD} + \overset{p_1}{\overline{D}E} + \overset{p_2}{D\overline{E}}$$

$$f_1 = \overset{p_3}{A\overline{B}} + \overset{p_4}{\overline{B}C\overline{D}E} + \overset{p_2}{D\overline{E}}$$

$$f_2 = \overset{p_0}{\overline{A}BCD} + \overset{p_4}{\overline{B}C\overline{D}E}$$

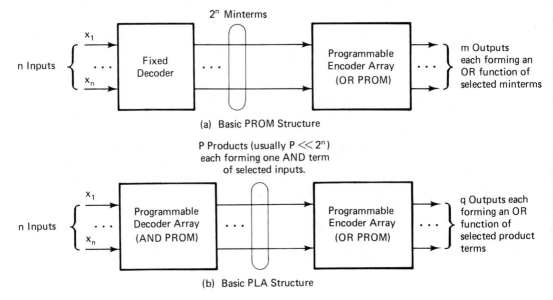

Figure 7.16 Two Contrasting Programmable Logic Structures.

The individual functions can be simplified or not depending on the designer's requirements; but in any case, the first implementation step is to label all unique product terms starting with P_0. Notice that the labeling was accomplished in the original function set by placing a subscripted P above each product term. For these functions there are five unique product terms out of a total of eight terms. In our system, each P term is implemented once in the AND PROM. Then, these terms are selected as needed in the OR PROM to produce the three desired functions. The realization of the above function set is shown in Figure 7.17. First consider the basic structure by identifying: the A through E input variable feeding the AND PROM, its output producing the five product terms P_0 through P_4, and finally the OR PROM, producing the three desired functions f_0 through f_2. On closer examination, it is seen that the PLA internally generates all complemented variables so that only uncomplemented quantities need be supplied as inputs. It should be understood that the basic PLA is manufactured with all structure finalized except the logic elements (diodes) represented by heavy dots in Figure 7.17, which by programming select the variables to be included in each product term and also select the product terms to be in each function. The "dot notation" employed here, of course, is essentially the same as that previously used in Figures 7.10 and 7.12 except that in those figures only an OR array is implemented. (Another small change found in Figure 7.17, and in all future PLA circuits, is that for simplification the vertical and horizontal line terminating resistors and power supply connections are omitted.)

As a partial check on the Figure 7.17 PLA circuit, consider the P_0 term and the f_1 function. The dot positions on the right-most vertical line of the AND array indicate that $\bar{A}$, B, C, and D are selected (ANDed), yielding $P_0 = \bar{A}BCD$. Next on the middle line of the OR array the P_2, P_3, and P_4 terms are selected (ORed), yielding $f_1 = P_2 + P_3 + P_4 = D\bar{E} + A\bar{B} + \bar{B}C\bar{D}E$. The reader should likewise check the other portions of our PLA designs.

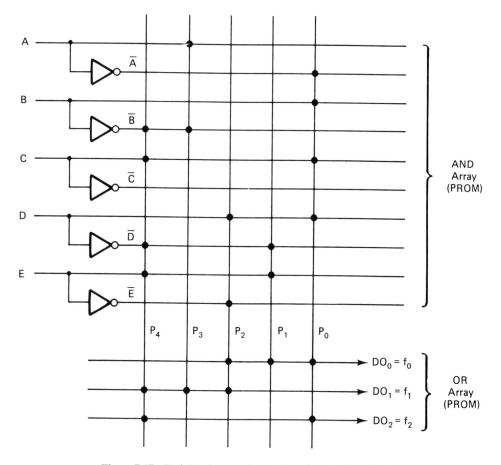

Figure 7.17 PLA Implementation of Functions f_0, f_1, and f_2.

From the general structure of Figure 7.17, it should be clear that the three keys to controlling PLA size (hence cost) are the required number of input variables, the number of outputs, and the number of product terms. Thus, it may be helpful to employ a Karnaugh map, or other simplification technique, to reduce the above parameters. On the other hand, it is easily seen that the size of a particular product term has no effect; the PLA cost is the same whether the product is a minterm containing all variables or merely one containing only a single variable.

7.4 APPLICATIONS TO STRUCTURED DESIGN

Now that you have an understanding of both fixed and programmable components, it is appropriate to consider some specific elementary examples of Structured Design. As should already be evident, structure is not a binary quantity, i.e., either present or absent (true or false); but it is a matter of degree—some systems are much more structured than others. Of course, all other things being equal, better structure leads to an improved design. Moreover, structure must always be judged in context. Cost or other reasons may lead to a design having a small portion of a circuit that appears to be less structured than it could be, but the complete circuit may actually be well-structured. Consider the function $f = \bar{A}_1\bar{A}_2A_3 + A_2\bar{A}_3 + A_1\bar{A}_3 + A_1A_2$. It is easy to show with a Karnaugh map that the function is already simplified. Assuming that both complemented and uncomplemented variables are available, the gate-type logic diagram in Figure 7.18(a) results. On the other hand, by using a multiplexer implementation, it is easy to show that the circuit in Figure 7.18(b) results. (Directly from the MUX, $f_{\text{MUX}} = \bar{A}_1\bar{A}_2A_3 + A_1\bar{A}_2\bar{A}_3 + \bar{A}_1A_2\bar{A}_3 + A_1A_2$, which may easily be reduced to f.) It is an elementary example, but if we assume that this is the complete system, one must agree that the circuit in part (b) of the figure is more regular and systematic in its appearance. (Also, it only requires one-half an IC package such as the CMOS dual, 1 of 4 MUX, 4555.) In addition, it is easier to understand and it employs a higher-level primitive than the circuit in part (a). (Incidentally, the circuit in part (a) may be implemented with two IC packages, such as the CMOS triple three-input NAND, 4023, and there would be one spare gate.) Thus, you must agree that the circuit in part (b) is better structured than that in part (a). If a second three variable function f_2 were required, another MUX from the same package could be added and the circuit would still be considered well-structured. If several complicated functions were required, however, a PLA would become cost competitive with the MUX and the better structured PLA version would be selected as the best design.

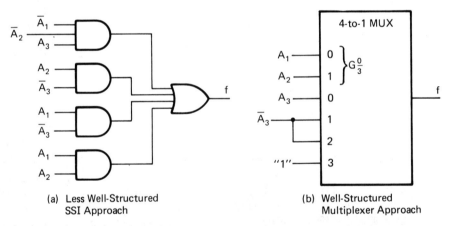

Figure 7.18 Two Implementations of $f = \bar{A}_1\bar{A}_2A_3 + A_2\bar{A}_3 + A_1\bar{A}_3 + A_1A_2$.

Design requires good engineering judgment, and there will be times when a solution will be selected as being better structurally even when it lacks one of the structural properties possessed by a competing design. Figure 7.19 is an example of this situation. Here it is assumed that the only system requirement is to develop the function $f = \bar{A}_1 + A_2 + A_3 + \bar{A}_4 + A_5 + \bar{A}_6 + A_7 + \bar{A}_8$. Part (a) of the figure shows a solution which may appear to be the best structurally—it does employ a higher-level primitive. The solution shown in part (b), however, is more regular, less costly, simpler, and employs fewer chips. Thus, it is not only cheaper, but it is better

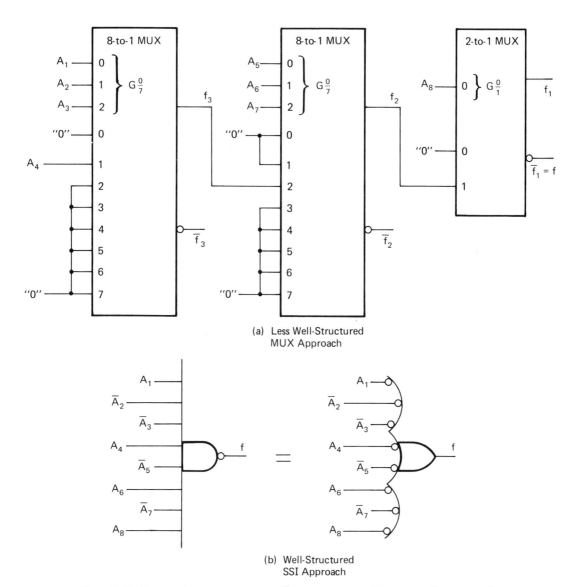

Figure 7.19 Two Implementations of $f = \bar{A}_1 + A_2 + A_3 + \bar{A}_4 + A_5 + \bar{A}_6 + A_7 + \bar{A}_8$.

structurally even though only SSI is employed. (Incidentally, the eight-input 7430 NAND IC would be a good chip to implement f. Similar OR functions with up to 13 variables can be implemented with a single 74133 NAND IC, but it would take four 8×1 MUX packages to accomplish the same task.)

Normally, a MUX provides a well-structured circuit because a complete set of minterms are available directly at its terminals, which adds to its power by providing a convenient, systematic method for developing complicated functions. Moreover, the essence of the difficulty and inefficiency of the MUX solution shown in Figure 7.19(a) is the fact that only one minterm is used from each MUX.

7.5 CONTROL MEMORIES

Although the memories in Section 7.3 were employed only in combinational design, they are certainly also applicable to sequential problems. The latter application is known as a *Control Memory*, CM, since the memory is used to control a sequence of digital circuit actions. The state generator, the logic and timing module, and some of the other circuitry of Chapter 6 could conveniently be replaced by means of a CM unit.

The CM idea was conceived in 1951 by M. V. Wilkes,[15] of the Cambridge University Mathematics Laboratory, as a means of adding systematic order to methods for designing digital computers. His original concept involved a ROM in which each word was divided into two parts. The first part specified the location in memory where the next instruction word could be found; the second part specified the states of certain signals which were employed to control individual logic circuits in the ALU and other parts of the computer. Because the control existed over very basic, low-level, ALU actions, the instructions became known as *microinstructions*. The instruction words stored in our CM will henceforth be called *microinstructions*. In addition to being employed here, they will also be considered in great detail in Chapter 9.

These ideas were soon further developed by Wilkes and other people; and one of the IBM series 7000 computers was microprogrammed in the early 1960s. Moreover, the IBM Systems 360 and 370, the Honeywell H4200, and several machines since have employed CM techniques. The current availability of inexpensive memories, however, has made this technique more popular than ever before. In addition to its use purely as a design method, it also possesses some very practical advantages. For example, the presence of a CM allows the configuration of the digital system to be easily changed simply by changing the contents of the memory instead of completely rewiring the control unit. In fact, a writeable store such as a RAM can be employed as the CM so that the configuration can be changed about as fast as a program is changed in a standard computer. (The term *Control Read-Only Memory*, CROM, is frequently used when the CM is contructed from a ROM rather than from another type of memory.)

An Elementary CM System

Figure 7.20 shows that the major components of a CM system are the memory itself, the memory address register, and the set of input/output signals. (In more complex systems, particularly those discussed in Chapter 9, another component called the *next-address logic* is added.) The response to each current address is the contents of that memory location which consists of an address output, $A_m \ldots A_1 A_0$, and a group of control signals, $C_k \ldots C_1 C_0$. The address output is combined with some external signals called *jump bits*, $J_R \ldots J_1 J_0$, to produce the next-address function, which for our simple system becomes the current address after the occurrence of the clock pulse. It is easy to visualize how a single jump bit, J_0, may be employed to produce conditional branching, much as was done with the machine language instruction APO in Chapter 3. When $J_0 = 0$, the next-address function becomes $0A_m \ldots A_1 A_0$; and when $J_0 = 1$, the function becomes $1A_m \ldots A_1 A_0$.

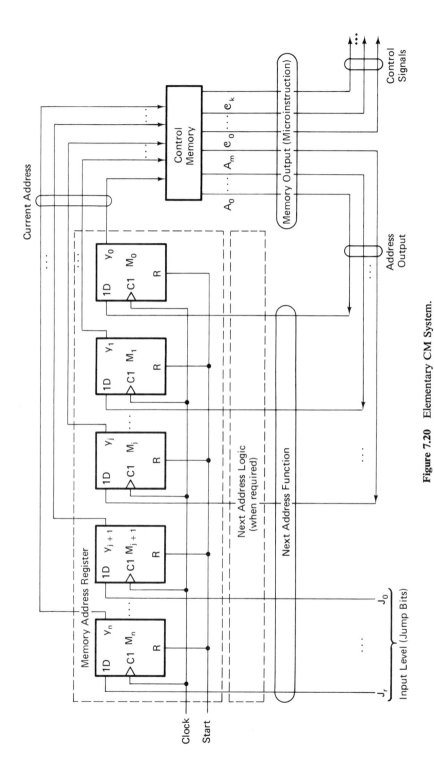

Figure 7.20 Elementary CM System.

Note that the start signal in Figure 7.20 (operating through the clear, R, input of each flip-flop) sets the memory address to zero. Thus, in this system, the first microinstruction is always taken from memory location zero.

CM implementation of an up-down counter and decoder. In order to demonstrate the CM method of design in a very simple way, consider the implementation of a two-bit up-down counter and its binary-to-decimal decoder. The CM required to satisfy this problem is shown in Table 7.4. As expected, each current

TABLE 7.4 CM DESIGN OF AN UP-DOWN COUNTER AND DECIMAL DECODER

	Current Memory Address, MA			Memory Contents (Microinstruction)					
	Input M_2 (J_0)	Counter M_1	M_0	Address Output A_1	A_0	Control Signals (Decimal Decoder Output) c_3	c_2	c_1	c_0
Up-Counter	0	0	0	0	1	0	0	0	1
	0	0	1	1	0	0	0	1	0
	0	1	0	1	1	0	1	0	0
	0	1	1	0	0	1	0	0	0
Down-Counter	1	0	0	1	1	0	0	0	1
	1	0	1	0	0	0	0	1	0
	1	1	0	0	1	0	1	0	0
	1	1	1	1	0	1	0	0	0

memory address has a corresponding memory contents, which consists of the address output and a set of control signals. While only a two-bit counter is implemented, the memory address, MA, requires an extra bit, M_2, to provide the up-down branching for the two counter modes. As an example, if the unit is in the state $MA = 001$, it is acting as an up-counter; and the address output is $A_1 A_0 = 10$, which means that following the clock pulse the next state would be $MA = 010$. Observe that the action taken by the counter at each clock pulse is determined by the current memory output (more specifically, the address output). The unit may be changed to a down-counter by the following sequence:

1. Between clock pulses, the J_0 level changes to $J_0 = 1$.
2. The clock pulse occurs and the counter increments since the memory address register is still set for the up-counter mode. At the same time, the M_2 bit is being loaded from J_0.
3. After a short delay, the current memory address becomes that of a down-counter, $M_2 = J_0 = 1$, and the address output shows a decrement.
4. The next clock pulse occurs and the counter performs the first decrement.

(In practice, the counter could be made to decrement immediately; but it would require a somewhat more complicated circuit than that shown in Figure 7.20.) The control signals are straightforward, and as required, they are a decimal translation of the counter's binary state, M_1M_0.

The CM, such as in Table 7.4 may be designed through the following steps:

1. From the statement of the problem, construct a CM table with the required number of memory addresses, including the jump bits.
2. Add the required address output columns and the control signal columns to the table.
3. Starting at a convenient memory address, fill in the corresponding address output as dicated by the problem statement. Using this output as the current memory address, fill in its corresponding address output. Continue in the natural sequence of the problem itself until the address output entries are complete for the entire table.
4. Review the entries generated in step 3 and be sure they correctly satisfy the requirements of the problem.
5. Complete the control signal columns of the table by determining what values are required by the problem statement for each memory address.

In order to get a clear and concise picture of the original problem requirements, it may be desirable to construct a flow diagram or write an RTL program before starting step 1 in the above procedure. The next example will demonstrate the value of an RTL program as a starting point for CM design.

CM implementation of a multiply algorithm. The multiply algorithm of Figure 7.21 is an elementary but interesting possibility for CM design. Here we consider the problem of multiplying two 6-bit positive numbers and obtaining the 12-bit product. Figure 7.21 shows the block diagram and RTL statements required to solve the problem. In the logic diagram, note that the multiplier basically consists of two shift registers, A^* and B, for the two input numbers, a result register, C, and a 12-bit full adder. The timing signals are indicated by means of quantities within parentheses. Since the algorithm involves six add/shift steps and each step consists of two parts, 12 microinstructions are required for the basic program. To these must be added the $T = 1$ initiation step for a total of 13 microinstructions. Table 7.5 shows these microinstructions and the other aspects of the CM program. The reader should now go through the entire program and be sure it does indeed correctly implement the multiply algorithm. We see that four control signals are employed; these are in a one-to-one correspondence, subscriptwise, with the T signals shown in Figure 7.21(a), e.g., $T = 2$ is the same as $\mathcal{C}_2$.

As shown in Figure 7.21(a), the $T = 2$ statement is a conditional one. This may be satisfied in two possible ways. One method is to use the condition, i.e., the B_0 signal, as a jump bit in the current address. The second method is to use *auxiliary logic* to realize the conditional statement. Since the first method would almost double the

---

*Since it must be shifted left without losing significant bits and since it operates with a 12-bit adder, observe that register A contains 12 bits.

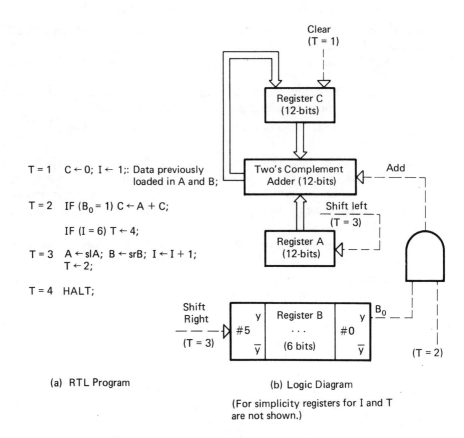

(a) RTL Program

$T = 1$ $C \leftarrow 0$; $I \leftarrow 1$;: Data previously loaded in A and B;

$T = 2$ IF $(B_0 = 1)$ $C \leftarrow A + C$;

IF $(I = 6)$ $T \leftarrow 4$;

$T = 3$ $A \leftarrow slA$; $B \leftarrow srB$; $I \leftarrow I + 1$;
$T \leftarrow 2$;

$T = 4$ HALT;

(b) Logic Diagram

(For simplicity registers for I and T are not shown.)

Figure 7.21 Multiply Algorithm ($C = B \times A$).

TABLE 7.5 CM IMPLEMENTATION FOR THE MULTIPLY ALGORITHM OF FIGURE 7.19

	Current Memory Address				Memory Contents (Microinstruction)							
					Address Output				Control Signals			
	M_3	M_2	M_1	M_0	A_3	A_2	A_1	A_0	c_4 (Halt)	c_3 (Shift)	c_2 (Add)	c_1 (Clear)
Start →	0	0	0	0	0	0	1	0	0	0	0	1
$I = 1$	0	0	1	0	0	0	1	1	0	0	1	0
	0	0	1	1	0	1	0	0	0	1	0	0
$I = 2$	0	1	0	0	0	1	0	1	0	0	1	0
	0	1	0	1	0	1	1	0	0	1	0	0
$I = 3$	0	1	1	0	0	1	1	1	0	0	1	0
	0	1	1	1	1	0	0	0	0	1	0	0
$I = 4$	1	0	0	0	1	0	0	1	0	0	1	0
	1	0	0	1	1	0	1	0	0	1	0	0
$I = 5$	1	0	1	0	1	0	1	1	0	0	1	0
	1	0	1	1	1	1	0	0	0	1	0	0
$I = 6$	1	1	0	0	1	1	0	1	0	0	1	0
Halt →	1	1	0	1	—	—	—	—	1	0	0	0

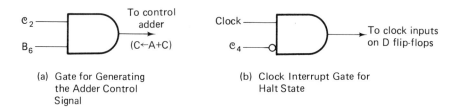

Figure 7.22 Auxiliary Logic for Use with CM of Table 7.5.

size of our CM program, we have elected to employ the second method with the conditional logic shown in Figure 7.22(a). There, the $\mathcal{C}_2$ state is ANDed with the B_0 signal to control the adder exactly as it is done in Figure 7.21(b). (The halt state is also accomplished by auxiliary logic—see Figure 7.22(b), but another method will be considered in the exercises.)

Although the two simple applications described above were of a tutorial nature and may not be sufficiently complex by themselves to justify the use of a CM, the current low price of semiconductor memories, particularly ROMs, have made them competitive with other types of circuitry for many relatively small digital systems. Certainly, the RTL programs for the CPU of a small computer, such as those shown in Table 6.6, could be directly implemented by means of a CM and several MSI packages composing the ALU, the various registers, and any needed auxiliary logic. An alternate to this approach will be considered in Chapter 9, where the CM and microinstruction techniques will be further developed to produce a so-called microprogrammable microprocessor, which is a very powerful computer.

The Sequential PLA Circuit

The CM concept can be implemented by several different methods—a ROM, an MSI circuit, a customized LSI chip, or a sequential PLA chip. For special counters and other moderate-size sequential circuits, the sequential PLA, shown in Figure 7.23, is a particularly attractive design choice. Note how well Figure 7.23 matches the CM diagram of Figure 7.20, the only real difference being that the Jump Bits of Figure 7.23 are not stored by the flip-flops but may immediately effect the control memory. For a counter, the difference is of minor importance because the register content is dependent on the arrival of the clock signal and the decoder output is based on the state of the register and not directly on the Jump Bits.

To gain insight into the detailed design of a sequential PLA circuit, consider the CM Up-Down counter previously described in Table 7.4. The basic PLA receives the current memory address consisting of bits M_0 through M_2 as inputs and develops the Address Output along with the Decimal Decoder signal. What is actually required in the PLA design is a translation of the information in Table 7.4 into a PLA program. Figure 7.24 shows the complete design. Notice that the PLA input lines form two groups—the up-down control, M_2, and the Flip-Flop Register outputs, M_0 and M_1. Similarly, the PLA output lines form two groups—the Address Output, which drives D_0 and D_1 of the Memory Address Register, and the Control Signals, which are the decimal decoder outputs. In comparing Figure 7.24 with the general configuration of

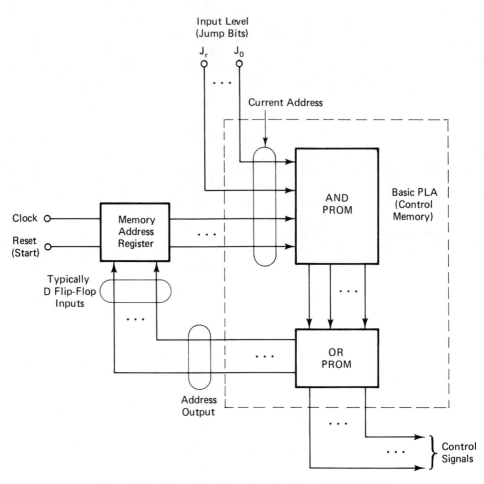

Figure 7.23 General Configuration of a Sequential PLA Chip.

Figure 7.23, the reader will recognize that as a convenience, some components have been rearranged. Similarly, the AND PROM is drawn with the M_0 and M_1 inputs from the right. Finally, the register is represented by Dependency Notation, which is very convenient for several flip-flops having a single control terminal. Thus the common reset control simultaneously drives both flip-flops of the register to the logical zero state, and the common clock provides a similar action. This notation, described in more detail in Section 5.10, has recently been extensively developed. It is particularly useful for concisely describing the detailed operation of large logic circuits (see Reference 13).

Conclusion

A wide range of components were considered in this chapter. Although each of these has merit as the solution to a certain class of problems, we must avoid stereotyped thinking. For example, in the case of a certain challenging requirement, maybe one

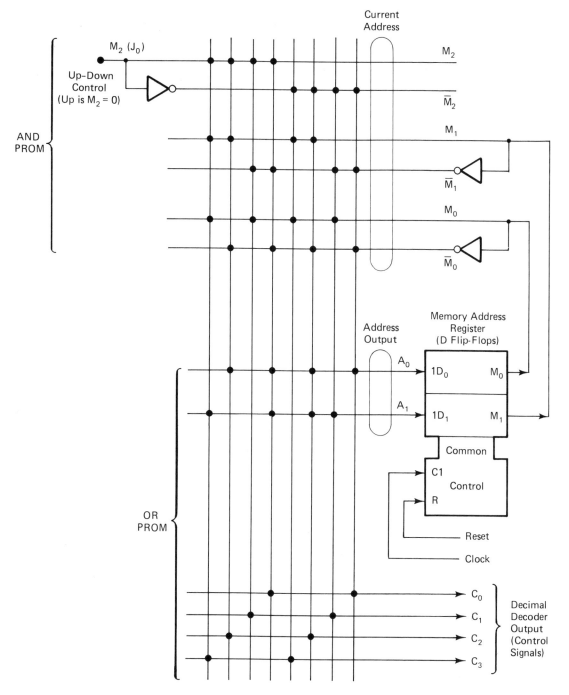

Figure 7.24 Sequential PLA Chip for the Up-Down Counter of Table 7.4.

of the following approaches would be better: a microprocessor system (discussed in the next two chapters), a combination of fixed and programmable components, a VLSI Circuit,[11] or perhaps some new system not discussed in this text. Although the designer does have considerable freedom (and responsibility) in selecting components, the philosophy of structured design *must* be paramount in his thinking; otherwise the system will likely fail on one or more of the following grounds: reliability, maintainability, cost, or documentation.

7.6 BIBLIOGRAPHY

1. Blakeslee, T., *Digital Design with Standard MSI and LSI* (2nd ed.). New York: John Wiley & Sons, Inc., 1979.
2. Boutin, N., "PSK Modulator Resolves Phase Shifts to 22.5°." *Electronics*, Vol. 53, April 24, 1980, p. 131.
3. Carr, W. N., and J. P. Mize, *MOS/LSI Design and Applications*. New York: McGraw-Hill Book Company, 1972.
4. *CMOS Databook*. Santa Clara, Calif.: National Semiconductor Corporation. (Consult most recent edition.)
5. Fletcher, W. I., *An Engineering Approach to Digital Design*. Englewood Cliffs, N.J.: Prentice-Hall, Inc., 1980.
6. Hnatek, E. R., *A User's Handbook of Semiconductor Memories*. New York: John Wiley & Sons, Inc., 1977.
7. Hope, G. S., *Integrated Devices in Digital Circuit Design*. New York: Wiley-Interscience, 1980.
8. Hurst, S. L., *The Logical Processing of Digital Signals*. New York: Crane, Russak, & Company, 1979.
9. Luecke, G., J. P. Mize, and W. N. Carr, *Semiconductor Memory Design and Application*. New York: McGraw-Hill Book Company, 1973.
10. Mano, M. M., *Digital Logic and Computer Design*. Englewood Cliffs, N.J.: Prentice-Hall, Inc., 1979.
11. Mead, C., and L. Conway, *Introduction to VLSI Systems*. Reading, Mass.: Addison-Wesley Publishing Company, Inc., 1980.
12. Muroga, S., *Logic Design and Switching Theory*. New York: John Wiley & Sons, Inc., 1979.
13. Peatman, J. B., *Digital Hardware Design*. New York: McGraw-Hill Book Company, 1980.
14. *TTL Data Book for Design Engineers*. Dallas, Tex.: Texas Instruments, Inc. (Consult most recent edition.)
15. Wilkes, M. V., "The Best Way to Design an Automatic Calculating Machine." *Manchester University Computer Inaugural Conference*, 1951, p. 16.
16. Winkel, D., and F. Prosser, *The Art of Digital Design—An Introduction to Top-Down Design*. Englewood Cliffs, N.J.: Prentice-Hall, Inc., 1980.

7.7 EXERCISES

7.1 Using the fewest possible multiplexers (no gates), draw a circuit to realize the expression $R = t^2 + 5$ shown in Table 7.1.

7.2 Using the fewest possible multiplexers (no gates), draw a circuit to realize the expression $E = x^2 + 3x + 1$, where x and E are the input and output binary numbers, and $0 \leq x \leq 3$.

7.3 Repeat Exercise 7.2, with $E = -2x^2 + 15x + 32$ and $0 \leq x \leq 7$.

7.4 Using 4-to-1 multiplexers only, draw the circuit for a 16-to-1 multiplexer. Provide an enable input for the new circuit.

7.5 Repeat Exercise 6.17, but employ a multiplexer approach instead of the decoder and tristate gates. If more than one multiplexer is required, show each separately.

7.6 At the lowest possible cost, design a circuit for the following function with multiplexers and inverters only. *No variables* are available directly in complemented form. Assume that an $n \times 1$ multiplexer costs $n/4$ dollars and inverters cost \$0.10 each. (Determine the cost of your design.)

$$f = \bar{A}\bar{C}D + AC + AB + \bar{A}C\bar{D} + \bar{A}\bar{C}E$$

7.7 Using the smallest possible multiplexer (no gates) realize a circuit for the following function:

$$f = B\bar{C}\bar{D} + \bar{A}B + A\bar{C}D + A\bar{B}$$

Both complemented and uncomplemented versions of the variable C are available.

7.8 Repeat Exercise 7.7 for the following function, but only uncomplemented variables are available.

$$f = \bar{A}\bar{B} + ABD + \bar{A}C$$

In the following five exercises (7.9–7.13) complete the first three steps in our design process—through Detailed Design. Also, be sure that all are well-structured solutions.

7.9 Design a variable-modulo frequency divider and employ a single-pole multiposition switch to select integer divider ratios from 1 to 16. The input signal is a symmetric square wave; but the output is required only to be a series of pulses of the proper frequency. (*Hint:* A counter with a synchronous load or clear feature may be helpful.)

7.10 For your design of Exercise 7.9, sketch a timing diagram with the circuit set to divide by three. Show both the input for a total of 10 clock pulses and the corresponding output on the diagram.

7.11 Repeat Exercise 7.9, but in this case assume that a symmetric square-wave output is required; but the divider need only be of even modulus (i.e., 2 to 16).

7.12 When determining the half-life (a measure of how quickly the radiation decays) of short-lived radioactive substances and in many other practical applications, it is desirable to sample pulse rates (counts per short time interval) from the source over several successive periods of time. Later the recorded data are examined at the operator's leisure.
 (a) Prepare the design for a system having the pulse source sampled during four 0.5-s intervals with each interval separated by a 0.5-s off-period.
 (b) Repeat part (a) but increase the number of sampling intervals from 4 to 64.

(c) Compare the two designs and explain why each employs a somewhat different philosophy.

7.13 A common asynchronous counter is driven by eight different input pulse sources. (The counter must reach a maximum of 255_{10}.)
  (a) Design a counter circuit where a constraint guarantees that none of the pulses from the sources will be coincident.
  (b) Repeat part (a) but with coincident pulses permitted between two or more sources.
  (c) What is the main limitation of the design in part (b), and how can this be overcome?

7.14 If a single ROM is employed, how many bits are required to multiply an n_1-bit number by an n_2-bit number?

7.15 Design a 2-bit binary multiplier showing both the equivalent logic diagram and ROM truth table.

7.16 How can Figure 7.13 be redrawn so that the bits from the middle two multipliers are not immediately combined?

7.17 Similar to Figure 7.12, draw the ROM circuit for a Gray-code-to-binary-code converter. (Employ the Gray-code version shown in Table 5.18.)

7.18 Similar to Figure 7.12, draw the ROM circuit for a code-X-to-excess-3 code converter. (These codes are listed in Table 2.10.)

7.19 Sketch the circuit similar to Figure 7.15 for a PLA which implements the following set of functions:

$$f_0 = B\bar{C}E + \bar{C}\bar{E} + \bar{A}DE$$
$$f_1 = C + ABDE + \bar{C}\bar{E}$$
$$f_2 = \bar{A}DE + B$$

7.20 An elementary PLA has the following maximum characteristics: four pins for input variables, four product terms (P_0 through P_3), and two pins for output variables. Which of the following sets of functions can be realized with a single elementary PLA? If any set cannot be realized, explain why. Draw a circuit similar to Figure 7.15 for each set of functions which is realizable.

Set (a): $f_0 = ADE + BCE + AC\bar{D}\bar{E} + A\bar{C}D\bar{E} + \bar{B}CE$
$f_1 = \bar{C}D\bar{E} + A\bar{C}DE$

Set (b): $f_0 = \bar{A}B\bar{C} + \bar{C}D + A\bar{B}, f_1 = \bar{A}B\bar{C} + \bar{B}D + BC + A\bar{B}$

7.21 A double-pole seven-position digital switch forms a complete system. Design (draw the circuit) the most well-*structured solution* you can find. Also, design the next best structured circuit and compare them.

7.22 A 9-line-decimal-to-4-line-BCD priority encoder is required. (If two or more input lines are asserted simultaneously, this decoder gives priority to the largest input line asserted.)
  (a) To gain familiarity with the circuit principles, first design a 3-line-decimal-to-2-line-BCD priority encoder with no input line asserted, yielding zero as an output.
  (b) For the full 9-line encoder, list the three most *structured design* approaches in rank order (only outline the circuit) and discuss the advantages, disadvantages, and flexibility of each to charges in the required function specifications.

7.23 Design the best structured logic for

$$f = (D\bar{E}H + BD\bar{E})(\bar{B} + \bar{G})(C + \bar{H})(\bar{A}\bar{H}J + \bar{A}\bar{C}J)(\bar{C} + G)$$

Assume that f is the only function required by the system and that both complemented and uncomplemented variables are available.

7.24 Repeat Exercise 7.23 for the function
$$f = A\bar{B}\bar{D} + \bar{A}E\bar{G} + G\bar{H} + \bar{A}B\bar{H} + B\bar{C}\bar{D} + A\bar{E}G + AK + BCDK$$

7.25 Repeat Exercise 7.23 for the function
$$f = \bar{A}B\bar{C}D + \bar{A}\bar{B}\bar{E} + A\bar{B}E + CDE + C\bar{D}\bar{E} + AB\bar{E} + \bar{C}D\bar{E} + \bar{C}\bar{D}E + \bar{A}BC\bar{D}$$

7.26 Repeat Exercise 7.23 for the function
$$f = (\bar{X}_3\bar{X}_4 + X_3X_4 + \bar{X}_3X_4X_7)(\bar{X}_1X_2 + X_1\bar{X}_2) + X_1X_2\bar{X}_5 + \bar{X}_1\bar{X}_2X_6$$

(*Hint:* Implement the function as it appears.)

7.27 Design a CM to implement the modulo-3 counter of Figure 5.7, with reset R, being achieved by means of the jump bit, i.e., $R = J_0$. (Present your results in a form similar to Table 7.4.)

7.28 Find and describe, in detail, another method for producing the halt state of Table 7.5 without employing auxiliary logic.

7.29 For the up-down counter of Table 7.4, find two ways to modify Figure 7.16 so that the counter decrements on the first pulse input, rather than the second, following the level change from $J_0 = 0$ to $J_0 = 1$. What are the advantages and disadvantages of each method?

7.30 If the auxiliary logic of Figure 7.18(a) is not employed, i.e., a jump bit is used instead, list the new entries for the CM of Table 7.5 for the $I = 2$ step. Assume $M_3M_2M_1M_0 = 0100$ at the end of the $I = 1$ step. (*Hint:* What is the advantage in connecting B_6 itself as the MSB in the memory address register rather than going through the intermediate operation of having the clock transfer $J_0 = B_6$ to a new M_4 flip-flop?)

7.31 Employing a control memory approach with a sequential PLA circuit and D flip-flops (similar to Figure 7.20), design a modulo-10 up-counter. The counter should be properly decoded for a decimal output.

7.32 Sketch a sequential PLA circuit employing clocked J-K flip-flops which satisfies the requirements of the Karnaugh maps of Figure 5.12 and thereby yields a BCD counter. (For simplification no output signals are required.)

7.33 Employing a sequential PLA circuit, with clocked J-K flip-flops, design a four-bit up-down counter (modulo-16), which requires the fewest possible PLA product terms. Provide an input, U/D, where $U/D = 1$ for counting up and $U/D = 0$ for counting down. The counter should be properly decoded for a binary output. (*Hint:* Review the process used in Exercise 7.26 to implement the BCD counter.)

7.34 Employing a sequential PLA circuit, with clocked R-S flip-flops, design a modulo-10 up-counter. Include a PLA section to convert the present count into decimal (10 lines) and a final section to decode these signals for driving a seven-segment display. The output lines to the display should be labeled 0, 1, 2, ..., 6, as shown in Figure 2.5(b). (*Hint:* Review the process used in Exercise 7.26 to implement the BCD counter.)

7.35 Employing a sequential PLA with your choice of flip-flops, design a four-bit ring counter that requires the fewest possible PLA product terms. The uncomplemented state of each flip-flop is to form the output signal. Provide a single input, IN, which allows any desired pattern to be formed in the counter after a sequence of clock pulses and explain your process for loading the counter.

Chapter 8

Logic Design with Microprocessor Systems

A microprocessor, μP, is a single LSI chip (or at most a small group of such chips) which implements the CPU of a digital computer; and as such it possesses a powerful array of arithmetic, logic, and control functions. In order to give you a "feeling" for a large modern μP, the 16-bit MC68000 is presented in Figure 8.1. The lower part of the picture shows the 64-pin package with the actual silicon chip exposed so that you can see how small a fraction of the total package the silicon actually occupies. The photomicrograph of the circuit with its approximately 68,000 transistors is seen in the upper part of the figure, and the various patches contain the functional units, such as the control unit, the internal memory, the ALU, etc. Sometimes the requirements of a processing task will be sufficiently simple that it can be adequately handled by the μP alone, without any additional chips. Other tasks, however, may be so extensive and complex that a multiboard microcomputer, μC, such as Motorola's VERSA module chassis, with its front panel and top removed, shown in Figure 8.2 is required. The 15-A power supply is easily identified at the rear of the chassis and the three circuit cards, from bottom up, are: the microcomputer, a floppy disk controller, and a 128K RAM. (Note the informal distinction between a μP and a μC is that the μC contains a larger memory and a wider range of capabilities that move its performance characteristics toward those of a larger-scale machine such as a minicomputer.)

The study of μP's is not an isolated topic but it is a continuation and extension of the material developed earlier, particularly in Chapters 3 and 6. Although it is necessary for this chapter to begin with the fundamentals of μP architecture and interfacing, its main emphasis is on the μP system as a method of solving logic design problems. Thus, we are interested in considering μP's as a design alternate to the

Figure 8.1 The MC68000: An Advanced Microprocessor with 16-bit Arithmetic Capabilities. (Courtesy of UPDATE, Motorola Semiconductor Products Inc.)

MSI/LSI components approach of Chapter 7. This chapter is intended to provide a solid basis for later detailed study of one or more specific μP families.[1,10,14]

Currently there are a large number of μP's available in the market place, many new ones are being developed, and advanced versions of successful older ones are being added every year. There is a temptation to have the bulk of this chapter consist of a detailed study of one or two currently popular μP's and then try to briefly generalize from there. Consistent with what is usually done when considering existing systems in other engineering areas, however, we prefer to develop an in-depth general theory, including practical examples, based on the principles of as many different μP's as possible but divorced from a detailed study of any one commercial type. Hopefully, then, the theory can be employed to understand and apply any current or future μP.

Most microprocessor manufacturers supply computer-aided design software

Figure 8.2 The VERSA Module Chassis: A Microcomputer with Extensive Capabilities. (Courtesy of UPDATE, Motorola Semiconductor Products Inc.)

packages, which can be very helpful in applying their products. For example, simulation programs allow a microprocessor system to be tested so that most design errors can be corrected even before a breadboard is actually wired. Moreover, programs called *cross assemblers* are available permitting the designer to employ the more extensive facilities of a larger scale computer for writing and debugging machine language programs, which are later to be run on a particular microprocessor. Since these software packages are tailored for a specific microprocessor, the reader is referred to the manufacturers' literature for details. Many of the required fundamentals, however, are developed in Chapters 3 and in the current chapter.

8.1 AN INTRODUCTION TO MICROPROCESSORS

We are interested in microprocessors for four reasons:

1. The microcomputers built from them are very useful for many types of data processing tasks previously requiring a large computer.

2. Microprocessors are an excellent model for teaching, studying, and experimenting with general problems in information processing.
3. A single microprocessor frequently can replace 100 or more lower-level SSI components (such as gates and flip-flops) in a system. Moreover, the new system will usually be more versatile and required engineering changes can be made much more quickly.
4. Since a μP has so much capability that is systematically and conveniently arranged, it is one of the major options in achieving structured design. Moreover, it allows the software to share the structure with the hardware, which is appropriate because of the added dimension of power and flexibility available from programming.

In previous chapters we have learned how to do logic design and even to build computers by combining SSI and MSI chips, each of which perform a fairly simple operation (e.g., multiplexing). Now we employ one or a small number of complex LSI μP chips to do the same tasks. Since μP's are so cheap and so easy to use, the trend is to do even fairly simple tasks with them. There is no general rule to say which problems must be solved with a μP, however, so that into the foreseeable future some designs will be done with μP's while others will continue to be done with SSI and MSI. The critical thing that the skillful designer will always do is to use his or her knowledge and good judgment in selecting the design philosophy appropriate to the project goals and constraints. He or she will continue to learn and experiment while being careful to avoid getting "locked into" current fads or into doing it a certain way just because that worked in the past.

In this chapter we first study microprocessor internal logic through the use of RTL; various methods for addressing memory are considered; an LSI chip for interfacing with peripherals is presented; and finally some methods for expressing designs are considered along with a comparison between μP and MSI designs.

8.2 MICROPROCESSOR ARCHITECTURE

Although the general digital computer system block diagram in Figure 3.1(a) is, of course, applicable to a microcomputer system, the redrawn diagram in Figure 8.3 emphasizes the structure which is typical in these systems. The μP, actually being a CPU, is the source of control signals, executes instructions, performs the ALU operations, and contains special registers such as the program counter, the accumulator, and the condition code register. Both RAM and ROM are present in Figure 8.3, as is common in many systems. Often the nonvolatile ROM contains the program and the RAM is used to store intermediate results. The peripheral interface unit, PIU, provides input/output ports for transmitting information to and from the peripherals. It should be understood that the PIU module(s) can provide both parallel and serial interface functions; thus, all types of peripherals may be serviced. An interesting feature of our typical system is that RAM, ROM, and PIU are all addressed in the same way; thus, the PIU merely looks like one or more memory locations. This *unified bus* feature has some significant advantages which will be discussed later.

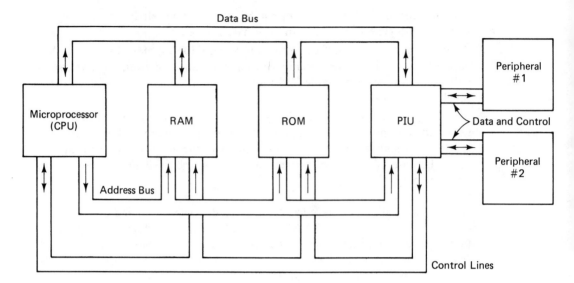

Figure 8.3 Typical Microcomputer System.

The reader should carefully examine the direction arrows for the various buses as they enter each unit; e.g., the data bus is bidirectional at all units except the ROM, which is expected since the ROM can only deliver but not receive data. The fact that the PIU is both a source and receiver for controls lines is explained by the fact that it can send signals back to the μP. A common control signal of this type is an interrupt generated by one of the peripherals—more details are given later in this chapter and in Chapter 11.

It has already been stated that a microprocessor is actually a CPU, but now let us look at it in more detail. A block diagram which is a simplified representation of the 6800, 8086, Z-8000 families and most of the other μP's in current use is shown in Figure 8.4. Again the features differ from those in the diagram for larger computers only in the details emphasized. Note how this figure is an extension of Figure 8.3. The system internal bus is the heart of the CPU—it carries both data and address information, but not simultaneously. An interesting aspect about the signals carried by the internal bus is that from the bits themselves there is no way to distinguish between data and addresses. The control circuitry, however, is aware of which instruction is being processed and on what part of the instruction or execution cycle it is currently operating. Thus, it is able to direct information properly between internal registers and between an internal register and one of the external buses. One feature of Figure 8.4 different from that of most larger machines is the fact that it employs an external address bus which is twice as wide as the data bus, 16 bits versus eight bits in a typical μP. This permits a relatively large number of memory words to be selected while achieving a savings in cost due to the lower number of bits per word. It does, however, lead to a certain amount of complication in μP internal design since some registers, e.g., the program counter and memory address register, must be 16 bits long, while others, e.g., the accumulator(s), are only eight bits long. The portions of the internal bus carrying the higher-order bits are marked with an H and those with the lower-

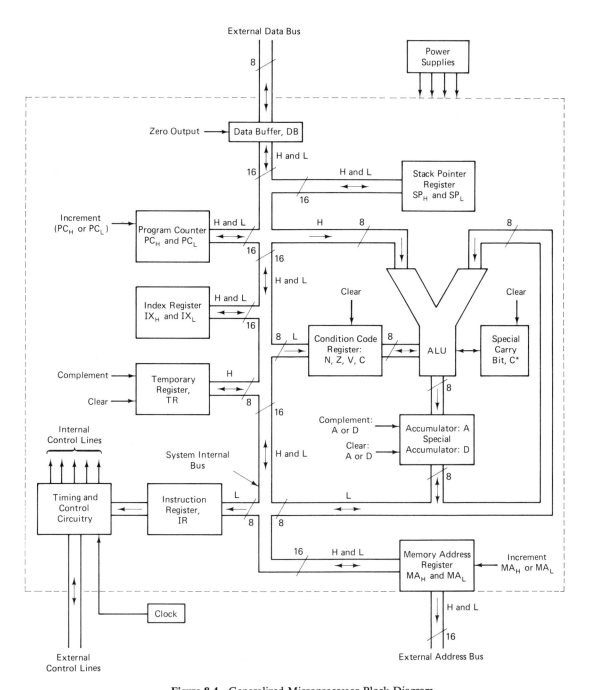

Figure 8.4 Generalized Microprocessor Block Diagram.

order bits are marked with an L in Figure 8.4. In addition, the number of bits carried by each signal is listed adjacent to the slash (/) on the bus. Consistent with this is the fact that registers like the program counter are divided into two sections, i.e., PC_H and PC_L. These may be loaded and incremented separately, e.g., $PC_L \leftarrow MA_L$; $PC_L \leftarrow PC_L + 1$. Alternatively, they may be loaded and incremented as a single register, with carry propagation in the case of the increment, e.g., $PC \leftarrow MA$; $PC \leftarrow PC + 1$.

While both the data buffer and the memory address register provide an interface function with the external buses, they are somewhat different in their method of operation. The MA is a true register and it points to the current memory (bus) address. The DB, however, is really a bilateral switch with the following six possible connections to the external data bus:

1. Receive on the internal lower bus [e.g., $PC_L \leftarrow M(MA)$].
2. Receive on the internal higher bus [e.g., $PC_H \leftarrow M(MA)$].
3. Transmit from the internal lower bus.
4. Transmit from the internal higher bus.
5. No connection between the internal and external buses, but the internal bus is released to carry data.
6. Zero output (i.e., transmit a logical zero to all bits on the external bus while the internal bus is released to carry internal signals).

One significant difference between microprocessor operation and that of larger-scale computers is in the number of bits available on the data bus. Although μP's range from 4- to 32-bit systems, the 8-bit variety is still the most popular. One consequence of such a narrow word is that the instructions often must be two or even three memory words long, which makes for more complicated instruction cycles. Thus, to clear an arbitrary memory location, CLR Extended, the μP must fetch the first word, all of whose bits are required to encode the clear instruction. Then the next two words must be fetched in sequence to determine the L and H halves of the memory address. The result is that several CPU cycles are required. To study this in more detail, consider the RTL program in Figure 8.5. At the end of the instruction cycle, $T = 4$ in this case, the memory address register is prepared to get the first word in the next instruction.

```
T = 1   IR ← M(MA); MA ← MA + 1; PC ← PC + 1;: Fetch Cycle;
T = 2   TR ← M(MA); MA ← MA + 1; PC ← PC + 1;: Start Execution Cycle;
T = 3   MA_L ← M(MA); MA_H ← TR; PC ← PC + 1;
T = 4   M(MA) ← 0; MA ← PC; N,V,C ← 0; Z ← 1; T ← 1;
```

Figure 8.5 RTL Implementation of the CLR Extended Instruction.

In comparing Figure 8.4 and 8.5 with the corresponding ones for the elementary computer (Figure 6.16, Figure 6.17, and Table 6.6), the following differences will be observed:

1. The data buffer is a switch, not the MB register.

2. Fetch is reduced to one clock cycle by the omission of the memory buffer register, and by doing a prefetch operation* (MA ← PC) at the end of the execution cycle.
3. The Fetch and Execution cycles form a continuous "T-sequence"; thus, an S flip-flop is not required.
4. To reduce instruction time, the number of clock cycles for the various instructions are of different lengths; e.g., Figure 8.5 requires four clock cycles, but for more complicated instructions, e.g., ADDA, five or more cycles may be needed.
5. Since one operand comes through the data buffer, the ALU has a slightly different bus configuration than has been seen previously.
6. The pertinent bits of the condition code register are changed at $T = 4$ to reflect the ALU status resulting from the current instruction. From the definition of the clear instruction, it is reasonable that the concatenated bits N, V, and C should be set to zero and that the Z bit should be set to 1.
7. Although for practical reasons most μP's employ a clock with two or more phases, for simplification only a single phase is employed here. Thus, in practice, the RTL statements shown for each clock pulse will merely be split between the clock phases.

Figure 8.6, the RTL program for the JMP instruction, demonstrates that by careful planning the number of clock cycles may sometimes be reduced, and it shows how data can be simultaneously read from the same memory location into two different μP registers. Part (a) of the figure is the longer program and makes it appear that the prefetch operation cannot be combined with the last part of the execution cycle. Part (b) of the figure, however, corrects the impression given by the original program by combining the productions in the original last two clock cycles. The reduced program also illustrates that different RTL statements can control the H and L sections of the bus.

$T = 1$ IR ← M(MA); MA ← MA + 1; PC ← PC + 1;
$T = 2$ PC$_H$ ← M(MA); MA ← MA + 1;
$T = 3$ PC$_L$ ← M(MA);
$T = 4$ MA ← PC; T ← 1;

(a) Original Program

$T = 1$ IR ← M(MA); MA ← MA + 1; PC ← PC + 1;
$T = 2$ PC$_H$ ← M(MA); MA ← MA + 1;
$T = 3$ PC$_L$ ← M(MA); MA$_L$ ← M(MA); MA$_H$ ← PC$_H$; T ← 1;

(b) Reduced Program

Figure 8.6 RTL Implementation of the JMP Extended Instruction.

Correction of a serious error is demonstrated in Figure 8.7. At $T = 4$ in the original figure, the $M(MA)$ operand to the ALU occupies the H-half of the internal bus while the MA ← PC statement places different information on the same bus.

*Sometimes physical constraints will force the prefetch operation to be done separately from an execution operation, resulting in fetch requiring two memory cycles.

$T = 1$ IR ← M(MA); MA ← MA + 1; PC ← PC + 1;
$T = 2$ TR ← M(MA); MA ← MA + 1; PC ← PC + 1;
$T = 3$ MA_L ← M(MA); MA_H ← TR; PC ← PC + 1;
$T = 4$ A ← A + M(MA); MA ← PC; N,Z,V,C ← (update); T ← 1;

(a) Original Program: Contains Bus-Overlap

$T = 1$ through $T = 3$ same as figure in (a).
$T = 4$ A ← A + M(MA); N,Z,V,C ← (update);
$T = 5$ MA ← PC; T ← 1;

(b) Corrected Program

Figure 8.7 RTL Implementation of the ADDA Extended Instruction.

This contradiction, or overlap of information, is defined as *bus-overlap* and is corrected in part (b) of the figure. Note that the addition of the $T = 5$ clock cycle, in part (b) of the figure, is actually a prefetch step and illustrates a case, mentioned previously, where prefetch cannot be combined with part of the execution cycle. The RTL statement N, Z, V, C ← (update), found at $T = 4$ in both parts of the figure, is a slight extension of our notation and indicates that the N, Z, V, and C bits which are concatenated to form the condition code register are updated according to the results of the production A ← $A + M(MA)$.

8.3 MEMORY ADDRESSING

When viewing Figure 8.3 from the unified bus concept, one realizes that the three units being addressed—RAM, ROM, and PIU—require some type of decoding scheme to enable the intended device properly. If only a small portion of the total available addressing space is to be employed, the external decoding can be very simple; but if a large fraction of the addressing space is to be utilized a correspondingly more complicated decoding scheme must be employed. There are three major methods for decoding memory address lines: full external decoding, one-of-N decoding, and multiple enables.

Full External Decoding

This method has been implied by the work we did in previous chapters and is represented by Figure 8.8.* In order to be specific, somewhat arbitrary but typical parameters are shown; e.g., each memory requires 10 lower address lines, A_0 through A_9. Note that the external chip enable, E, pins which are not required are properly tied to a voltage to select the chip.

The following characteristics of this method should be observed:

1. No addresses are wasted since all possible $2^{16} = 65K$ memory locations are available.

*The decoding principles are emphasized by including only the enable circuitry (for Figures 8.8 through 8.11) in all except one component, which is complete to allow the full system to be visualized.

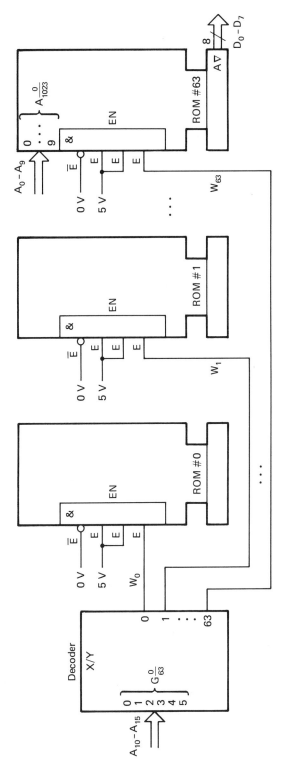

Figure 8.8 Memory Chip Selection by Full External Decoding.

2. An extra chip(s) is needed for the decoder, yet three of the enables provided with each memory are wasted.

One-of-N Decoding

In small systems this is the most popular method. As shown in Figure 8.9, each of the upper address lines is employed to select one of the memory chips. For comparison with Figure 8.8 the same type of memory and address line configuration is employed.

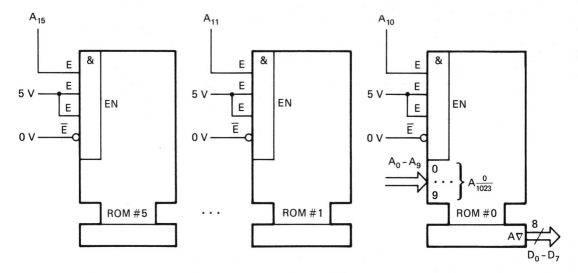

Figure 8.9 Memory Chip Selection by One-of-N Decoding.

There are some significant differences between the decoding in the two figures. The method in Figure 8.9 possesses the following characteristics:

1. Only $6(2^{10}) = 6K$ memory locations are actually available; thus, there is a significant waste. In a small system, however, not even the full 6K address location will normally be used.
2. Again all but one of each memory's enable pins are wasted, but no external decoding chips are required, which results in a significant savings and also simplification in wiring.

To clarify which memory addresses are available for programming it is helpful to construct the memory map shown in Table 8.1. Observe that the memory is contiguous between ROM 0 and ROM 1, but there are holes of various sizes between all other ROMs. Consider ROM 1; when bit 11 equals one the ROM is on and the location addressed is determined by the condition of bits 0 through 9. Although the other ROMs receive the same bits 0 through 9, there is not a bus conflict because Table 8.1 shows that bit 10, and bits 12 through 15, equal zero, which disables all the other ROMs.

TABLE 8.1 MEMORY MAP FOR ONE-OF-N DECODING

Memory Section (ROM Number)	Address Bits*				Hex Range	Decimal Range
	15 14 13 12	11 10 9 8	7 6 5 4	3 2 1 0		
0	0 0 0 0	0 1 X X	X X X X	X X X X	0400–07FF	1024–2047
1	0 0 0 0	1 0 X X	X X X X	X X X X	0800–0BFF	2048–3071
2	0 0 0 1	0 0 X X	X X X X	X X X X	1000–13FF	4096–5120
3–4			etc.			
5	1 0 0 0	0 0 X X	X X X X	X X X X	8000–83FF	32768–33792

*X ≡ variable bit, either 0 or 1 is allowed.

Decoding with Multiple Enables

The final method which we will consider involves the simultaneous use of more than one of the enable pins on each memory. ROMs provide particular flexibility with this scheme because usually each enable pin may be configured in either normal or complemented form when the programming mask is applied. This method is also applicable, however, with other memory types either in combination with ROMs or separately. Assuming that all enable pin combinations are available, the memory selection system of Figure 8.10 results.

Again there are significant differences between this method and the two previous methods. The following points in particular should be observed:

1. Here $16(2^{10}) = 16K$ addresses are available, which is considerably more memory locations than can be reached with the method of Figure 8.7.

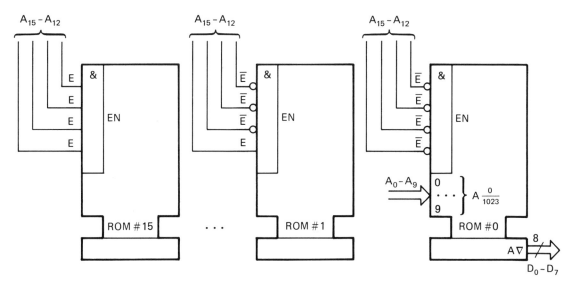

Figure 8.10 Memory Chip Selection by Decoding with Multiple Enables.

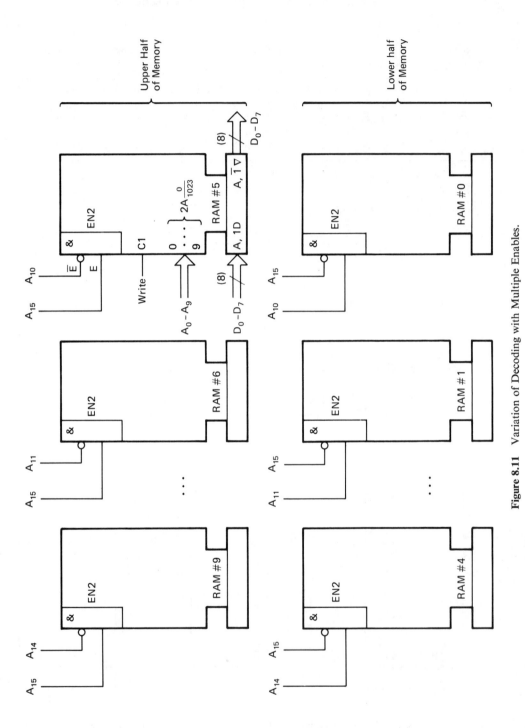

Figure 8.11 Variation of Decoding with Multiple Enables.

292

2. None of the memory enable pins are wasted and no external decoding chips are required. On the other hand, approximately 75% of the memory space is lost when this exact scheme is employed.

Several variations of the scheme of Figure 8.10 are possible. Since RAM and PIU modules frequently have a fixed group of E and $\bar{E}$ pins, one interesting multiple enable technique is shown in Figure 8.11. Here it is assumed that each RAM has only one E and one $\bar{E}$ pin. One way to visualize this technique is to employ A_{15} and either E or $\bar{E}$ to select either the top or bottom half of memory and then to employ the remaining $N - 1$ address lines to provide one of $N - 1$ decoding. Figure 8.11 has been drawn to emphasize the above visualization. The reader should study the figure and be sure to determine how the process can be generalized to cases where more enable pins are available.

8.4 INTERFACING PRINCIPLES

A μP is not of any value until it is interfaced to one or more external circuits—one cannot even read data out of it without at least interfacing the μP with a set of LEDs, light-emitting diodes. Some μP's have special output lines and special instructions for I/O; but for simplicity here we assume that I/O read and write is handled in the same way we do memory. Often we employ a special I/O chip, i.e., a PIU. A specialized chip, however, is not always necessary and simple tasks are often accomplished without them. The main value of a PIU is that it provides in a single chip many functions normally required to solve rather complicated interfacing problems.

Elementary Interfacing with Small-Scale Integration

As an example of an elementary, SSI interface, assume that a μP program depends on the condition of a single switch in the outside world. This switch is to be read under program control and if it is closed the program takes a certain action; otherwise it takes an alternative action. Since the complete system will require some areas in memory for the program, fixed constants, and other interfaces, assume that the memory map shown in Table 8.2 is valid for the system and that all other memory locations are available for our use.

TABLE 8.2 SYSTEM MEMORY MAP

Address Range (Hex)	0000–00FF	C000–C001	D000–D001	FC00–FFFF
Device	RAM #1	PIU #1	PIU #2	ROM #1

Following the concepts of Section 8.3, let us assign the switch a position on the memory map which makes its decoding very simple. There is a large empty area in the middle of the map, e.g., 0800 is typical. This location is particularly suitable because it may be satisfied by $A_{15} = 0$ and $A_{11} = 1$; all of the other bits can be don't cares and not decoded. Examination of the map reveals that $A_{15} = 0$ eliminates everything except RAM 1 and $A_{11} = 1$ eliminates that.

Since there are several other devices connected to the data bus, our switch must be isolated from these; otherwise bus conflicts will develop. The simplest way to accomplish this is place a tristate buffer amplifier between the switch and the data bus as shown in Figure 8.12(a). The tristate amplifiers are represented by the standard

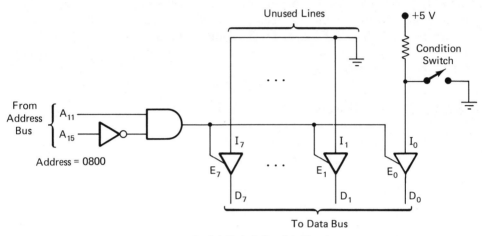

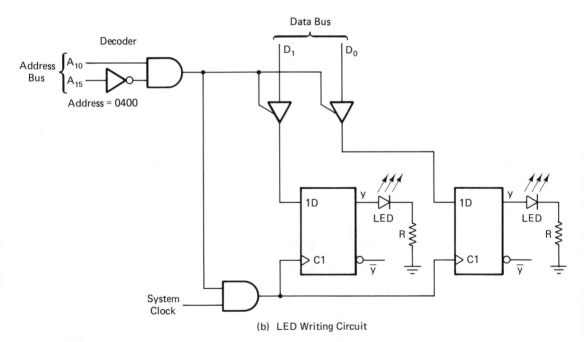

Figure 8.12 μP Read and Write Interfacing with SSI.

amplifier symbol, but with the signal entering from the side being the one which controls whether the amplifier has a connection between input and output or whether the amplifier places an open circuit on the data bus. The above description for the amplifiers in Figure 8.12(a) is concisely represented by Table 8.3.

TABLE 8.3 TRISTATE AMPLIFIER LOGIC

E_0	I_0	D_0
0	—	(Open circuit, ∞ resistance)
1	0	0
1	1	1

Because the address of our switch is only partially decoded, it should be clear that many other numbers on the address lines such as 0803, 0F00, etc., would also read the switch position. The important facts are, however, that there is no interference with other devices in the memory map and that a relatively simple interface was developed.

Figure 8.12(b) shows that a basic circuit permitting a μP to write information into a peripheral is only slightly more complicated than one permitting data to be read. A typical application of the circuit is to provide a data display for the results of information processed by the μP. The D-type flip-flops, of course, hold the output signal so that the μP can devote itself to other tasks instead of periodically refreshing the LEDs.

Since this circuit has an address of 0400, which is not overlapping with that of the switch in part (a) of the figure, both interfaces may be employed in the same system. Note that when the LED interface is deselected (the decoder at 0400 is off) the system clock is disconnected. This prevents an erroneous signal from being clocked through the D-input of the flip-flop. The resistor R in series with each LED prevents excessive current from being drawn by the flip-flop. (Normally, R is in the neighborhood of a few hundred ohms.)

Interfacing with SSI can lead to interesting and challenging logic design issues. Some of these possibilities, which are straightforward extensions of the current work, are explored in Section 8.8 starting with Exercise 8.15. At this point many readers will want to consider a few of these exercises in order to understand SSI interfacing more fully.

The Peripheral Interface Unit, PIU

The SSI circuits of Figure 8.12 are practical for fairly simple tasks requiring only a few bits of data. When many bits, interrupt functions, or the ability to reconfigure the interface electrically are required, however, a specially designed LSI chip is usually the best solution. Many manufacturers currently sell such chips at a very nominal cost, and the following discussion should provide the fundamentals for understanding them. (Our development most closely parallels the 6821 chip, which is the PIU for the 6800 system.)

Shortly we will describe the PIU-1, which will be one of the standard chips to be employed later in application examples and sample designs. To gain insight into PIU operation, however, we first consider a simplified read-only PIU—one that allows data to be read, from a peripheral to the μP, but the reverse, write, path does not exist. Figure 8.13 presents this PIU and the following features should be observed:

1. The data path from the peripheral through the data register to the μP.
2. The method for addressing the PIU through partial decoding by the chip select pins, CS_i, and the tristate buffer.
3. The way the load data line provides a signal for transferring information into the data register; but it is not in synchronization with the μP clock. (Note that the unclocked R-S flip-flop discussed in Section 5.6 is employed.)
4. The method for setting the read flip-flop by the Load Data line with the resulting level becoming the interrupt signal fed to the μP.
5. The way the level coming off of the read flip-flop's y output tells the peripheral that the μP has received the data (when $y = 0$).
6. That the read flip-flop is reset in synchronization with the μP clock once the

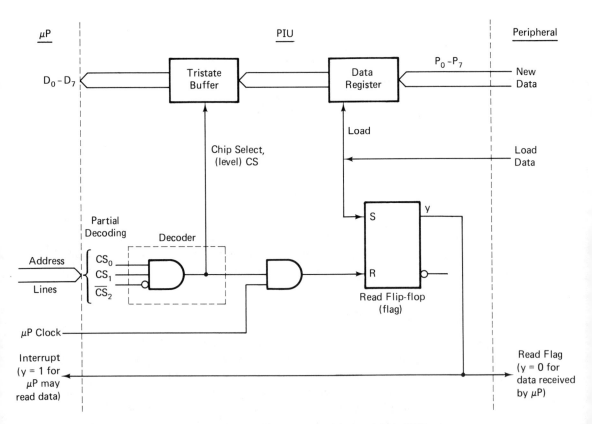

Figure 8.13 Basic Read-Only PIU.

chip is selected. Note that the μP and the peripheral logic can prevent the simultaneous occurrence of S and R signals by monitoring the flip-flop's y output.

The read flip-flop is the main component in a very important digital operation called *handshaking*, which is a process for supervising the interchange of data between two computer units—the μP and the peripheral in this case. Handshaking, through the interrupt signal, tells the μP when new data are available to be read; and through the Read Flag signal, it tells the peripheral that the μP has copied the information from the data register.

The counterpart of the above device is the write-only PIU shown in Figure 8.14. These two PIUs have many similarities except that they interchange data in opposite directions. Again there is a handshaking feature in this, but here the interrupt is employed in a somewhat unorthodox manner in that it acknowledges receipt of the data by the peripheral. This signal is important in that it informs the μP that the PIU is ready for the next byte to be placed into the data register. Note in Figure 8.14 that the data register is loaded in synchronization with the μP clock because it is that signal which will have the proper timing relationship to the other μP signals, e.g., the address and the data bits. On the other hand, the Data Received signal sets the write flip-flop without being synchronized by the clock (it is asynchronous).

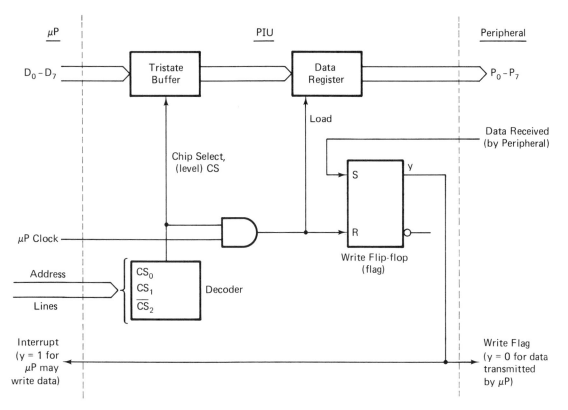

Figure 8.14 Simple Write-Only PIU.

It is highly desirable in a practical PIU to have the features of both the read and the write units in the same package. Moreover, it is also desirable to have some convenience features not previously discussed, e.g., the ability to have some of the bits in the data register to be in the read mode while others are in the write mode, to modify the flag flip-flops so that the interrupts may be inhibited yet the flag set to store the information but on a less urgent basis. To demonstrate the above principles a circuit called the PIU-1 is established following the logic and notation shown in Figures 8.13 and 8.14; e.g., the μP Read and Write flags have essentially the same inputs and outputs as in the previously separate diagrams. In order to permit the desired conve-

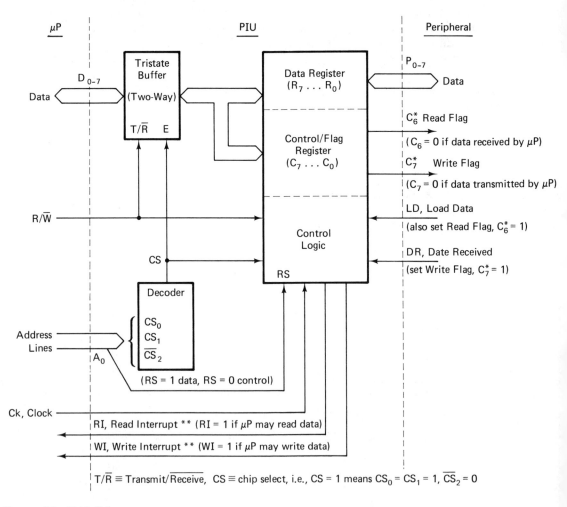

Notes: *See Table 8.4.
      **See Table 8.5 for enable conditions.

Figure 8.15 Read/Write PIU-1 Circuit.

nience and flexibility, the PIU-1 (shown in Figure 8.15) contains a second register called the control/flag register. This has its own address and is read and written in the same manner as the data register. It contains two bits (C_0 and C_1) for determining which data register bits are in the read mode and which are in the write mode, a pair of bits (C_2 and C_3) to determine the way the Load Data signal controls the Read Interrupt and its flag, yet another pair of bits (C_4 and C_5) that perform the same function for the Data Received signal. Finally, there is a pair of flag bits with C_6 being the Read flag and C_7 being the Write flag. This scheme is summarized in Table 8.4.

TABLE 8.4 PIU-1 CONTROL/FLAG REGISTER FORMAT

C_7^*	C_6^*	C_5	C_4	C_3	C_2	C_1	C_0
Write Flag	Read Flag	Data Received Function		Load Data Function		Data Direction Control	

*All the bits in this register, except these, may be written and read in the same way as the data register. C_6 and C_7, however, may be read in the normal way, but they are each written separately by logic such as that shown in Figures 8.13 and 8.14.

In the process of combining the read and write functions in a single PIU, it is not surprising that a few extra control lines need to be added. One of these is the register select line, RS, which in one state, $RS = 1$, selects the data register and in the other state selects the control/flag register. On system startup the program initializes C_0 through C_5; then frequently these bits are not changed during system operation. The program, however, will normally read the flags in bits C_6 and C_7 many times during system operation to monitor peripheral requests.

In order to understand the function of the various bits in the control/flag register, consider Table 8.5. Part (a) of the table shows that in addition to all read and all write

TABLE 8.5 CONTROL/FLAG REGISTER BIT CODING*

(a) Data Direction Control

C_1	C_0	Function
0	0	All bits are μP read.
0	1	P_4–P_7 μP read; P_0–P_3 μP write.
1	0	P_2–P_7 μP write; P_0–P_1 μP read.
1	1	All bits are μP write.

(b) Read Data from PIU

C_3	C_2	Load Data Pin (Active Transition)	Read Interrupt Enabled?
0	0	↓	No
0	1	↓	Yes
1	0	↑	No
1	1	↑	Yes

(c) Write Data to PIU

C_5	C_4	Data Received Pin (Active Transition)	Write Interrupt Enabled?
0	0	↓	No
0	1	↓	Yes
1	0	↑	No
1	1	↑	Yes

*The Read flag is always set on active transition by the Load Data input. The flag is reset (cleared) by the μP reading the data register. The Write flag is always set on active transition by the data received input. The flag is reset (cleared) by the μP writing into the data register.

the register can be divided in two mixed ways. By allowing either two or four data bits for μP reading and the remainder for μP writing it is possible to do a small I/O task with a single PIU. Of course, during a μP write to the data register the "read bits" are not effected in any way and the converse also holds. In a similar vein, the Load Data signal will only cause those P_0 through P_7 bits to be transferred to the data register which agree with the $C_1 C_0$ specification.

Parts (b) and (c) of the table basically describe the on/off control of each interrupt; they also describe selection of either negative- and positive-going transitions as the active signal. Note that the flag is set by the active Load Data signal independent of the state of its interrupt; further, once the flag is set if the interrupt is later enabled, it will immediately be passed to the μP. As Figures 8.13 and 8.14 show, the flags are reset by the chip select signal and the μP clock. In the case of Read, reset means that the μP has read the data register. On the other hand, in the case of Write, reset means that new information from the μP has been deposited in the data register.

Example: Reading Data from a Peripheral through the PIU-1 into the μP

To clarify the above principles consider the timing diagram shown in Figure 8.16, which basically describes the reading of data from a peripheral into the μP, $D_{0-7} \leftarrow P_{0-7}$. You should carefully follow the data transfer steps starting at the peripheral with the Load Data signal causing the information to be transferred first into the data register, R. Then observe that the PIU-1 is selected and information is sent through the tristate buffer out on the data bus, D, where it is finally transferred into the μP by the clock pulse. (It will probably be helpful to refer to Figure 8.15 and Table 8.5 while studying the timing diagram.) Note that we employ some new notation on the lines representing bus and register signals; e.g., for $t_1 \leq t \leq t_2$ the timing diagram shows that the data register, R, carries the word $R_7 R_6 \ldots R_0 = 0110\ 1100$. The large dot on the diagram represents a transition point for data; but the data word is always constant between dots. The symbol "X" is employed for any bit of a signal that is unknown; e.g., before t_1 all bits of the data register are in that category. As an aid in following the sequence of signals, large groups of bits that are unchanged from the previous period are labeled with the abbreviation "unch." The information below the timing diagram should help in understanding the sequence of signals that occur when the μP reads data through the PIU-1.

Example: Writing Data into the Peripheral and Changing the Control Function

Figure 8.17 presents another example in which the PIU-1 serves as a vehicle for the transfer of data between a μP and its peripheral. It then changes part of the control word so that the nature of the data received function is modified; and finally, by employing the new control information, it writes the second data word out to the peripheral. Again, detailed information about the main features of the timing diagram is given at the bottom of the figure. In this case, however, most of the features emphasized are ones not appearing in Figure 8.16.

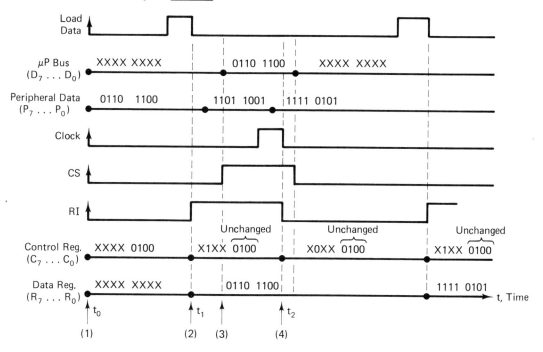

Major Events:

1. At t_0, the registers show the following starting conditions:
   a. From Table 8.5, a $C_1 C_0 = 00$ indicates that all PIU bits are in the μP read mode.
   b. $C_3 C_2 = 01$ indicates that the Load Data Signal, LD, is active on the negative going transition and RI is enabled.
   c. The Peripheral Bus has good data on it.
   d. The Tristate Buffer had disconnected the PIU from the D_{0-7} Bus; thus $D_7 \ldots D_0 = $ XXXX XXXX.
2. The negative going edge of LD causes RI = 1, $C_6 = 1$, and the peripheral data to be loaded into the Data Register.
3. The information from the Data Register is placed on the D_{0-7} Bus by the proper address performing the chip select, CS = 1.
4. The data on the D_{0-7} Bus is transferred to the μP, and the peripheral is informed from the line $C_6 = 0$. Between t_1 and t_2, $P_7 \ldots P_0 = $ 1101 1001, but this data is lost because a Load Data signal does not occur until after the P_{0-7} signal is changed.

Notation:
X ≡ logic value of this bit unknown
Logic levels:

Figure 8.16 Timing Signals for Reading Data from a Peripheral ($D_{0-7} \longleftarrow P_{0-7}$).

Constants: R/$\overline{W}$ = 0, Not used: RI, LD.

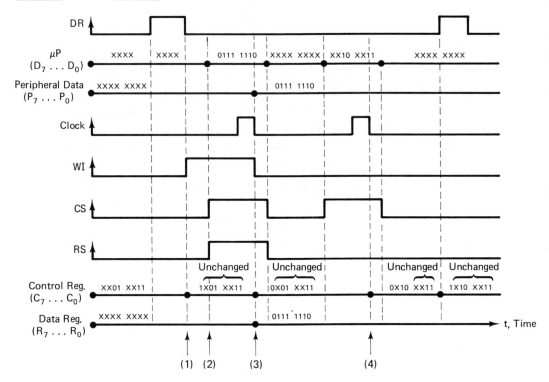

Figure 8.17 Timing Signals for Writing Data from a μP ($P_{0-7} \leftarrow D_{0-7}$ and $C_{0-7} \leftarrow D_{0-7}$).

Major Events:

1. Since C_{54} = 01, the active transition of DR is on this negative going edge, which is also the instant that the Write Interrupt occurs and that the flag, C_7, is set.
2. The data is placed on the Data Bus coincident with the onset of the address producing the chip select, CS, and the register select.
3. Although CS = 1 considerably earlier, it is not until down clock that the incoming data are transformed into the Data Register. Note that at this same time C_7 = 0 indicating to the peripheral that the data has been transferred and is available on the P_{0-7} Bus.
4. At this point, bits C_4 and C_5 of the Control Register are changed giving the Data Received function its active transition on the positive going edge, and disabling the interrupt.

Notation:
See Figure 8.16.

Example: Program to Service the PIU

As a contrasting example to the above, consider the following requirement: Construct a 6800 program fragment that causes the μP to pause until the peripheral acknowledges receipt of the previous data word, then writes the hex number 2F into the bits P_0 through P_7. The Data Received signal is to be active on its negative-going transition and although not used, the Load Data signal is to be active on its positive-going transition. Neither interrupt is to be active. The program is to be placed anywhere in a 256-word ROM having A_0 through A_7 connected in the usual way and the only two enable inputs connected with $\bar{E} = A_{15}$ and $E = A_{14}$; in addition, the PIU-1 has the connection $CS_0 = A_{15}$, $CS_1 = +5\text{ V}$, $\overline{CS_2} = A_{14}$, and $RS = A_0$. The solution to this problem first involves recognition of the fact that the hex address range for the PIU is 8000–8001 and the range for the ROM is 4000–40FF. The program for this problem is shown in Table 8.6. The first instruction sets all bits in the PIU to write with $C_0 = C_1 = 1$; it sets the no interrupt and the Load Data signal active on the positive-going transition with $C_3 = 1$, $C_2 = 0$; and it sets the Data Received signal active on the negative-going transition with $C_5 = 0$, $C_4 = 0$.

TABLE 8.6 PROGRAM FRAGMENT TO TRANSMIT THE WORD 2F TO A PERIPHERAL

Starting Memory Address	Label	Instructions	Comments
4000		• • •	
		LDAA #$0B	Initialize: load control/flag reg.
		STAA $8000	
	LOOP	LDAA $8000	Test Write flag and wait until it is set.
		BPL LOOP	
		LDAA #$2F	Place 2F in data register.
		STAA $8001	
		• • •	

8.5 STRUCTURED MICROSYSTEM DESIGN

A μP is a universal data processor which contains some fixed elements (e.g., arithmetic and control functions), but it has great power for algorithm construction through its software. Thus, the design techniques for a μP system are somewhat different from the ones we developed in Chapter 7 for the MSI/LSI components implementation. In both cases, however, we employ structured design, which includes the top-down approach as a main part of its philosophy. Thus, we proceed first to a broad analysis of the system requirements, then to a partitioning of the design into compatible sections (components), and finally to the detailed logic (circuit wiring or software coding). We now intend to employ our new understanding of μP's in the design of practical digital systems. The first task is to develop a *Design Outline Technique* for presenting basic system concepts as opposed to complete working drawings. (Although done on a smaller scale, this includes most features of a total structured design except detailed logic.) The information given, however, should be sufficiently complete so that a computer engineer would be able to use it to complete all the details of both the hardware and software design. The three steps to be employed in preparing a design outline are as follows:

1. Draw a hardware block diagram, normally containing 12 blocks or less, which emphasizes those aspects of the design not common to most other systems. Since μP, ROM, and RAM are usually employed in a conventional way, they may be combined into a single block. Careful attention should be given to special circuitry required by the peripherals themselves and in connecting the peripherals to the PIUs.
2. Construct a general flow diagram, normally 15 blocks or less, which emphasizes those aspects of the software most important to understanding the system under consideration. To justify program concepts not previously discussed or simply for clarity, it is often necessary to supplement the general flow diagram either with one or more program fragments or with a detailed flow diagram(s) for the critical section(s) of the general diagram.
3. Present a verbal explanation of operation and emphasize special features of the design that may not be obvious to the reader.

Example: Data Processor Which Displays the Average

In order to see how easy it is to outline the design of a small data processing system with μP's, consider the following problem. A μP system is to be built to find the mode (the most frequently occurring value, which is one measure of the average) in a set of data consisting of eight-bit binary words whose values are all larger than zero. The data will be transferred to the μP under interrupt control and the end of data will be marked by the contents of the last word being zero. After the last word, the value of the mode is to be displayed in binary on a set of eight LEDs. Any important constraints or information not given above are to be specified in the best way by the designer and they are to be explicitly listed and discussed.

Solution: Since this problem states that the information is transferred under interrupt control, we are automatically guaranteed that the peripheral will supply the data on eight lines (these become P_{0-7}) and that when they are ready a pulse is supplied on a separate line, which becomes the Load Data signal. Further, the peripheral will pause until the read flag, C_6, resets to zero before issuing the next Load Data signal. Figure 8.18 is a block diagram of the required μP system and it satisfies the needs of step 1 of the system design outline. In agreement with our philosophy of only showing the skeleton of those items common to most systems the μP address bus, data bus, and control lines have been combined into a single bus feeding the two PIUs. Also note that since PIU 2 contains an eight-bit register, it stores the output and drives the LEDs directly without requiring an additional chip. Certain other displays, e.g., the gas discharge type, may require a buffer amplifier chip to provide the required power, but unless specifically requested the Design Outline Technique will not reach this level of detail.

There are several points concerning our system that were not completely specified in the original problem statement. This lack of completeness is typical of almost all specifications and it is part of the challenge of design. The main criterion for completing the specifications was to include the most probable cases and to have a reasonably simple system. Lower-probability cases could be included but usually at a cost of more

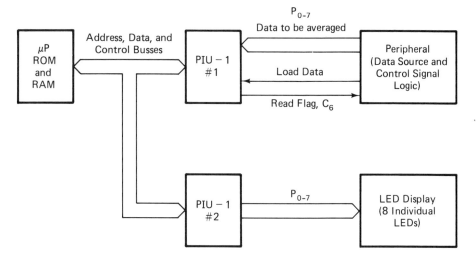

Figure 8.18 Data Averaging and Display System.

complicated software. The following list contains the major points which were decided by the designer:

1. After the first mode is displayed, the system will accept data to perform a new calculation but will display the old value until the new value is ready.
2. In cases where data come in short rapid groups, the rate may be too high for the observer to follow satisfactorily. A software delay would be one solution to the problem; but it is not thought to be of a high enough priority to be included in this program.
3. A single mode value is to be selected for each set of input numbers. If two or more inputs occur with the same highest frequency, the one corresponding to the lowest input will be displayed.
4. Zero will be displayed until the first mode value is ready.
5. If a particular number entering into a mode calculation occurs often enough ($>2^8 - 1$ times), the storage table will overflow. The resulting fault could be detected with software and corrective action taken; but it is assumed that overflow is very unlikely and does not justify the additional complication in the program.

By employing the above assumptions, the software flow diagram in Figure 8.19 was prepared.

The algorithm is based on a "table-lookup" procedure and consists of three major program loops. In the first loop, which included block 1 in the figure, the data are read stored sequentially in the table until the presence of a zero signals the end of the current data set. Although not detailed in Figure 8.19, a loop is then employed in block 3 to search through the table to find the mode. Finally, the outer loop is closed by returning to read more data.

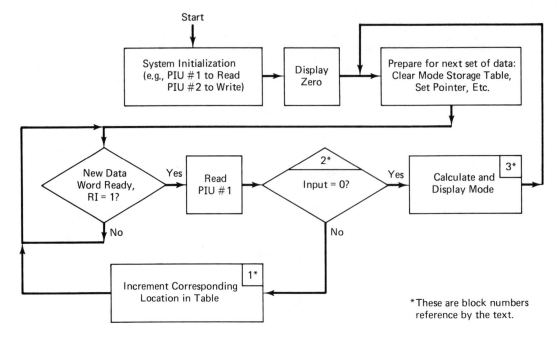

Figure 8.19 Flow Diagram for the Mode Calculating System.

Each number read from PIU 1 becomes the relative address in the table and its presence is recorded by incrementing the contents of that address. Assuming that the number read from the data source is in accumulator A, the brief program fragment in Table 8.7 will properly achieve the increment operation. The principle here is to store A's contents, which becomes the relative table address, into the third byte of the INC instruction. This instruction, theorefore, will increment the desired location in the table. The base address of the table, given here as 2000_{16}, may be changed to any convenient value within the RAM's range.

TABLE 8.7 PROGRAM FRAGMENT FOR INCREMENTING THE TABLE CONTENTS

RAM Address	Mnemonic	Remarks
	• • •	
150	STAA $155	Both are extended,
153	INC $2000	three-byte instructions.
	• • •	

Rather than employ a program fragment to "prove" the mode calculation method of block 3 in Figure 8.19, we employ the flow diagram of Figure 8.20. The heart of this algorithm is to initially set HPC = 0; thus, by sequentially comparing HPC with each table entry, a larger value will eventually be reached and a possible mode

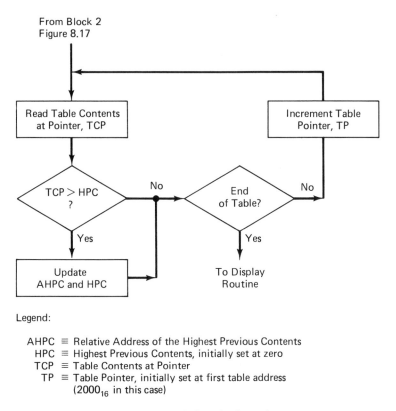

Figure 8.20 Mode Search Flow Diagram.

candidate will be established. The search for larger HPC values will continue, with higher candidates being stored in HPC and their relative address in AHPC. When the end of the table is reached the relative address stored in the AHPC location will be the mode of the set of data being considered.

‡8.6 MICROPROCESSOR VERSUS RANDOM LOGIC DESIGN

There is no single design technique, microprocessor system, or MSI/LSI components implementation which is optimum for all situations—each specification must be considered and its requirements carefully analyzed before a method is chosen. When trying to decide between a μP system and components, the following facts should be considered.

1. μP's are being employed in an increasingly large percentage of the designs. There are many reasons for this, but two of the main ones are that μP capabilities are continually increasing and cost is decreasing. Increased capabilities means not only that μP's can perform a wider variety of tasks and that they are faster, but also that more memory and PIU functions are being placed in the

same package as the µP itself, resulting in a very compact system with a very low chip count, which leads to even further decreases in cost.
2. Limitations associated with the µP should be appreciated. Typical ones are:
   (a) They are basically serial devices and are relatively slow. Although a typical µP instruction may require only about 2 µs, many of these instructions are often required to complete a relatively simple task that can be accomplished by components in a small fraction of a microsecond. Moreover, many component circuits can be operating in parallel to accomplish a complicated task rapidly.
   (b) Fairly strict limits exist on the amount of power µP's can supply. Thus, buffers or power amplifiers may be required even in modest systems.
   (c) Design documentation and debugging for µP software can be even more expensive than it is for a corresponding task done exclusively by component circuits.
3. The following are heuristics that may influence the designer's use of µP's:
   (a) For a purely combinational logic problem (such as a large decoder) or for a major combinational segment of a sequential problem, a cleaner design will frequently result if a ROM or a PLA is employed rather than a µP.
   (b) Where specialized MSI or LSI chips* exist for complicated tasks, a better design will often result if the specialized chip is employed as an auxiliary circuit for the µP or even to replace the µP entirely.
4. On the other hand, typical µP strengths are:
   (a) It is easier to make field changes in a system by replacing a program in ROM than by changing wiring and circuit components.
   (b) Particularly for large production runs, it is cheaper to produce µP systems than other methods. The reason is that, to a great extent, wiring is being exchanged for software deposited in a ROM and the software can be copied into the ROM at a much faster and cheaper rate than the best automatic systems can do circuit wiring.
   (c) All other things being equal, the more complicated a task is, the more likely it is that a µP will be a good solution; and the "break point" is continuing to move toward the use of µP's in simpler systems.
   (d) Unless there are strong reasons to the contrary, such as discussed above in paragraphs 2 and 3, a good heuristic is first to consider the use of a µP in satisfying problem specifications.

Example: Maximum Recording Digital Thermometer

We now study competing designs for microprocessor and component versions of a digital thermometer. This problem is somewhere in the gray region between a system so simple that components are obviously the optimum approach and a more complicated problem with required expansion capabilities where a µP is the obvious choice. The details of the problem are: Design a digital celsius thermometer, based on an analog sensing element, that remembers the maximum temperature occurring

---

*For obvious reasons, these are often called *peripheral chips*.

during any desired interval of time. It is to have the following pushbutton controls:

1. Display the present indoor temperature, at least within the range 15 to 35°C.
2. Display the maximum temperature, which has been recorded during the current interval, but continue sampling and recording the present temperature.
3. Clear the maximum temperature storage and start a new recording interval.

Only one button is to be pushed at a time and after the button for one type of display is pushed, that display is to remain active until the other display is requested.

Microprocessor version. Using the Design Outline Technique, the block diagram for the µP version is given in Figure 8.21. Three PIUs are employed—the first to drive the display, the second to monitor the system controls, and the third to take data from the thermal sensor. This sensor produces an analog signal which must be converted to digital for the µP's use by a special IC called an *analog-to-digital, A/D, converter*. It periodically samples the analog signal; stores it; and converts its magnitude to a binary number of sufficient bits to meet the system accuracy specifications. The key to communication between a µP and much of the outside world is through the A/D converter and its inverse, the *digital-to-analog, D/A, converter*. (See Reference 3 for a detailed discussion of all facets of A/D and D/A conversion.) Note that the specified temperature range does not require a sign, which simplifies the procedure for handling data. It is assumed that the A/D converter contains its own timer, which causes the temperature to be automatically sampled at approximately 1-s intervals.

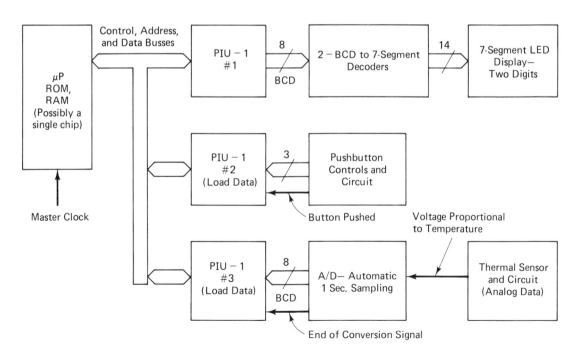

Figure 8.21 Digital Thermometer System—µP Hardware.

Sec. 8.6   Microprocessor versus Random Logic Design

The end of the conversion signal then causes the BCD result to be loaded into PIU 3 and the flag set so that the µP can update the current temperature and check to determine if a new maximum has occured.

A flag is employed, rather than an interrupt, to request service from the microprocessor by both the A/D input and by the controls. Since the data rates in this system are quite low, the flag approach rather than use of interrupts is entirely satisfactory. Moreover, the use of flags makes the software somewhat simpler.

The three pushbuttons are wired so that each establishes the logic level on a single bit line feeding PIU 2. In addition, the button circuitry simultaneously provides a logic level on a separate line when any of the buttons are depressed. This line feeds the Load Data input and sets the flag flip-flop.

An interesting hardware/software trade-off exists in the display logic. Figure 8.21 shows a hardware decoder transforming each of the BCD digits into seven-segment code for the LEDs. The alternative would be to do the decoding in software and present the seven bits to each LED through two separate PIUs. Thus, the trade-off is two decoders for an extra PIU and the decoding software. The original Figure 8.21 is considered to be somewhat superior to the alternative because the extra PIU would probably be more expensive than two decoders. Also, in production runs of only a few systems, the extra software would add to the cost of the alternative. Addi-

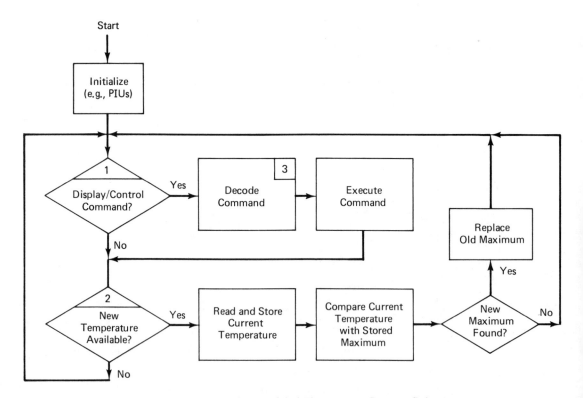

Figure 8.22 Digital Thermometer System—Software.

tional decoding software costs would, however, be negligible for even moderate-size production runs.

The flow diagram for this system's software is shown in Figure 8.22. Blocks 1 and 2 are places where the above-mentioned flags are checked to determine if a control button has been pushed or if the A/D converter is ready with a new temperature. Until one of the flags is set, the μP continues to loop through blocks 1 and 2. If the control button for maximum temperature is pushed, the flag in PIU 2 will be set. At block 1 the μP will sense this condition and in block 3 it will determine that the active command is maximum temperature. The decoding of this command will be straightforward since only one of the three pushbutton command lines will be high at a time. The μP will next retrieve the maximum temperature from its storage location and write the BCD number into the data register in PIU 1. This register will continue to hold the new temperature until another display command is given and meanwhile the data register's two BCD digits will pass through a seven-segment decoder and be displayed on the LEDs. The reader should continue to study both Figures 8.21 and 8.22 and be sure to understand how all the other aspects of the original specifications are satisfied by the system.

MSI/LSI components version. Following the MSI/LSI design process of Section 7.2, the Detailed Design step yields the circuit of Figure 8.23. First consider the similarities between this and the μP solution. Both employ the same thermal sensor and the same A/D converter. Also, both employ the same display circuitry: a BCD-to-seven segment decoder followed by an LED display. Instead of using a μP to implement the system logic in software, Figure 8.23 shows an eight-bit comparator at the heart of the components version. The current temperature register and the maximum temperature register each feed the comparator; and if the current temperature is larger, the $A > B$ signal coincident with the master clock pulse causes the new maximum temperature to be loaded into the maximum temperature register. Whether the maximum temperature or the current temperature is displayed depends on the signal from the pushbutton controls. The reader should study the differences in the way the pushbutton control signals are employed in the two systems.

Summary and conclusions. The two solutions to our digital thermometer problem, Figures 8.21 and 8.23, require a comparable number of blocks; and if each block required one IC, the two solutions would be equivalent from an IC count standpoint. Although this type of comparison is somewhat simplistic, our two solutions do appear to be quite competitive from many practical viewpoints. This standoff, however, is sensitive to relatively small changes in the system specifications. For example, if the minimum temperature were also required to be stored and displayed, the balance would begin to swing in favor of the μP solution. Moreover, if a day-by-day time history of maximum and minimum temperatures were required to be stored and displayed, the μP system would have a definite advantage over the components approach. The design shown in Figure 8.21 could be extended to the Component Finalization stage and a similar extension made in the system of Figure 8.21. Although this would permit a more exact comparison, which is often demanded in practice, it would not contribute

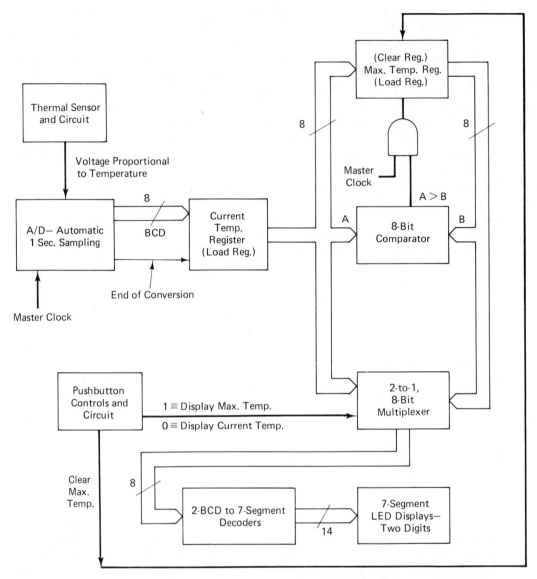

Figure 8.23 Digital Thermometer System—Detailed Design (components version).

anything to our understanding of digital design, nor would it help in understanding the sensitivity of the two approaches to changes in specifications.

It is our conclusion that while a larger and larger percentage of digital systems will be designed with μP's, SSI and MSI components will remain important to the engineer into the future for the following reasons:

1. Interfacing a μP to memories, peripherals, and other devices is often facilitated if some SSI and MSI components are employed.

2. To understand fully the operations and limitations of μP's, a detailed knowledge of gates, registers, and other primitives is required.
3. System operating speed and other special requirements sometimes prevent conventional μP's from being employed.

The discussion of Structured Design continues in the next chapter, where microprogrammable microprocessors are considered. For more details on certain topics of this chapter, however, you may want to study: Winkel and Prosser's book[15] on top-down design, and Peatman's book[12] on self-test, signature analysis, and dependency notation. The bibliography also lists several excellent texts that consider specific μP's and that emphasize other useful topics such as 16-bit machines.

8.7 BIBLIOGRAPHY

1. Bishop, R., *Basic Microprocessors and the 6800*. Rochelle Park, N.J.: Hayden Book Company, Inc., 1979.
2. Dollhoff, T., *16-Bit Microprocessor Architecture*. Reston, Va.: Reston Publishing Company, Inc., 1979.
3. Hnatek, E. R., *A User's Handbook of D/A and A/D Converters*. New York: John Wiley & Sons, Inc., 1976.
4. Klingman, E. K., *Microprocessor Systems Design*. Englewood Cliffs, N.J.: Prentice-Hall, Inc., 1977.
5. Krutz, R. L., *Microprocessors and Logic Design*. New York: John Wiley & Sons, Inc., 1980.
6. Leventhal, L. A., *Introduction to Microprocessors: Software, Hardware, Programming*. Englewood Cliffs, N.J.: Prentice-Hall, Inc., 1978.
7. McGlynn, D. R., *Microprocessors: Technology, Architecture, and Applications*. New York: John Wiley & Sons, Inc., 1976.
8. McGlynn, D. R., *Modern Microprocessor System Design: Sixteen-Bit and Bit-Slice Architecture*. New York: Wiley-Interscience, 1980.
9. Osborne, A., *An Introduction to Microcomputers*, Vol. 1: *Basic Concepts*. Berkeley, Calif.: Adam Osborne and Associates, Inc., 1976.
10. Osborne, A., *An Introduction to Microcomputers*, Vol. 2: *Some Real Microprocessors*. Berkeley, Calif.: Adam Osborne and Associates, Inc., 1978.
11. Peatman, J. B., *Microcomputer-Based Design*. New York: McGraw-Hill Book Company, 1977.
12. Peatman, J. B., *Digital Hardware Design*. New York: McGraw-Hill Book Company, 1980.
13. Soucek, B., *Microprocessors and Microcomputers*. New York: John Wiley & Sons, Inc., 1976.
14. *The 8080–8085 Microprocessor Book*. New York: John Wiley & Sons, Inc., 1980.
15. Winkel, D., and F. Prosser, *The Art of Digital Design—An Introduction to Top-Down Design*. Englewood Cliffs, N.J.: Prentice-Hall, Inc., 1980.

8.8 EXERCISES

8.1 Write an RTL program for the CLRA instruction.

8.2 Write an RTL program for the STAA instruction.

8.3 A new instruction rotate left extended [C, M ← rlC, M] is to be added to Table 3.10. Its operation is the same as ROLA except that the operand address is specified by the second and third bytes as is done with the other extended instructions. As indicated in the notation above, the carry bit, C, acts as the most significant bit, effectively forming a nine-bit operand. Write an RTL program for the new instruction.

8.4 Using ROMs *only* (no auxiliary logic allowed) with the same general properties and requirements as those discussed in Section 8.3, except with each ROM having *512 8-bit words* and three enable inputs, in any combination of E and $\bar{E}$ forms, design the largest possible *continuous* memory, starting with address zero. Be sure your block diagram makes it clear what address lines go to which part of each ROM. How many words are in the complete memory?

8.5 For the single memory chip shown in Figure E8.5, determine the addressing range in hex. (If more than one range is possible, give both the lowest and highest ranges.)

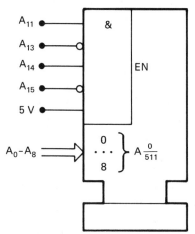

Figure E8.5

8.6 The PIU-1 control/flag register contains the number $E6_{16}$. Give a detailed explanation concerning the operating state of the PIU-1.

8.7 The PIU-1 is to be in the following state:

1. All data bits are μP write.
2. The write interrupt is enabled.
3. The active transition of the data received pin is on the positive-going edge.

What hex number should be placed in the control/flag register? (Fix all bits at "zero" which cannot be determined or which are arbitrary.)

8.8 Assuming a power supply pin for +5 V and pin for ground, how many total pins are required for our PIU-1 chip? Name them.

8.9 Using the Design Outline Technique, produce a well-structured μP system for the following specification:

1. If the input signal to the system is in the true state (true = +5 V, false = 0 V), the square wave shown in Figure E8.9 is to appear at the output terminal or else the complement is to be produced. (The complement has the high and low voltages reversed from the original signal.)

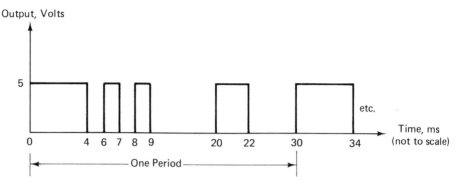

Figure E8.9

2. Switching between the signal and its complement must occur only at the end of a full period of the wave.
3. Waveform accuracy of within 5% of the specification is required.
4. Emphasize the description of the method for generating the waveform timing and the method of switching between the waveform and its complement in your documentation.
5. A square-wave (0 and 5 V) oscillator circuit with outputs of 0.5, 1, and 10 kHz is available as a timing source.

8.10 Using the design outline approach, develop a well-structured μP system to be placed

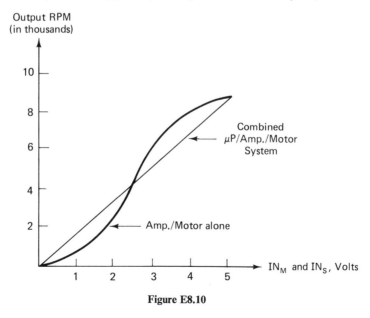

Figure E8.10

in series with a motor and the power amplifier driving it, for producing a linear combined system response (output versus IN_S) from the nonlinear amplifier/motor characteristics (output versus IN_M); both graphs are shown on Figure E8.10, where IN_M is the input voltage to the amplifier/motor, IN_S is the input voltage to the combined system, and the output RPM of the motor is common to both graphs. In your solution, emphasize the method for converting from the system input, IN_S, to the voltage required by the amplifier/motor to yield the desired output RPM. Since a ROM as large as a few hundred words is cheap and programming time is expensive, make your software as simple as possible at the expense of ROM words, if necessary.

8.11 At $t \geq 0$ a PIU system has the following information on its peripheral data bus: $(P_7-P_0) = 10001101$; also $R/\bar{W} = 1$, $R_S = A_0 = 1$. Employing all our usual notation for the PIU, fill in (mark heavy) *as much information as possible* into the timing diagram (Figure E8.11). (Even x, unknown, is considered information—if that is all that can be determined.)

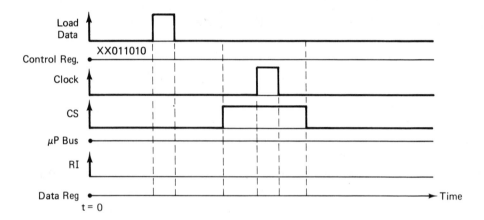

Figure E8.11

8.12 A μP system (not the 6800) has 18 address bits. The RAM chips present have address inputs A_0 through A_7, with only one E and one $\bar{E}$ input each. Using as many of these RAM chips as needed, but no external decoding, determine the maximum possible actual memory locations (words) that may be made available. Sketch a block diagram of your circuit.

8.13 Study the circuit of Figure 8.9 and prepare a memory map for address bits A_9 through A_{15}.

8.14 Which of the following addresses also reads the switch position for the interface of Figure 8.12: 7312, 7C12, 9E31, C753, 1849, 5E31, AAAA, and 7AAA?

8.15 Using the smallest number of IC packages, design (sketch the circuit) an interface to read information from an eight-bit register onto the μP data bus. The address of the register is to be E100 and the system memory map of Figure 8.11 applies. NAND and NOR packages are available with each containing two four-input gates of the same type. The tristate packages each contain four gates. PIUs are not available. Minimize the loading placed on the address lines.

8.16 If the AND gate in Figure 8.12 is replaced by this circuit (Figure E8.16), list the highest and lowest addresses which will select the switch.

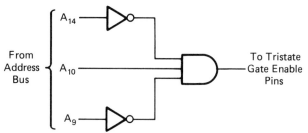

Figure E8.16

8.17 Write a program, starting at 0100, for a μP which is to transmit the sequence A3, 15, CF to a peripheral through a PIU whose address starts at 7A30. Assume that the peripheral is very fast so that it is ready to receive data as soon as the μP can send it both at the beginning and during transmission of the sequence. Minimize the amount of memory required for your program and make it run as far as possible. If any bit in the control/flag register is arbitrary or is unknown, set it to zero in your program.

8.18 Repeat Exercise 8.17 but in this case the peripheral is slower and the μP must wait for an interrupt before each member of the sequency is sent.

8.19 Construct a program fragment, starting at 0200, to poll three PIUs sequentially; determine the first one that is ready to deliver a data word; and store that data at address 21FE. If initially none is ready, continue polling until the data from one PIU have been stored; then immediately halt. The three PIUs start at addresses 5010, 5020, and 5030.

8.20 Repeat Exercise 8.11 using Figure E8.20, except employ the following bus and control line information: μP bus $(D_7-D_0) = 10111101$, $R/\bar{W} = 0$, and $R_S = A_0 = 1$. (The above values are actually valid only during $CS = 1$.)

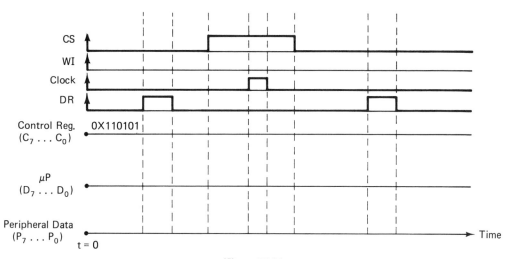

Figure E8.20

8.21 A μP system (not the 6800) has 17 address bits. For its single memory chip shown in Figure E8.21, determine the addressing range in hex. (If more than one range is possible, give both the lowest and highest ranges.)

Sec. 8.8   Exercises

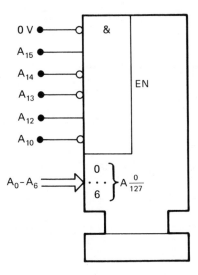

Figure E8.21

8.22 Design a μP write-only interface for an 18-bit LED display. The *only* address constraints are the map of Figure 8.11 and the requirement that the write address(es) be as close as possible to 0200_{16}. The lowest-numbered display bit should be at the lowest address and the bits should run in sequence. Where there is a choice, employ the lowest address lines in the decoder. Do not show all 18 bits in your circuit but only enough to *clearly* indicate your logic.

8.23 Design a μP bidirectional interface that exchanges information between the D_4 bit of the data bus and a clocked *J-K* flip-flop. The only address constraints are the map of Figure 8.11, the requirement that the fewest possible address lines should be employed by the decoder, and within these constraints the interface should be at the highest available address. The clock and other control signals normally available from the μP for a PIU are available but, of course, not a PIU itself.

8.24 Large incandescent lamp arrays have become very popular as animated displays for baseball scoreboards and many other applications. To get a better understanding of how such systems are implemented, design the SSI interface for a 16×16 LED prototype display controlled by a μP. The LEDs are to operate only from the D_0 data line with an address range of 8000 to $80FF_{16}$. In addition to the usual SSI components, you may employ MSI, 4-to-16 line decoders. (These contain a single enable, E, input.)

8.25 A circuit to implement the arithmetic equation $R = \frac{1}{35}(t^2 - 100t + 5)$ forms a complete system where R is a 2's complement number, $0 \le t \le 127$, and R is to be rounded to the nearest integer. Outline the designs for the best and the next best *structured solutions* you can find and compare them.

Chapter 9

Microprogrammable Microprocessors

Standard μP's were considered in Chapter 8, where they were employed in several logic design applications. Although very powerful, these machines do have limitations in that they are somewhat slow and their architecture is fixed. Thus, it is not possible, for example, to add new operations to their instruction set. Another class, called microprogrammable microprocessors, however, does have a great amount of flexibility in its instruction set and general architecture. Moreover, this new component class, considered here, provides many opportunities for Structured Design; and as mentioned in Chapter 7, one of the earliest applications of Structured Design was with microprogramming. (At the end of the current chapter we will give a comparison summary for all the component classes.)

Here we distinguish between a *macroprogram* and a *microprogram*. A macroprogram is written with ordinary machine-level instructions, *macroinstructions*, such as those presented in Table 3.10. On the other hand, a microprogram contains more primitive instructions, *microinstructions* (each equivalent to one or two register transfer operations), from which macroinstructions can be implemented. For example, the multiply macroinstruction is implemented by means of a series of add and shift microinstructions.

9.1 MICROPROCESSOR CLASSES

Except for the fact that the CPU is implemented with LSI circuits, the architecture of a microprocessor is essentially the same as we have previously discussed in Figure 6.16. Thus, the programming concepts of Chapter 3 and the principles developed in the other chapters are applicable. Since it will be particularly valuable later and since it provides an application of the ROM methods of Chapter 7, the elementary microprocessor shown in Figure 9.1(a) is based on the control memory concept.

Although it may not be obvious at first, Figure 9.1(a) is essentially the same as Figure 3.1 except that more details are provided concerning the way control signals are generated. The control memory is the source of signals (microinstructions) which actuate the various basic operations of the ALU. It also provides signals to other microprocessor units, e.g., the main memory receives its read/write signal from the CM. Of course, the control memory is addressed in the same way as in Chapter 7.

Figure 9.1(a) includes several classes of microprocessors.

1. The *nonprogrammable* class is one in which the user is given a fixed processor structure such as is present in most pocket calculators. About the only freedom the user has is in the selection of the available keyboard operations and the data. Since the unit is not programmable, the main memory and the macroinstruction bus is omitted in this type of processor. (The ALU in most microprocessors contains a few registers for the storage of intermediate results.) In addition, the control memory is loaded at the factory and the end-user will not be able to modify it. Part of the microinstruction supplies the next-address function. Another part supplies the ALU signals for the various mathematical operations and for control of the input/output devices. For example, in a pocket calculator, the input is a small keyboard and the output is usually a liquid crystal or LED, light-emitting diode, display. Surprisingly, a scientific calculator usually requires less than 1000 microinstructions for complete implementation including the keyboard decoding, arithmetic, and display functions. Within this space, it employs some intriguing algorithms to perform trigonometric calculations, logarithms, and other functions. In addition, the hardware contains some very interesting logic. The reader is directed to the *Hewlett-Packard Journal*, particularly Reference 13 in the Bibliography, for details which are beyond the scope of this text.
2. The *macroprogrammable* class is the standard μP we discussed in Chapter 8, with the main memory being user-programmable with conventional machine language instructions. As shown in Figure 9.1(a), each macroinstruction is decoded by means of the next-address logic; this usually results in a sequence of several microinstructions from the CM. (Details of this process will be discussed in Section 9.3.) Again the CM may not be modified by the end-user; but there usually is considerable freedom to interface various input/output devices with the processor. In fact, several microprocessor manufacturers have a series of LSI chips which have been especially designed for interfacing between their processor and external devices. Tutorial versions of a very popular member of this class, the 6800, were previously considered in Chapters 3 and 8.
3. The *microprogrammable* class, also called a bit-slice μP since the ALU is divided into chips typically containing from two to four bits each, is the most general and powerful of the three. Here the user is free to program the control memory (or course, the main memory also) which in turn allows direct management of the individual operations of the ALU (and of other parts of the processor).

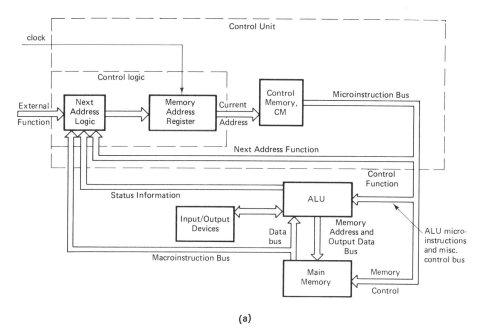

(a)

(b)

Figure 9.1 (a) A Microprogrammable Microprocessor. (b) Photomicrograph of the 2901 Four-Bit Microprocessor ALU Slice. (Photo Courtesy of Advanced Micro Devices Inc.)

Through this method new macroinstructions may be defined, which means that the user has complete control of the instruction set employed in programming the main memory. Although we are concentrating on the application of microprogramming to microprocessors in this chapter, the basic ideas are applicable to all scales of digital computers. For example, extensive use was made of microprogramming in the implementation of both the IBM 360 and 370 families of computers and several minicomputers may be microprogrammed by the end-user. (Again, hardware details will be discussed in Section 9.3.)

Figure 9.1(b) shows a photomicrograph of the Advanced Micro Devices 2901B four-bit processor slice, which is designed to serve as a fast, cascadable ALU chip. Along with other members of the 2900 family, such as the control logic chip, it can easily be employed to produce a complete 16-bit or large microprocessor. In addition to their capability to emulate (imitate the performance of) a wide variety of computers, systems constructed from the 2900 family also have numerous applications in building various other types of data processors. The actual 2901B circuit is quite small (0.117 inches by 0.128 inches), yet it contains the equivalent of 538 gates. At room temperature this chip permits a minimum clock period of 60 ns. (maximum frequency of 16.7 MHz), which is a rough measure of the speed at which primitive operations such as a register shift may be performed.

9.2 MICROINSTRUCTIONS AND MICROPROGRAMMING

In Chapter 7 we saw how a control memory could be employed to implement sequential circuits (thus, also register transfer logic). From this, it is only a short step to visualize how a microprogram in the control memory of Figure 9.1(a) can implement the macroinstructions required by the processor. The ALU and the other units need a large number of control signals to manage the various register transfer operations. Over 200 separate signals is not an uncommon requirement for a large-scale computer (the IBM 360/65 employs about 250). The most straightforward method for realizing these signals is to assign one bit in the control memory's word for each needed signal. While this is a very general approach, it is extravagant in the number of bits required per instruction word. Moreover, many control signals are mutually exclusive, e.g., the same register would not be simultaneously shifted and cleared. Thus, the control word may be coded so that, in the extreme, n bits may manage 2^n microinstructions. In the first method, n operations are controlled simultaneously (parallel processing) but with a high cost in bits per word. In the second method, only one operation is controlled at a time, resulting in a low bit-per-word cost; but a decoder is required and the desired function may consume a series of several control memory cycles. For obvious reasons, the first approach is called *horizontal microprogramming* while the second is called *vertical microprogramming*. Most processors employ a combination approach, which is between these two extremes. As a simple example, if five bits are reserved for control, with three bits for vertical and two bits for horizontal microprogramming, one of eight ($2^3 = 8$) operations could be executed, through a decoder, with two others always available. This is a total of three simultaneous operations from a repertoire of $2^3 + 2 = 10$ instructions. Because it follows the trend of current microprocessors and for tutorial reasons, ours will be a combination approach leaning toward vertical microprogramming.

Establishing the Control Memory Address

Now let us consider the method of achieving the next control memory address. In Chapter 7, an elementary branching method was described in which a bit (or group of bits) was employed to control branching and the next-address logic was trivial. Here, a more powerful technique is achieved by employing an actual jump instruction which, together with some auxiliary conditions, establishes the memory address through a next-address logic circuit.

The control memory word format employed by our microinstructions is shown in Figure 9.2. As in the discussion in Chapter 7, the control word is broken into two parts—the next-address function and the control function. In order to achieve both conditional and unconditional branching with the next-address function, bits 8 through 11 code several different jump (branching) possibilities. These include jumps on flag conditions, jumps on status of groups of bits from the main memory, and random jumps. For convenience, the control memory is visualized as a square array of memory locations with bits 0 through 3 specifying the column and bits 4 through 7 specifying the row.

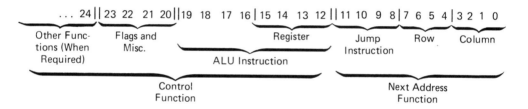

(Bit numbers are shown divided into hexadecimal digits.)

Figure 9.2 Microinstruction Format.

Note that bits 12 through 19 specify the ALU microinstruction, bits 20 through 23 specify flags, and bits 24 and above control other functions when required. These will be explained in more detail shortly.

Details of the next-address function are shown in Table 9.1. Note that each row of the table (particular address control function) contains the binary code of the next-address function (as read from the CM) and the resulting actual next memory address. Consider the Jump Random instruction, JR. In this case, a jump may be made to any place in the CM since there is a one-to-one correspondence between the next row-column in the function and the actual next memory address row-column. For example, the mnemonic instruction which is to cause an unconditional branch to row 5, column 7 would be JR 5, 7. It would have the hexadecimal code 157_{16} for microinstruction bits 0 through 11.

The Jump Flag Column function, FC, has the same one-to-one row-column correspondence except for the fact that the least significant bit in the next column

TABLE 9.1 ADDRESS CONTROLS

Branch Mnemonic	Next-Address Function												Next Memory Address	
	11	10	9	8	7	6	5	4	3	2	1	0	Row	Column
JR ≡ Jump Random	0	0	0	1	d_7	d_6	d_5	d_4	d_3	d_2	d_1	d_0	$d_7 d_6 d_5 d_4$	$d_3 d_2 d_1 d_0$
FC ≡ Flag Column	0	0	1	0	d_7	d_6	d_5	d_4	d_3	d_2	d_1	–	$d_7 d_6 d_5 d_4$	$d_3 d_2 d_1 f$
MA ≡ Multiplexer Segment A	1	0	1	0	d_7	d_6	d_5	d_4	–	–	–	–	$d_7 d_6 d_5 d_4$	$A_3 A_2 A_1 A_0$
MB ≡ Multiplexer Segment B	1	0	1	1	d_7	d_6	d_5	d_4	–	–	–	–	$d_7 d_6 d_5 d_4$	$B_3 B_2 B_1 B_0$
etc.														
MF ≡ Multiplexer Segment F	1	1	1	1	d_7	d_6	d_5	d_4	–	–	–	–	$d_7 d_6 d_5 d_4$	$F_3 F_2 F_1 F_0$

Abbreviations: d_i ≡ bit i of data word.
A_i ≡ bit i of macroinstruction word (main memory bus) segment A (similar for $B_i, C_i, \ldots F_i$).
f ≡ current state of flag flip-flop.
– ≡ bit is a don't care.

address takes on the value of the flag, f. If the flag's value is established by the sign of some register, a Branch on Register Sign instruction would result. On the other hand, the Jump on <u>M</u>ultiplexer Segment <u>A</u>, MA, instruction is somewhat different from the previous jump instructions. Here the next memory row address is the same as in the next-address function; but the column address is obtained from a four-bit segment of the main memory bus (part of the macroinstruction) selected by a multiplexer in the next-address logic. The various four-bit segments of the input bus are labeled A, B, C, etc. (with A being the most significant four-bit segment, B the next four-bit segment, etc.) resulting in multiplexer instructions with mnemonics MA, MB, MC, etc.

One major use of the multiplexer instructions is to decode macroinstructions from the main memory. The actual decoding process causes a jump to the proper starting location in CM for the realization of the macroinstruction by means of a microinstruction program.

As an example of address control, assume that the next-address function is $B53_{16}$, the main memory bus contains $9E8C_{16}$, and the multiplexer segments correspond, in sequence, with the hexadecimal digits starting with the most significant digit being segment A. Here, B, the most significant digit of the function, indicates that the branch operation is Jump on Multiplexer Segment B, MB; the next digit, 5, specifies the next memory address row; finally the next digit, 3, is neglected in this example. The resulting next memory address becomes 5E, where E is the digit found in the multiplexer Segment B.

ALU Microinstructions

As previously indicated, the ALU microinstructions are the lowest-level, most primitive operations defined for a computer. From a small group of these instructions the macroinstructions for *any* computer can be constructed. It is one of the major reasons for our interest in these primitives—they are the key to simulation (the more specific word in this context is *emulation*) of one computer by another. Emulation is employed by computer manufacturers to test a computer design while the actual hardware for the new machine is still under construction. In addition, it is frequently employed to represent an old machine by a newer one so that programs using the old instruction set can be run without any changes.

The ALU instruction set has some features in common with machine language instructions such as those discussed in Chapter 3. Again, the instruction is broken into operation and operand parts, with certain instructions like clear having the accumulator as an implied operand. Although the instructions under consideration refer only to the ALU, units such as the memory and input/output devices are controlled indirectly through registers manipulated by these instructions.

Table 9.2 summarizes the ALU microinstructions which are employed in this chapter. The reader should study the table, taking note of the RTL equivalence and the several other features listed there which aid in understanding the instructions. The ALU instructions have been given two-letter mnemonics to emphasize their more primitive character and to distinguish them from the macroinstructions of Tables 3.2 and 3.10. The following other features should be observed concerning the entries in Table 9.2:

1. Bits 16 through 19 in the control memory bit format indicate the operation, and bits 12 through 15 indicate the operand. Note that each part is expressed as a hexadecimal number; and to emphasize its base-16 nature, the bit pattern is divided into two groups by a vertical line.
2. There are eight registers within the ALU, which are denoted by 000 through 111 in bits 12 through 14. These are used to hold one operand during certain instructions and as temporary storage—R^i represents register i, $0 \leq i \leq 7$.
3. When bit 15 is zero, the register class of instructions is being considered. AD, AN, OR, LR, and several others belong to this class.
4. For bits 12 through 15, the following definitions are employed: $1000 \equiv$ accumulator and $1001 \equiv$ memory address register.
5. In the AD, CM,* and several other instructions, the quantity CI is the Carry Input. (A similar quantity is the Left Input, LI, found in the SR instruction.) It forms an LSB input for various functions much as the carry-in signal, C_{i-1}, does for the full adders shown in Figure 6.31. On the other hand, the SL and TR as well as some other instructions have a Carry Output, CO, signal. (A similar quantity is the Right Output, RO, found in the SR instruction.) This is analogous to the carry generated by the MSB in a parallel full-adder circuit. The exact role of the *CI, LI, RO,* and *CO* quantities for all instructions, where they are applicable, should be clarified by the RTL column in the table.

As an example of microinstruction operation, assume that the contents of register 3 and the accumulator for a six-bit processor are, respectively, 011011 and 110001. After execution of the microcode: 10100011 (bits 12–19), the contents of register 3 and the accumulator would be 011011 and 010001. These results may be checked from

*Any possible confusion between this mnemonic and the abbreviation for control memory should be resolved by the context.

Sec. 9.2 Microinstructions and Microprogramming

TABLE 9.2 ALU MICROINSTRUCTIONS

Mnemonic	Meaning	RTL Equivalent	Control Memory Bit Format 19 18 17 16 (Operation)	15 14 13 12 (Operand)
1. CL	Clear accumulator	$A \leftarrow 0$	0000	$1000 = 08_{16}$
2. NO	No operation	—	0000	$1111 = 0F_{16}$
3. RR	Rotate Acc. right one bit (similar to LINC ROR)	$A \leftarrow rrA$	0010	$1000 = 28_{16}$
4. SR	Shift Acc. right one bit	$A \leftarrow srA$; $A_{n-1} \leftarrow LI$; $RO \leftarrow A_0$	0011	$1000 = 38_{16}$
5. SL	Shift Acc. left one bit	$A \leftarrow slA$; $CO \leftarrow A_{n-1}$; $A_0 \leftarrow CI$	0100	$1000 = 48_{16}$
6. AD R^i	Add $C(R^i)$ to Acc. results in Acc. R^i unchanged	$A \leftarrow A + R^i + CI$ (1's comp. result)	0101	$OR_2^i R_1^i R_0^i = 5i_{16}$
7. ST R^i	Store Acc. in register R^i; Acc. is unchanged	$R^i \leftarrow A$	0110	$OR_2^i R_1^i R_0^i = 6i_{16}$
8. DR R^i	Decrement register R^i	$R^i \leftarrow R^i - 1$	0111	$OR_2^i R_1^i R_0^i = 7i_{16}$
9. IR R^i	Increment register R^i	$R^i \leftarrow R^i + CI$ ($CI = 1$)	1000	$OR_2^i R_1^i R_0^i = 8i_{16}$
10. IM	Increment memory address register	$MA \leftarrow MA + CI$ ($CI = 1$)	1000	$1001 = 89_{16}$
11. LR R^i	Load Acc. from register R^i	$A \leftarrow R^i$	1001	$OR_2^i R_1^i R_0^i = 9i_{16}$
12. LM	Load memory address register, MA, from Acc.	$MA \leftarrow A$	1001	$1001 = 99_{16}$
13. LD	Load Acc. from data input bus	$A \leftarrow (Data\ Bus)$	1001	$1101 = 9D_{16}$
14. AN R^i	AND $C(R^i)$ with Acc. leave results in Acc. $C(R^i)$ unchanged	$A \leftarrow A \wedge R^i$	1010	$OR_2^i R_1^i R_0^i = Ai_{16}$
15. OR R^i	OR $C(R^i)$ WITH Acc. leave result in Acc. $C(R^i)$ is unchanged	$A \leftarrow A \vee R^i$	1011	$OR_2^i R_1^i R_0^i = Bi_{16}$
16. CM	Complement Acc.	$A \leftarrow \bar{A} + CI$ $\begin{cases} CI=0 \text{ is 1's comp.} \\ CI=1 \text{ is 2's comp.} \end{cases}$	1100	$1000 = C8_{16}$
17. TR R^i	Test Register R^i for zero	IF ($R^i = 0$) $CO \leftarrow 0$ ELSE $CO \leftarrow 1$	1111	$OR_2^i R_1^i R_0^i = Fi_{16}$

Table 9.2 which indicates that the mnemonic AN 3 corresponds to the above microcode. It requires the AND function between register 3 and the accumulator, with the results placed in the accumulator but leaving the register unchanged. (More extensive microinstruction examples will follow in the next section.)

9.3 AN ELEMENTARY MICROPROCESSOR SYSTEM

The microword format, the ALU microinstructions of Table 9.2, and the system examples to follow have been somewhat idealized for tutorial purposes; but they are similar to those for existing microprocessors (e.g., the Advanced Micro Devices 2900 family[3] and the Intel 3000 series[10]). Once the material presented here is understood, the reader should be in a good position to master any of the commercial units.

As one gains knowledge of this versatile new processor, one should think of application possibilities in terms of Structured Design. Most of the same structured properties are available to a microprogrammable microprocessor and it is actually more powerful and flexible than a standard μP. There is a cost, however, in that the new flexibility is gained through extra chips and extra complexity, which in turn makes it more difficult to actually achieve the same degree of Structured Design in hardware and particularly in software as may be obtained with a standard microprocessor.

Although the basic processor which we will consider in this section has the capabilities of handling both macroinstruction and microinstruction, for simplicity an elementary application requiring only microinstructions will be considered first. In this example, pressure and temperature are measured for a chemical plant. Next the hypothetical performance function

$$F = T + \frac{P}{2} + 105$$

(where T is temperature and P is pressure) is calculated and displayed. Then the process is repeated. A flow chart for the problem is shown in Figure 9.3. It faithfully follows the

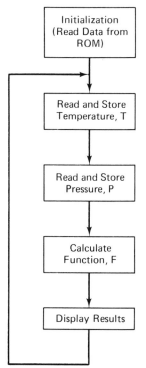

Figure 9.3 Method for Calculating the Performance Function.

above problem statement, but the need for the initialization step will be justified shortly.

The hardware to implement the flow chart for this problem is shown in Figure 9.4. It is a more specific version of the block diagram of Figure 9.1. While more details will be added to several blocks as the requirements develop, the basic form shown will be employed through the remainder of the text in examples and in the exercises. The control logic, control memory, and ALU are the same major units that were discussed previously. Note that the carry-input (CI) terminal and the left-input (LI) terminal are both explicitly shown on the ALU; these signals are required by several of the ALU instructions of Table 9.2, e.g., AD and IM. Since only one of the two terminals receives information at one time, CI and LI share a control line. For the same reason, they also share a bit in the microinstruction.

The constants needed by the processor are stored in the data ROM, Figure 9.4, and the bracketed quantity at the ROM input is employed to represent the addresses accepted. If an address between 0 and $2F_{16}$ appears at the ROM input, the contents of that address are immediately placed on the data bus. On the other hand, when an address outside 0 to $2F_{16}$ appears at the ROM input, the output will be effectively disconnected from the bus. In all cases, the data bus contents will be determined

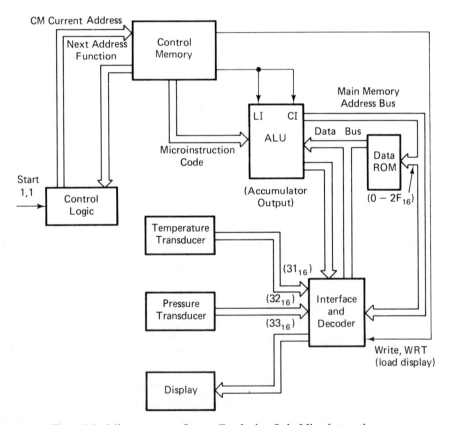

Figure 9.4 Microprocessor System Employing Only Microinstructions.

only by the device whose address is on the address bus. The transducers and the display have addresses 31, 32, and 33, respectively, so that information may be exchanged with them much as is done in a RAM. (Note that the accumulator output serves as the source of data to be written, or loaded, by the display.)

Now consider Table 9.3 which contains the microprogram to implement the flow diagram of Figure 9.3. (The accompanying main memory contents are shown in Table 9.4.) The table is constructed in a manner very similar to those for the machine language programs given in Chapter 3. The leftmost column contains the control memory address, the next group of columns contains the microinstruction mnemonic, and the third group of columns contains the actual microinstruction code. The microinstruction mnemonic format is the same as previously shown in Figure 9.2. The address controls follow the scheme of Table 9.1, and the ALU Instructions follow the scheme of Table 9.2. In the present case, only two bits of the miscellaneous function are employed: $\begin{pmatrix} W \\ R \\ T \end{pmatrix}$, which loads the interface display from the accumulator when

TABLE 9.3 PERFORMANCE FUNCTION PROGRAM

CM Address (Row, Col.)	( )	( )	$\begin{pmatrix} W \\ R \\ T \end{pmatrix}$	$\begin{pmatrix} CI \\ LI \end{pmatrix}$	ALU Instruction	Next-Address Function	Microinstruction Code (Base 16)	Remarks
1,1 (Start)	0	0	0	0	CL	JR 0,5	008105	
0,5	0	0	0	0	LM	JR 1,5	099115	
1,5	0	0	0	0	LD	JR 2,5	09D125	$A \leftarrow C(\text{Bus Add} \cdot 0)$ = 105_{10}
2,5	0	0	0	0	ST 2	JR 3,5	062135	$R2 \leftarrow 105_{10}$
3,5	0	0	0	1	IM	JR 4,5	189145	
4,5	0	0	0	0	LD	JR 5,5	09D155	$A \leftarrow C(\text{Bus Add} \cdot 1)$ = 31_{16}
5,5	0	0	0	0	ST 3	JR 6,5	063165	$R3 \leftarrow 31_{16}$
6,5	0	0	0	0	LM	JR 0,6	099106	
0,6	0	0	0	0	LD	JR 1,6	09D116	$A \leftarrow T; :T \equiv \text{Temperature}$
1,6	0	0	0	0	ST 4	JR 2,6	064126	$R4 \leftarrow T$
2,6	0	0	0	1	IM	JR 3,6	189136	
3,6	0	0	0	0	LD	JR 4,6	09D146	$A \leftarrow P; :P \equiv \text{Pressure}$
4,6	0	0	0	0	SR	JR 5,6	038156	$A \leftarrow \dfrac{P}{2}$
5,6	0	0	0	0	AD 4	JR 6,6	054166	$A \leftarrow T + \dfrac{P}{2}$
6,6	0	0	0	0	AD 2	JR 7,6	052176	$A \leftarrow T + \dfrac{P}{2} + 105_{10}$
7,6	0	0	0	1	IM	JR 8,6	189186	
8,6	0	0	1	0	NO	JR 9,6	20F196	Numeric Display ← Acc.
9,6	0	0	0	0	LR 3	JR 6,5	093165	Now repeat calculation

(Rows 1,1 through 5,5 are bracketed as "Initialization")

TABLE 9.4 DATA CONTROLLED BY THE MAIN MEMORY BUS

Main Memory Address (Base 16)	Contents	Remarks	
0	105_{10}	Constant for performance function, F	⎫
1	31_{16}	Temperature transducer address	⎬ Data ROM
2 through 2F	—	Not used	⎭
31	(Temperature values)	⎫	
32	(Pressure values)	⎬ Interface	
33	(Display)	⎭	

$\begin{pmatrix} W \\ R \\ T \end{pmatrix} = 1$, and $\begin{pmatrix} CI \\ LI \end{pmatrix}$, which is the carry-input/left-input bit. The reader should now study the program. First consider the mnemonic group of columns and verify that the flow chart of Figure 9.3 is properly implemented; then check that the microcode column is a correct representation of the corresponding mnemonics. (The comments in the Remarks column should help to clarify the various steps of the program.)

‡9.4 A MACROINSTRUCTION APPLICATION

As pointed out in the previous section, our basic hardware is capable of supporting both macroinstructions and microinstructions. This facility will now be demonstrated by a design for implementing a portion of the LINC computer, discussed in Chapter 3. Recognize that the proposed system is a member of the third microprocessor class, a microprogrammable processor. Here the LINC program (macroinstructions) and data will be placed in the main memory. The control unit, through the control logic and CM, will decode the macroinstructions and realize the necessary register transfer logic to implement each instruction. This will include the realization of the fetch and execution cycles and the maintenance (in the ALU) of such items as the memory address register and the program counter. The above process of realizing one computer by means of another computer is known as *emulation*.

Processor Configuration

The block diagram for the current application is shown in Figure 9.5. It follows the basic pattern established by the general block diagram in Figure 9.1, but it emphasizes those details which are most appropriate in explaining the macroinstruction implementation. For example:

1. The macroinstruction word and the multiplexer are octal oriented while hexadecimal is employed throughout the rest of the system.
2. The multiplexer connections are defined so that it is easy to see which macroinstruction bits are sampled by each multiplexer segment, Seg. (Note that a 12-bit octal macroinstruction word is employed.)

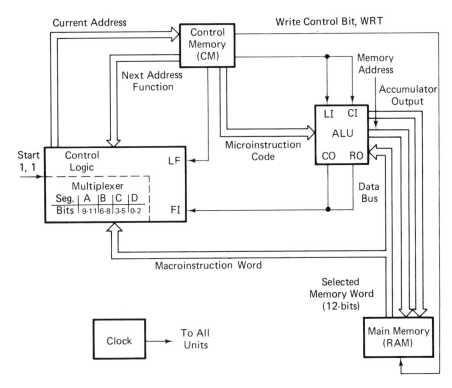

Figure 9.5 Microprocessor Having Both Macroinstruction as Well as Microinstruction Capabilities.

3. The accumulator output is supplied directly to the RAM so that data can be stored (written) in the memory. Since the same memory address lines are employed both for data being retrieved from as well as for data being stored in memory, the write control bit is necessary. (When the write bit is 1, data are stored; when it is 0, data may only be read.)

4. The carry-output and right-output (CO and RO) bits from the ALU are explicitly shown. The carry-output signal is familiar from Chapter 6—it goes to 1 whenever a carry would be produced from the MSB, e.g., when the accumulator's MSB is a 1 and a shift-left microinstruction is executed. The right-output signal is produced in a similar way, e.g., when the LSB is 1 and a shift-right microinstruction is executed. The presence of the CO or RO signal may be stored by setting the Flag Input bit to 1 in the control logic (through the FI terminal).

5. It is assumed that the Load Flag logic is such that, when $LF = 1$, the flag flip-flop, is connected to the ALU through the FI and CO/RO terminals in such a way that it looks like a bit added to one end of the accumulator. For example, when the SL instruction is executed and $LF = 1$ in the microinstruction, the flag will look like it is a bit added to the left end of the accumulator, forming a large shift register.

The flag must be set on a previous step to effect a branch instruction like

FC. (It is the state of the flag before the clock pulse which is transmitted to the FC logic and which determines the next memory address.)

Hardware similar to that required to implement Figure 9.5 is found in Figure 9.6. This single card contains the HEX-29, which is a complete 16-bit computer design by Advanced Micro Devices engineers employing the AM 2900 family of bit-slice components. The four long packages at the center of the card are AM 2901 chips with each implementing a four-bit slice of the ALU. The three large packages on the left edge form a major part of the control logic and the bulk of the remaining packages form the memory.

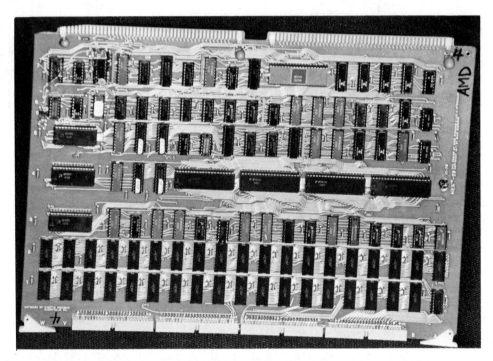

Figure 9.6 The HEX-29: A Microprogrammable Microprocessor on a Single Printed Circuit Card. (Courtesy of Advanced Micro Devices Inc.)

CM Programs for the LINC Instructions

Once the basic units of the microprocessor, shown in Figure 9.5, have been properly connected, all that is required to complete the LINC design is to deposit the required programs in the control memory. Before actually writing any of the microprograms, however, it is desirable to construct a map of the CM and plan the arrangement of the various programs within the memory. (As programs are completed, it can be updated to provide a map of available memory locations.) Figure 9.7 shows the map for our emulation of the LINC. Since we do not know beforehand exactly how many locations are required to implement each macroinstruction, a line with an arrowhead

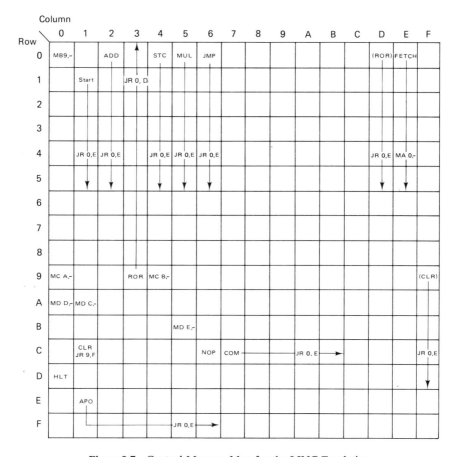

Figure 9.7 Control Memory Map for the LINC Emulation.

is employed to represent the intended sequence of locations. The line is broken with an unconditional branch symbol, e.g., JR, to indicate the approximate location of the jump to the next program. When no arrow is associated with an unconditional branch, such as MA or MB, the instruction will be placed in that exact location. Parentheses around a macroinstruction. e.g., (ROR), denote the continuation of a program.

Figure 9.7 shows that the start signal initiates the CM at address 1,1. A small initialization program begins at that point, and it is followed by an unconditional branch to begin the first fetch-cycle program, at address 0,E. The fetch program ends each time with the next-address function MA 0,–. The purpose of this instruction is to cause a jump to row 0, with the column determined by the multiplexer sampling of segment A (bits 10 through 12) of the macroinstruction word. Thus, if the macroinstruction word is 6213_8, the new CM address will be 0,6, which is indicated by the LINC mnemonic JMP in the corresponding position on the CM map. (Once a multiplexer branch instruction entry is made in a row, we reserve the remaining columns, 0 through 7, in that row so that other macroinstructions, with different machine codes, can be detected there.)

The reader should now verify that ADD, STC, and MUL are in their proper positions on the map. If segment A of the multiplexer yields 0_8, the 0,0 location shows that the jump MB 9,– would be executed, which means that the next instruction would come from row 9. From Table 3.2 it can be seen that, for our pseudo-LINC instructions, the only possibilities for segment B are 0, 3, and 4. If it is 3, the ROR instruction sequence begins. If it is 4, an APO instruction is probably required, but the other bits will also need to be checked. Finally, if it is a 0, a branch to row A is indicated for further decoding. The reader should continue this process and verify that the programs for the other instructions start at the proper locations as indicated on the map. (Note that the only code allowed for the instruction NOP is 0016; but the YYYY form of Table 3.2, will be considered in the exercises.)

The implementation of fetch, STC, and ROR are now presented since they are typical of a large class of microprograms. After studying them, you should be able to justify all the entries in each column. (The RTL programs in Figure 6.17 and Table 6.6 may serve as a useful guide to the required algorithms.) The fetch program is shown in Table 9.5. As predicted, the MA 0,– next-address function is contained in the program as a first step in decoding the macroinstruction. In this and the following two programs, it will be observed that register 1 is employed for the program counter, and register 2 is employed to store the macroaccumulator. The term *macroaccumulator* emphasizes the fact that the accumulator for the machine being emulated is different from the microaccumulator, or accumulator for short, of the microprocessor itself. Because this distinction is made, the microaccumulator can be employed for other tasks without disturbing the macroaccumulator.

TABLE 9.5 MICROPROGRAM FOR THE FETCH CYCLE

CM Address (Row, Col.)	Microinstruction Mnemonic						Microinstruction Code (Base 16)	Remarks
	( )	( )	$\begin{pmatrix}W\\R\\T\end{pmatrix}$	$\begin{pmatrix}CI\\LI\end{pmatrix}$	ALU Instruction	Next-Address Function		
0,E	0	0	0	0	LR 1	JR 1,E	09111E	A←PC
1,E	0	0	0	0	LM	JR 2,E	09912E	MA←A
2,E	0	0	0	1	IR 1	MA 0,–	181A0–	PC←PC+1 ; then to execute cycle

*Context allows the memory address register, MA, to be distinguished from MUXA.

Next consider the microprogram for the STC instructions, shown in Table 9.6. A small problem develops here when the first instruction loads the accumulator from the memory. Only the three least significant digits of the octal number are valid address information—the fourth digit is the code of STC, which is a 4. The MSB of this digit, however, is the only bit differing from zero, and the SL and SR microinstruction

TABLE 9.6 MICROPROGRAM FOR THE STC INSTRUCTION

CM Address (Row, Col.)	Misc. Function		$\begin{pmatrix}W\\R\\T\end{pmatrix}$	$\begin{pmatrix}CI\\LI\end{pmatrix}$	ALU Instruction	Next-Address Function	Microinstruction Code (Base 16)	Remarks
	( )	( )						
0,4	0	0	0	0	LD	JR 1,4	09D114	$A \leftarrow (4XXX_8)$
1,4	0	0	0	0	SL	JR 2,4	048124	Remove leading bit to eliminate "4" in $4XXX_8$
2,4	0	0	0	0	SR	JR 3,4	038134	Restore address
3,4	0	0	0	0	LM	JR 4,4	099144	$MA \leftarrow XXX_8$
4,4	0	0	0	0	LR 2	JR 5,4	092154	$A \leftarrow (Macroacc.)$
5,4	0	0	1	0	NO	JR 6,4	20F164	Store data
6,4	0	0	0	0	CL	JR 7,4	008174	$A \leftarrow 0$
7,4	0	0	0	0	ST 2	JR 0,E	06210E	$(Macroacc.) \leftarrow A$; then return to Fetch cycle

sequence nicely eliminates the objectionable bit and restores the desired address. Another feature of the program worthy of mention is the way the data are stored in the main memory by the $\begin{pmatrix}W\\R\\T\end{pmatrix} = 1$ microinstruction at address 5,4.

Finally, we consider the ROR microprogram shown in Table 9.7. The machine code itself, i.e., the 3 in $0300 + n$, is eliminated this time by what is called *masking*. Since 0 through 5 are the only valid bits, a mask formed by the number 0077_8 when ANDed with $0300 + n$ will yield n, the required number of rotate steps. Another point to emphasize here is the loop implemented by the instructions at addresses 3,D through 7,D. Of particular interest is the way the loop counter is decremented at address 4,D and the way the test for completion is made at address 5,D.

Summary

The hypothetical bit-slice microprocessor we have been discussing is similar to several commercial units,[10] but it has been simplified for instructional reasons. The objective has been to present the basic principles of microprocessor hardware and some of the main design techniques for their use without getting tied up in the idiosyncrasies of particular types. In addition, an effort has been made to minimize routine details, e.g., both the CM and the microinstruction set are much smaller than most commercial units. For generality, the emphasis here has been on user-microprogrammable units. It should be understood, however, that microprocessors which do not have this capability are also quite popular, and as a matter of fact, are often easier to apply. By having an understanding of the concepts upon which all microprocessors are built, it is felt that the reader should be able to learn more quickly how to apply any of the commercial types to practical design problems.

TABLE 9.7 MICROPROGRAM FOR THE ROR INSTRUCTION

CM Address (Row, Col.)	( )	$\begin{pmatrix}L\\F\end{pmatrix}$	$\begin{pmatrix}W\\R\\T\end{pmatrix}$	$\begin{pmatrix}CI\\LI\end{pmatrix}$	ALU Instruction	Next-Address Function	Remarks
9,3	0	0	0	0	CL	JR 8,3	⎫
8,3	0	0	0	1	SL	JR 7,3	⎪
7,3	0	0	0	1	SL	JR 6,3	⎪
6,3	0	0	0	1	SL	JR 5,3	⎬ Create mask, which is 0077_8
5,3	0	0	0	1	SL	JR 4,3	⎪
4,3	0	0	0	1	SL	JR 3,3	⎪
3,3	0	0	0	1	SL	JR 2,3	⎭
2,3	0	0	0	0	ST 3	JR 1,3	$R3 \leftarrow A$
1,3	0	0	0	0	LD	JR 0,D	$A \leftarrow (0300+n)$
0,D	0	0	0	0	AN 3	JR 1,D	$A \leftarrow n$; by masking
1,D	0	0	0	0	ST 3	JR 2,D	$R3 \leftarrow n$
2,D	0	0	0	0	LR 2	JR 3,D	$A \leftarrow (Macroacc.)$
3,D	0	0	0	0	RR	JR 4,D	$A \leftarrow rrA$
4,D	0	0	0	0	DR 3	JR 5,D	$R3 \leftarrow R3 - 1$
5,D	0	1	0	0	TR 3	JR 6,D	Test Reg. 3 for 0 (finished)
6,D	0	0	0	0	NO	FC 7,D	If finished jump 7C otherwise 7D
7,C	0	0	0	0	ST 2	JR 0,E	Finished; $(Macroacc.) \leftarrow A$; then to execute cycle.
7,D	0	0	0	0	NO	JR 3,D	Not finished; return to 3,D

9.5 STRUCTURED DESIGN REVISITED

These last three chapters have considered various ways of producing Structured Hardware Design, and it now should be obvious that there is no universal component or class of components best for all applications. A conventional (macroprogrammable) microprocessor is good for many general situations, but it may need to be supplemented by, or even replaced entirely by, a PLA if the system requires many simultaneous (within tens of nanoseconds) combinational functions. In a similar way, one can fairly easily make a strong case, in special situations, for any group from SSI NANDs to the microprogrammable components of the current chapter. To classify a wide range of components according to their capabilities for structured design, Table 9.8 has been prepared. This table does not relieve the engineer of the responsibility for careful analysis of the system before components are selected; and it certainly will need to be augmented as details of the specifications become available; but it does present a concise summary that can be employed as a guide for making decisions.

Note that the potential for well-structured design varies from low for SSI components to very high for a μP and VLSI. The word "potential," for example in the case of a μP, indicates that the component has very high capabilities; but the problem here is that the designer must indeed correctly use these capabilities if a well-structured design is to result. A poor design can occur, for example, if the μP employs

TABLE 9.8 COMPONENT STRUCTURED DESIGN CHARACTERISTICS

Implementation Method (Component Class)	Potential for a Well-Structured Design	Nature of Structure	Advantage	Disadvantage
1. μP	Very high	Programmable operations (instructions); data and address buses; PIU for interfacing; specialized internal registers, e.g., Accumulator, CCR, etc.	Economical for all but very small tasks; program easily modified; powerful algorithms possible	Somewhat slow; low drive capability; basically serial in processing capabilities
2. Programmable component, e.g., PLA	High	Large regular array; general sum-of-products terms	Flexible; parallel nature	Sequential problems may require extra chips
3. Microprogrammable microprocessor	High	Similar to μP, but bit-slice ALU	More flexible and faster than μP; good parallel aspects	System is usually much more expensive than μP; low-level instructions may hide structure
4. MSI, e.g., multiplexer	Moderate	Higher-level primitives contribute to modularity and structure	Many powerful higher-level primitives; parallel and fast	System may require many clips; limited or no programmability
5. SSI, e.g., NAND gates and flip-flops	Low	Structure must be entirely developed by the designer	Parallel fast operations possible; cheap for small tasks	Very many clips often required for a complete system
6. VLSI	Very high	Complete system or subsystem built into component	Chip may be tailored to exact requirements; rapid parallel operations; very economical in high volume production	Very expensive for low-volume systems; chip development time may be long
7. Bus structure	High	Standard data and other information lines	Economical communication paths	May be slow; often has serial aspects
8. Special-purpose LSI chips, e.g., PIU	High	Standard well-developed functions often easily interfaced to bus or similar data path	Cheap, well-documented functions and algorithms	General use may be limited by μP or other chip family requirement

a very unstructured program. On the other hand, a well-structured system can result from the use of SSI if the designer is very careful to faithfully employ the structured design properties, e.g., if a top-down process is followed, and if all logic functions and algorithms implemented are simple in form, systematic, and well documented. In another way of viewing it, the word "potential" indicates how easy (and often how economical) it is to implement a typical design. Thus, if all other factors were equal, one would strongly prefer to do a μP rather than an SSI design.

Douglas Fairbairn in his article[5] on VLSI makes an interesting comparison between several methods for implementing a complex data encryption algorithm requiring several hundred multiplications of very large binary numbers. A conventional microprocessor solution is only able to achieve an encryption rate of 10 bits/s, and a TTL method costing about $3000 (160 MSI/LSI chips) might be able to reach 6000 bits/s. On the other hand, a VLSI solution has been designed by Ron Rivest and his colleagues at MIT to operate at 1200 bits/s and can be produced in volume for about $50. This chip has dimensions of 5.5 mm by 8 mm and contains 40,000 transistors.

The three right-hand columns of Table 9.8 should help to amplify special characteristics and limitations of each component. You should carefully study all parts of the table, and when necessary, refer back to earlier sections of the book for additional insight concerning the contents of any of the columns in the table.

9.6 BIBLIOGRAPHY

1. Agrawala, A. K., and T. G. Rauscher, *Foundations of Microprogramming.* New York: Academic Press, Inc., 1976.
2. Alexandridis, N. A., "Bit-Slice Microprocessor Architecture." *Computer,* Vol. 11, No. 6, June 1978, pp. 56–80.
3. *Am2900 Bipolar Microprocessor Family Data Book.* Sunnyvale, Calif.: Advanced Micro Devices, Inc., 1978.
4. Chu, Y., *Computer Organization and Microprogramming.* Englewood Cliffs, N.J.: Prentice-Hall, Inc., 1972.
5. Fairbairn, D., "VLSI: A New Frontier for System Designers," *Computer,* Vol. 15, No. 1, January 1982, pp. 87–96.
6. Husson, S. S., *Microprogramming—Principles and Practices.* Englewood Cliffs, N.J.: Prentice-Hall, Inc., 1970.
7. McGlynn, D. R., *Modern Microprocessor System Design: Sixteen-Bit and Bit-Slice Architecture.* New York: Wiley-Interscience, 1980.
8. Mick, J. R., and J. Brick, *Microprogramming Handbook.* Sunnyvale, Calif.: Advanced Micro Devices, Inc., 1977.
9. Mick, J., and J. Brick, *Bit-Slice Microprocessor Design.* New York: McGraw-Hill Book Company, 1980.
10. Myers, G. J., *Digital System Design with LSI Bit-Slice Logic.* New York: John Wiley & Sons, Inc., 1980.
11. Rauscher, T. G., and P. N. Adams, "Microprogramming: A Tutorial and Survey of Recent Developments." *IEEE Transactions on Computers,* Vol. C-29, 1980, pp. 2–20.
12. "Special Issue on Microprogramming." *IEEE Transactions on Computers,* Vol. C-23, 1974, pp. 753–837.

13. "Special Issue on the HP-65 Programmable Pocket Calculator." *Hewlett-Packard Journal*, Vol. 25, May 1974, pp. 2–17.

9.7 EXERCISES

9.1 A microprogrammable microprocessor does not have direct capabilities for loading constants, which may be required by its microinstructions. List, and discuss in a couple of sentences each, at least three good ways to solve this problem.

9.2 It is desired to create a new next-address function which jumps to a new row on the basis of the contents of the flag flip-flop. Following the scheme of Section 9.2 *as closely as possible*, prepare the required information for a new line in Table 9.1 which describes this instruction.

9.3 It is desired to create a new ALU microinstruction which increments the accumulator. Following the scheme of Section 9.2 *as closely as possible*, prepare the required information for a new line in Table 9.2 which describes this instruction.

9.4 In addition to the formally defined NO instruction ($0F_{16}$), list at least two other instructions which will perform no operation under certain conditions.

9.5 Create a new column for Table 9.7 which shows the microinstruction codes (in base 16) corresponding to the microinstruction mnemonics.

9.6 (a) In constructing Figure 9.6, it was assumed that only the LINC machine code 0016 would be employed as the NOP instruction. If the additional YYYY code of Table 3.2 is also allowed, insert NOP in the resulting locations on the CM map.
 (b) What microinstruction should be placed at these locations?

9.7 Using a format like Table 9.5, write the start (initialization) microprogram, which has its first location at CM address 1,1 (Figure 9.6). Assume that all macroprograms start at location 20 in the main memory.

9.8 Write a microprogram to implement the APO instruction of Table 3.2. (Present your solution in a form similar to Table 9.6.)

9.9 Repeat Exercise 9.8, but implement the JMP instruction.

9.10 The μP system of Figure 9.5 has all CM and register locations, not specified below, initially set to zero. Assume a 12-bit ALU; and any instruction code not defined by Table 9.2 is a NO operation, except that a halt instruction is added with mnemonic HT and code 00_{16}, which halts the processor at the current CM location. After the program below halts determine:
 (a) The address where it halts
 (b) Contents of the μP's accumulator
 (c) The number of times the instruction in 1, 4 is executed
 (d) The contents of register 2

CM Address (Row, Col.)	Microinstruction Code (Base 16)	CM Address (Row, Col.)	Microinstruction Code (Base 16)
(Start) 1, 1	153132	3, 2	062133
1, 3	2D3151	3, 3	00811B
1, 4	4F2161	4, 3	1C8113
1, B	148122	5, 1	062124
2, 2	048143	6, 1	06227A
2, 4	038114	7, B	00F124

(Continuation)

9.11 Employing the program below, repeat Exercise 9.10; but replace part (c) with:
(c) The number of times the instruction in 6, 3 is executed.

CM Address (Row, Col.)	Microinstruction Code (Base 16)	(Continuation) CM Address (Row, Col.)	Microinstruction Code (Base 16)
(Start) 1, 1	008152	6, 2	074163
5, 2	15415A	6, 3	4F416A
5, 3	148157	6, A	00F26C
5, 7	064162	6, C	00F16F
5, 9	148153	6, D	00F162
5, A	019159		

9.12 Employing the program below, repeat Exercise 9.10. [Assume that the main memory has 12-bit words and that locations 000 through 777_8 initially contain (+0). Give *all* results in octal.]

CM Address (Row, Col.)	Microinstruction Code (Base 16)	(Continuation) CM Address (Row, Col.)	Microinstruction Code (Base 16)
(Start) 1, 1	182115	1, 7	148118
1, 2	162113	1, 8	1C823A
1, 3	548117	3, A	063123
1, 4	128116	3, B	082A58
1, 5	092114		
1, 6	218112		

9.13 Employing the system configuration of Figure 9.4, including the data ROM of Table 9.4 and the program below, repeat Exercise 9.10. (Give all results in octal.)

CM Address (Row, Col.)	Microinstruction Code (Base 16)
(Start) 1, 1	008117
1, 2	062118
1, 3	028112
1, 4	1C8116
1, 5	148113
1, 6	06211A
1, 7	148115

9.14 Your supervisor at a large steel company has a list of general requirements for a new 16-bit computer system to control a rolling mill for which he wants you to write a design proposal. (D/A converters will be employed in driving motors and other devices, but exact details are not needed here.) List the major steps, about one page, that you would follow preparing the proposal for a well-structured design. (*Hint:* Be careful about the assumptions you make.)

Part III

ADVANCED TOPICS

Chapter 10

Circuits for Computer Arithmetic

Having studied the principles of digital systems and modern methods for logic design in Parts I and II, we are now ready to consider some advanced topics, which have many practical applications. Although each is highly dependent on the earlier parts, the chapters of Part III themselves are only loosely coupled to each other. Thus, one can read these later chapters in any arbitrary order.

The principles of ALU design will be developed here using the tools of Chapter 4, 5, and 6; and high-performance circuits for addition, multiplication, and division will be studied. Excellent opportunities to produce innovative ideas and to appreciate the original work of others exist in the ALU area. Examples showing the trade-offs between cost, processing speed, and circuit complexity will be presented.

10.1 FAST ADDERS

In Section 6.6 basic hardware for addition and subtraction was considered. This material provides insights into ALU operation, but it does not give the reader much depth in understanding logic circuits, for computer arithmetic. The study of fast adders and other special arithmetic circuits are not only important in their direct applications to advanced digital computer design; but they are probably even more important in the practical vehicle provided for the study of general high-performance digital logic.

The main disadvantage of the basic parallel adder circuit in Figure 6.31 is that a carry originating at bit 0 may propagate (ripple) down the adder as far as the nth bit. (This occurs, for example, when $+1$ is added to -0 in the 1's complement system.) Now each carry stage is a two-level circuit; thus the ripple time, T, can be as long as

$T = 2n\,\Delta t$ seconds, where Δt is the propagation time for one level. If $n = 32$ and $\Delta t = 10$ ns, $T = (2)(32)(10) = 640$ ns, which is much too long for most applications. Of course, Δt can be reduced considerably by the use of ECL* or other fast circuitry, but there is a definite physical limit to how small Δt can be made. Moreover, if ECL is employed throughout the computer, there will still be a serious difference between the operating time of the adder and most other units. For these reasons, the basic adder is augmented in most computers by some additional circuitry.

One of the most popular adder improvements is known as the *carry look-ahead principle*. This method reduces the ripple time by employing fewer than $2n$ levels between the first and the nth bits. Since any Boolean function can be represented by a two-level circuit, it is possible to reduce the ripple delay to $T = 2\,\Delta t$; but for reasonable values of n, the cost is prohibitive. Therefore, the current approach seeks a tradeoff between cost and adder speed.

An adder stage can produce a carry by two mechanisms—either the carry is *generated* within the stage i, or it was generated previously and stage i is serving to *propagate* it. Both of these conditions are illustrated in Table 10.1. Observe that those cases where the carry-out depends on the carry-in are called *propagation stages*, and the case where $X_i = Y_i = 1$ is called a *generation stage* because a carry-out will be generated independently of the carry-in.

TABLE 10.1 CARRY CONDITIONS

Input		Carry Output	
X_i	Y_i	C_i	Remarks
0	0	0	
0	1	C_{i-1}	Propagation, P_i
1	0	C_{i-1}	Propagation, P_i
1	1	1	Generation, G_i

From Table 10.1, the equation that describes carry generation, G_i, is

$$G_i = X_i Y_i \tag{10.1}$$

and the equation that describes carry propagation, P_i, is

$$P_i = \bar{X}_i Y_i + X_i \bar{Y}_i = X_i \oplus Y_i \tag{10.2}$$

Therefore, the carry-out of state i becomes

$$C_i = G_i + P_i C_{i-1} \tag{10.3}$$

It is also possible to express the sum output in terms of P_i and C_{i-1} by rearranging the equation in Figure 6.30 and then substituting Equation (10.2):

$$\begin{aligned} S_i &= X_i \bar{Y}_i \bar{C}_{i-1} + \bar{X}_i \bar{Y}_i C_{i-1} + X_i Y_i C_{i-1} + \bar{X}_i Y_i \bar{C}_{i-1} \\ &= (X_i \bar{Y}_i + \bar{X}_i Y_i)\bar{C}_{i-1} + (\bar{X}_i \bar{Y}_i + X_i Y_i)C_{i-1} \\ &= P_i \bar{C}_{i-1} + \overline{(X_i \bar{Y}_i + \bar{X}_i Y_i)}C_{i-1} \\ &= P_i \bar{C}_{i-1} + \bar{P}_i C_{i-1} = P_i \oplus C_{i-1} \end{aligned} \tag{10.4}$$

*Emitter coupled logic—see Chapter 12 for a detailed description.

Returning to Equation (10.3), we see that this equation defines the carry output recursively. Thus, the equation for C_i may be rewritten in terms of the parameters of previous stages as follows:

$$C_i = G_i + P_i(G_{i-1} + P_{i-1}C_{i-2})$$
$$= G_i + P_iG_{i-1} + P_iP_{i-1}(G_{i-2} + P_{i-2}C_{i-3})$$

Thus, for m bits the result is

$$C_i = G_i + P_iG_{i-1} + P_iP_{i-1}G_{i-2} + P_iP_{i-1}P_{i-2}G_{i-3} + \ldots \quad (10.5)$$
$$+ P_iP_{i-1}P_{i-2}\ldots P_{i-m+1}C_{i-m}$$

Equation (10.5) can be expressed in the more concise form:

$$C_i = G_i + \left(\prod_{j=i}^{i} P_j\right)G_{i-1} + \left(\prod_{j=i-1}^{i} P_j\right)G_{i-2} + \ldots + \left(\prod_{j=i-m+1}^{i} P_j\right)C_{i-m} \quad (10.6)$$

Finally, by defining the carry-in as

$$G_{i-m} = C_{i-m}$$

we can express Equation (10.6) in the still more concise form:

$$C_i = \sum_{k=i-m}^{i}\left(\prod_{j=k+1}^{i} P_jG_k\right) \quad (10.7)$$

where $0 \le m \le i+1$. (Recall that $\prod_{j=i+1}^{i} P_j = 1$.)

For the case $m = i+1$, Equation (10.7) produces the carry for any bit i in terms of the previous bits all the way down to bit 0, and in this case $G_{-1} = C_{-1} =$ carry into bit 0. Since P_i is a two-level function, Equation (10.5) reveals that C_i is a four-level expression. Consequently, the carry delay is $T = 4\,\Delta t$, which is extremely fast compared to the ripple carry. For example, for $n = 32$ and $\Delta t = 10$ ns, it was previously determined that $T = 640$ ns; but in this case, it would only require $T = (4)(10) = 40$ ns. On the other hand, the logic gates to mechanize Equation (10.7) for all 32 bits would lead to prohibitive costs in terms of numbers of gates and the number of inputs per gate.

A good compromise in a situation like this is to construct carry look-ahead, CLA, logic stages consisting of l bits each, where l is usually a number not more than 6. (To avoid overly complicated diagrams, yet to bring out the basic principles, a stage consisting of four bits will be considered here.) First we will show, in Figure 10.1, how a modified form of the full adder can be produced having the necessary G_i, P_i, and S_i outputs. Now gate 2 generates the G_i output directly from Equation (10.1); the inverted output of gate 3 yields

$$P_i = \overline{X_iY_i + \bar{X}_i\bar{Y}_i} = (\bar{X}_i + \bar{Y}_i)(X_i + Y_i) = X_i\bar{Y}_i + \bar{X}_iY_i$$
$$= X_i \oplus Y_i$$

as expected from Equation (10.2), and the output of gate 6 produces the S_i term directly from Equation (10.4).

Having completed the modified full adder, MFA, it is now necessary to show how the carry signal is generated. Since we have agreed to use CLA stages consisting of four bits each, stage 2 representing bits four through seven is arbitrarily selected as

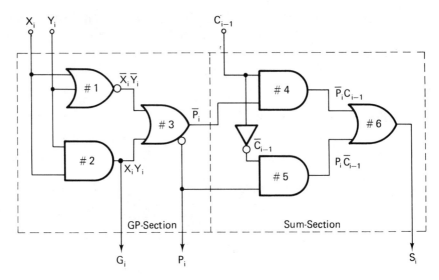

Figure 10.1 Modified Full-Adder Circuit.

a typical circuit. Here, $i = 7$, $m = 4$, and Equation (10.7) becomes

$$C_7 = \sum_{k=7-4}^{7} \left(\prod_{j=k+1}^{7} P_j G_k \right) = \prod_{j=4}^{7} P_j G_3 + \prod_{j=5}^{7} P_j G_4 + \prod_{j=6}^{7} P_j G_5$$
$$+ \prod_{j=7}^{7} P_j G_6 + \prod_{j=8}^{7} P_j G_7 \quad (10.8)$$
$$= P_4 P_5 P_6 P_7 G_3 + P_5 P_6 P_7 G_4 + P_6 P_7 G_5 + P_7 G_6 + G_7$$

Similarly, C_6 is found with $i = 6$, $m = 3$:

$$C_6 = \sum_{k=3}^{6} \left(\prod_{j=k+1}^{6} P_j G_k \right) = P_4 P_5 P_6 G_3 + P_5 P_6 G_4 + P_6 G_5 + G_6 \quad (10.9)$$

C_5 is found with $i = 5$, $m = 2$:

$$C_5 = \sum_{k=3}^{5} \left(\prod_{j=k+1}^{5} P_j G_k \right) = P_4 P_5 G_3 + P_5 G_4 + G_5 \quad (10.10)$$

Finally, C_4 is found with $i = 4$, $m = 1$:

$$C_4 = \sum_{k=3}^{4} \left(\prod_{j=k+1}^{4} P_j G_k \right) = P_4 G_3 + G_4 \quad (10.11)$$

(In all of the above equations, it should be noted that $G_3 = C_3 =$ carry into the stage.)

By employing Equations (10.8) through (10.11), and by representing the modified full adder with two blocks, as implied by the dashed sections of Figure 10.1, we can represent the complete CLA circuit for stage 2 as shown in Figure 10.2.* (It should be emphasized that the *GP* section of each MFA generates its outputs independently of the carry into that bit—of course, this is the key to the success of the CLA method.) There are seven other stages which are exactly the same as the one shown in Figure

*There are several commercial CLA adder units which employ the principles indicated in Figure 10.2, e.g., the 8260 MSI package manufactured by Signetics Corporation.

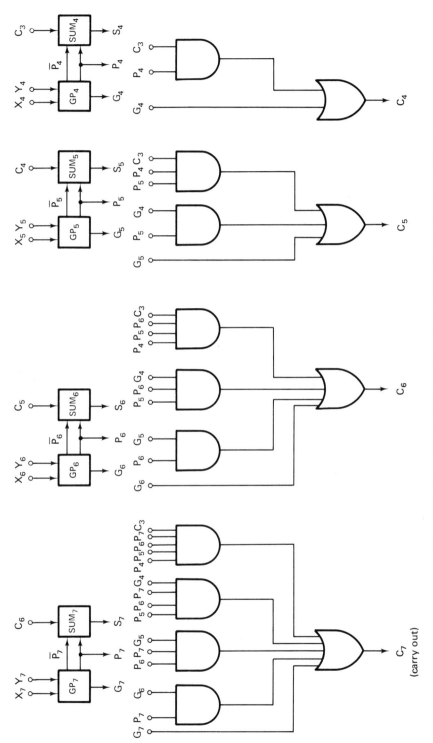

Figure 10.2 Typical Four-Bit Carry Look-Ahead Adder Stage. (Note that GP_i is the generate-propagate half and SUM_i is the sum half of the ith MFA.)

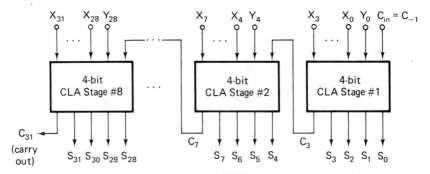

Figure 10.3 32-Bit Adder Employing Eight CLA Stages with Ripple Carry between Stages.

10.2 except that each subscript has a multiple of 4 either added to or subtracted from it. The complete CLA adder, with all eight stages, is shown in Figure 10.3.

How does the 32-bit CLA adder compare with the ripple adder? It was previously established that each CLA stage is a four-level circuit, and there is ripple carry between stages. Thus, each stage must wait for the previous one to settle before developing the carry output for the next stage. The total propagation time of the adder is therefore

$$T_{CLA} = (8)(4) \, \Delta t = 32 \, \Delta t$$

and this is to be compared with the ripple adder,

$$T_{ripple} = (32)(2) \, \Delta t = 64 \, \Delta t$$

which means a factor of 2 in speed has been gained with the CLA adder circuit. However, by employing more bits per stage, a larger improvement is possible at the cost of a more complicated circuit, e.g., with 32 bits and 8 bits per stage (a value which may not be practical) the ratio is

$$\frac{T_{ripple}}{T_{CLA}} = \frac{64 \, \Delta t}{(4)(4) \, \Delta t} = 4$$

There are other variations of the basic CLA circuit which can achieve improvements of four or more over the ripple method without costing as much as the 8-bit per stage unit. Some involve grouping CLA stages on two or three levels instead of the single level employed in Figure 10.2. These and other techniques are extensively discussed in the literature.[4,5,6,9]

10.2 MULTIPLICATION [5,7,9]

There are many alternative methods for performing multiplication. It is our purpose here not to catalog the various possible multiplication algorithms, but to present fundamental principles, to develop certain basic circuits, and, most important of all, to convey design techniques so that the reader can learn to arrive at a circuit satisfying a given specification. In the design process, the RTL method of Chapter 6 will assume a very prominent role.

Multiplication of Positive Numbers

Many multiply algorithms are based on the conventional manual method of successive additions and shifting of the multiplicand. Since all of our work will be with binary numbers, some simplification is produced; and if both numbers are positive, a particularly simple case arises. Even here, however, there are alternative possibilities which may lead to better and less expensive circuits.

Consider the problem of multiplying two 6-bit positive numbers and obtaining the 12-bit product. Figure 10.4 shows the block diagram and RTL statements required to solve the problem. In the logic diagram, note that the multiplier basically consists of two shift registers, A and B, for the two input numbers, a result register, C, and a 12-bit full adder. The timing signals are indicated by means of quantities within parentheses. Since it must be shifted left without losing significant bits and since it operates with a 12-bit adder, observe that register A contains 12 bits.

Unless the extra components are required for another task, a significant savings can be achieved by employing the logic diagram shown in Figure 10.5. The major difference between this circuit and the previous one is that register B serves the dual purpose of storing the multiplier at the beginning of the process and storing the right

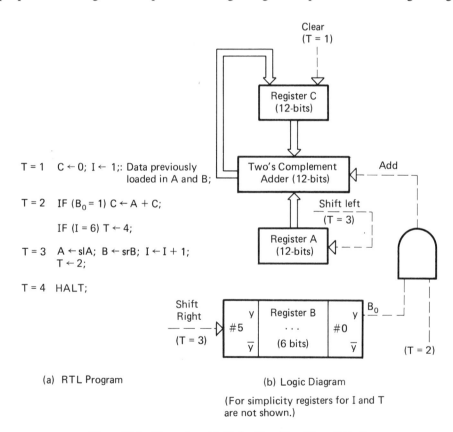

Figure 10.4 Elementary Multiply Algorithm ($C = B \times A$).

Sec. 10.2 Multiplication

six bits of the result at the end of the process. The method works because each time register B is shifted right to obtain a new LSB for testing, there is room at the left end of the register for another bit of the result. Due to this circuit modification, registers A and C, as well as the adder, can each be reduced from 12- to 6-bit units. The corresponding RTL program is shown in part (a) of Figure 10.5. Note the differences, at $T = 2$ and 3, between the program in Figure 10.4(a) and the one in Figure 10.5(a). These differences are due to the structural changes made in going to the new circuit.

$T = 1$ $C \leftarrow 0; I \leftarrow 0; A \leftarrow$ (Multiplicand); $B \leftarrow$ (Multiplier);
$T = 2$ IF $(B_0 = 1)$ $C \leftarrow A + C$; $I \leftarrow I + 1$;
$T = 3$ $C,B \leftarrow$ srC,B; IF $(I \neq 6)$ $T \leftarrow 2$;
$T = 4$ HALT;
(*Note:* Bit 0 is the LSD for all registers.)

(a) RTL Program

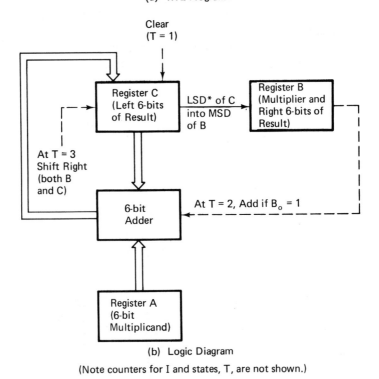

(b) Logic Diagram
(Note counters for I and states, T, are not shown.)

Figure 10.5 Alternative Multiply Algorithm.

Sign-and-Magnitude Multiplication

Again by analogy with the conventional manual process, the most obvious method for mutliplying signed numbers is to handle the sign and magnitude separately. There are four possible cases for the signs of the multiplicand and the multiplier: (1) $++$, (2) $+-$, (3) $-+$, and (4) $--$. Assuming 1's complement numbers, the four cases can be detected; both numbers can be made positive by complementation; if either case 2 or 3 occurs, a memory bit, M, can be set; positive-number multiplication can be performed; and finally, any correction for the proper sign can be made. Actually, the only modifications needed in Figure 10.4 to accomplish sign-magnitude multiplication are: add the memory bit, M, and provide a means for complementing registers A, B, and C. Figure 10.6 is an RTL program for this mode of multiplication, and the following points should be observed when studying the figure:

1. The IF statements, at $T = 2$, detect which of the four sign cases have occurred, and M is changed to remember the sign of the result.
2. The IF statement and the complement operation, at $T = 5$, establish the proper form for the result.
3. Since two n-bit numbers (magnitudes) generate a $2n$ ($2 \times 5 = 10$) bit product and since the MSD of register C, B is employed for the sign, there is one ($12 - 10 - 1 = 1$) unused bit, which is C_4.
4. Registers B and C must be shifted six times to get the LSB of the result into B_0 of the 12-bit C, B register.

$T = 1$ $C \leftarrow 0$; $I \leftarrow 0$; $A \leftarrow$ (Multiplicand); $B \leftarrow$ (Multiplier);
$T = 2$ IF ($\bar{A}_5 \wedge \bar{B}_5$) $M \leftarrow 0$;
         IF ($\bar{A}_5 \wedge B_5$) $B \leftarrow \bar{B}$, $M \leftarrow 1$; IF ($A_5 \wedge \bar{B}_5$) $A \leftarrow \bar{A}$, $M \leftarrow 1$;
         IF ($A_5 \wedge B_5$) $A \leftarrow \bar{A}$, $B \leftarrow \bar{B}$, $M \leftarrow 0$;
$T = 3$ IF ($B_0 = 1$) $C \leftarrow A + C$; $I \leftarrow I + 1$;
$T = 4$ $C,B \leftarrow$ srC,B; IF ($I \neq 6$) $T \leftarrow 3$;
$T = 5$ IF ($M = 1$) $C,B \leftarrow \bar{C},\bar{B}$;
$T = 6$ HALT;

Figure 10.6 RTL Program for Sign-Magnitude, 1's Complement Multiplication.

‡1's and 2's Complement Multiplication

The multiplication of (3) × (−5) in the 1's complement system, employing the usual convention, is shown in Figure 10.7. Several points should be observed concerning this example:

1. The symbol ⌋ is used to separate the sign bit from the others.
2. Since the product of two four-bit numbers yields an eight-bit result, the partial products are also shown in eight bits just as they would appear in a register.

$$-5 = -(101) = 1\rfloor 010$$
$$3 = +(011) = 0\rfloor 011$$

```
        11111010
        11110100
        00000000
        00000000
      ⎯⎯⎯⎯⎯⎯⎯⎯
    ① 11101110
        → + 1
      ⎯⎯⎯⎯⎯⎯⎯⎯
      1⌋1101111  = − (0010000) = − 16 ≠ − 15
```

Figure 10.7 Conventional 1's Complement Multiplication.

3. Since the first two partial products are negative, it is not surprising to find that the leading bits are 1's (For example, the first partial product is $11111010 = -5$, as it should be, not $00001010 = +10$.)
4. Although it appears that the correct operations were performed, the final result is unexpected.

This counter example is meant to illustrate that one must not make any assumptions, in performing binary arithmetic, that have not been proved.

In order to examine the above problem in detail, a general representation for the multiplication process will be presented. First, consider the conventional multiplication algorithm for two positive integers. The product, c, can be represented as

$$c = \sum_{i=0}^{n-1} b_i a 2^i \tag{10.12a}$$

This may be analyzed as follows:

1. (a) is the multiplicand.
2. $a2^i$ is the multiplicand shifted i bits to the left.
3. b_i is the ith multiplier bit ($b_i = 0$, or 1)
   (a) If $b_i = 1$, $a2^i$ is added to the partial result.
   (b) If $b_i = 0$, $a2^i$ is not added.
4. The n partial results are combined by carrying the sum from $i = 0$ to $i = n-1$.

Equation (10.12a) may be simplified as follows:

$$c = \sum_i b_i a 2^i = a \sum_{i=0}^{n-1} b_i 2^i$$

where, by definition,

$$\sum_{i=0}^{n-1} b_i 2^i = b_0 2^0 + b_1 2^1 + \ldots + b_{n-1} 2^{n-1} = b$$

Therefore, $c = ba$ as expected.

Now, examine the following 2's complement product of two, nonzero, n-bit numbers:

$$(b) \times (-a) = \sum_{i=0}^{n-1} b_i(2^{2n} - a2^i) \tag{10.12b}$$

The equation can be analyzed as follows:

1. The partial results contain $2n$ bits; therefore, by definition, the 2's complement of the multiplicand is $2^{2n} - a$.
2. The multiplicand is shifted i bits to the left by the 2^i term to form the 2's complement $2^{2n} - a2^i$.
3. Again, the multiplier for the ith bit, b_i, controls whether or not a particular partial result is included in the sum.

Equation (10.12b) may be written

$$(b) \times (-a) = 2^{2n} \sum_i b_i - a \sum_i b_i 2^i$$
$$= 2^{2n} \sum_i b_i - ab \tag{10.13}$$

where by definition $b = \sum_i b_i 2^i$. Consider the terms $2^{2n} \sum_i b_i$; since $b \neq 0$, there are two possible cases:

1. $b_i = 1$ for only one i. Here, $2^{2n} \sum_i b_i = 2^{2n}$.

2. $b_i = 1$ for k values of i. Here, addition of the partial products will result in $k - 1$ carry-outs, but these are neglected in the 2's complement system; and again $2^{2n} \sum_i b_i = 2^{2n}$.

Therefore, Equation (10.13) becomes

$$(b) \times (-a) = 2^{2n} - ba \qquad (10.14)*$$

which is the desired product. Now Equation (10.14) proves that direct multiplication is acceptable for $(b) \times (-a)$ in the 2's complement system.

In order to try this proof, we consider the product $(3) \times (-5)$ in Figure 10.8. As expected, the correct result of -15 was obtained.

```
     -5 = -(101) = 1|010 + 1 = 1|011
      3 = +(011)              = 0|011
                          ─────────────
                              11111011
                              11110110
                              00000000
                              00000000
                          ─────────────
                        ① 1|1110001 = -15
```

Figure 10.8 2's Complement Multiplication Employing the Algorithm of Equation (10.12b).

Now let us return to the problem which arose in considering Figure 10.7. The product of $(b) \times (-a)$ in the 1's complement system may be expressed in general as

$$(b) \times (-a) = \sum_{i=0}^{n-1} b_i[2^{2n} - (a+1)2^i]\dagger$$

$$= 2^{2n} \sum_i b_i - (a+1) \sum_i b_i 2^i = 2^{2n} \sum_i b_i - ba - b \qquad (10.15)$$

Since in the 1's complement system carry-outs are added to the LSB, the first term in Equation (10.15) becomes

$$2^{2n} \sum_i b_i = 2^{2n} + (k-1)$$

---

*A more rigorous way to obtain Equation (10.14) follows. By definition, $\sum_i b_i = k$. Thus, Equation (10.13) becomes

$$b \times (-a) = 2^{2n}k - ba = 2^{2n} + 2^{2n}(k-1) - ba$$
$$= \underbrace{2^{2n} - ba}_{\text{2's complement number}} + \underbrace{2^{2n}(k-1)}_{\substack{(k-1) \text{ carry-outs} \\ \text{which are neglected}}} = 2^{2n} - ba$$

†By definition, the 1's complement of $(-a)$ is $2^{2n} - a - 1 = 2^{2n} - (a+1)$. The quantity 2^i, again provides the required shift of i bits.

where $(k - 1)$ is the number of carry-outs. Therefore, Equation (10.15) becomes

$$(b) \times (-a) = (2^{2n} - ab - 1) + k - b$$
$$= \text{(correct result)} + k - b \qquad (10.16)*$$

In the example of Figure 10.7, $k = 2$ and $b = 3$; therefore, Equation (10.16) yields

$$(b) \times (-a) = (-15) + 2 - 3 = -16$$

which is the exact result obtained.

One way to do $(b) \times (-a)$ 1's complement multiplication is to proceed as we did in the $(3) \times (-5)$ example and to employ a correction of $k - b$. (That process is actually followed in some computers.) The other alternative is to find a modified algorithm which does not require a correction. To pursue this latter alternative, consider the result of multiplying 1's complement number, $-a$, by 2; from the definition, $-a = (2^n - a - 1)$. Now the conventional shift-left procedure will be employed; a correction factor, X, will be added; and the result will be equated to $(2^{n+1} - 2a - 1)$, which is the correct result. Thus,

$$-2a = 2(2^n - a - 1) + X = 2^{n+1} - 2a - 1$$

Simplifying:

$$2^{n+1} - 2a - 2 + X = 2^{n+1} - 2a - 1$$

Therefore, $X = 1$, and we see that to multiply a 1's complement negative number by 2, the usual shift left is accomplished, but a 1, not a 0, must be used to fill the vacated bit.† In a similar way, the reader should verify the following multiplication rules:

1. To multiply a 1's complement number by 2, shift left and replicate the sign bit in the vacant position.
2. To multiply a 2's complement number by 2, shift left and fill the vacant position with a 0.

Returning to the general case of $(b) \times (-a)$ in the 1's complement system, in place of Equation (10.15) we write the new algorithm:

$$(b) \times (-a) \equiv \sum_{i=0}^{n-1} b_i(2^{2n} - a2^i - 1) \qquad (10.17)$$
$$= 2^{2n} \sum b_i - a \sum b_i 2^i - \sum b_i$$

where the term within the parentheses should be recognized as the multiplicand $(-a)$ properly shifted i bits to the left. Again

$$\sum_i b_i = k \quad \text{and} \quad 2^{2n} \sum b_i = 2^{2n} + (k - 1)$$

*Again $\sum_i b_i = k$; therefore, Equation (10.15) can be expanded as follows:

$$b \times (-a) = 2^{2n}k - ba - b = \underbrace{[2^{2n} - (ba + b - 1) - 1]}_{\text{1's complement term}} + \underbrace{2^{2n}(k - 1)}_{(k - 1) \text{ carry-outs to be added to LSB}}$$

$$= [2^{2n} - (ba + b - 1) - 1] + k - 1$$

which agrees with Equation (10.16).

†In general, when a 1's complement number is shifted left i places, multiplied by 2^i, the result is $-2^i a = 2^{n+i} - 2^i a - 1$; since leading 1's may be eliminated in a 1's complemented number, this reduces to $-2^i a = 2^n - 2^i a - 1$.

Therefore, Equation (10.17) becomes

$$(b) \times (-a) = 2^{2n} + (k-1) - ba - k = 2^{2n} - ba - 1$$

which indicates that the correct product is obtained.

All of this means that Figure 10.7 can be corrected merely by shifting a 1 into the vacancy left in the second partial product, which would be 11110101. (Showing that the correct product now results in Figure 10.7 is left as an exercise.)

We must be careful in finding not only the product $(b) \times (-a)$, but also $(-b) \times (a)$ and $(-b) \times (-a)$. For the 2's complement, consider $(-b) \times (a)$:

$$(-b) \times (a) = (\bar{b} + 1)a = \left(\sum_{i=0}^{n-1} \bar{b}_i a 2^i\right) + (1)a \qquad (10.18)$$

In Equation (10.18), the 2's complement of (b) is formed by complementing the individual bits and adding 1. Thus, the first term is the partial sum created by shifting $\bar{b}$ the proper number of times; the second term accounts for the 1 which must be added to the 1's complement to get the 2's complement. Now,

$$\bar{b}_i = 1 - b_i \qquad (10.19)$$

This may be seen as follows:

If $b_i = 1$, $\bar{b}_i = 1 - b_i = 1 - 1 = 0$
If $b_i = 0$, $\bar{b}_i = 1 - 0 = 1$

Thus, the relation is true in general since it holds for all possible cases.

Now we substitute Equation (10.19) into Equation (10.18) and simplify:

$$(-b) \times (a) = \left(\sum_i (1 - b_i)a 2^i\right) + a$$

$$= a \sum_{i=0}^{n-1} 2^i - a \sum_i b_i 2^i + a = a(2^n - 1) - ba + a \qquad (10.20)$$

$$= a2^n - ba \neq 2^{2n} - ba$$

Since the desired result, $2^{2n} - ba$, is not obtained, *the straightforward method of Equation (10.18) is not a correct procedure for 2's complement multiplication.*

A fairly simple alternative procedure which does produce the correct result is shown in Equation (10.21):

$$(-b) \times (a) = [-(-b)] \times [-(a)] = (b) \times (-a) \qquad (10.21)$$

Equation (10.21) indicates that by multiplying both numbers by (-1), which does not upset the equality, a form is obtained matching an algorithm already developed. Changing the sign of the two numbers, of course, requires the 2's complement operation. At first it appears that this operation would require additional time. However, the multiplicand, (a), can be 2's complemented by adding 1 to the 1's complement. The actual process is accomplished by gating $\bar{a}$ into the adder forming the partial sum and at the same time employing a carry-in of 1 to the LSB. In the multiplication process, the multiplier bits are tested in sequence to determine whether or not the multiplicand is added to the partial product. Thus, the 2's complement of the multiplier can be determined sequentially, and no additional time is required to change the sign of either number.

The following algorithm for the sequential conversion of a number (a), with bits (a_i), to its 2's complement (a^*) can be established:

1. Examine the (a_i) bits from right to left, and place a 0 in each (a_i^*) bit until the first $a_i = 1$ bit is found.
2. Corresponding to this first $a_i = 1$ bit, make $a_i^* = 1$.
3. Continue to examine the a_i bits, and in each case produce the following transfer:
$a_i^* \leftarrow \bar{a}_i$.

(As an exercise, show that the above algorithm produces 10100 as the correct 2's complement of 01100.)

The only multiplication possibility that has not been covered is $(-b) \times (-a)$; but this case follows the same pattern as $(-b) \times (a)$. Similar to Equation (10.21), we have

$$(-b) \times (-a) = [-(-b)] \times [-(-a)] = (b) \times (a) \quad (10.22)$$

Again, both numbers must be complemented, but the same process as employed for $(-b) \times (a)$ can be used.

Figure 10.9 is the RTL program that forms the product of 2's complement numbers. Examination of the figure reveals that it treats both positive and negative numbers by means of the algorithms just discussed. At T_4 and T_7 note how the scale, instead of the shift, operation preserves the sign bit. The block diagram for this system is similar to Figure 10.4, but to avoid overflow* of the $n - 1$ magnitude bit into the sign bit during intermediate stages of multiplication for large numbers, registers A and C are made $n + 1$ bits wide. The multiplicand is transferred to A with the sign in both bits $n + 1$ and n; thus, as usual, bit $n - 1$ is actually the first magnitude bit.

$T = 1$ $C \leftarrow 0; I \leftarrow 0; A \leftarrow$ (Multiplicand); $B \leftarrow$ (Multiplier);
$T = 2$ IF $(B = 0)$ $T \leftarrow 9;:$ Test for multiplier equal to zero;
           IF $(B_{n-1} = 1)$ $T \leftarrow 5;:$ Test for multiplier negative;
$T = 3$ IF $(B_0 = 1)$ $C \leftarrow A + C; I \leftarrow I + 1;:$ Positive multiplier employed;
$T = 4$ $C,B \leftarrow scC,B;$ IF $(I \neq n)$ $T \leftarrow 3$ ELSE $T \leftarrow 9;$
$T = 5$ IF $(B_0 = 0)$ $C,B \leftarrow srC,B, T \leftarrow 5; I \leftarrow I + 1;:$ Continue executing until $B_0 \neq 0;$
$T = 6$ $C \leftarrow \bar{A} + C + 1;$ IF $(I = n)$ $T \leftarrow 9;:$ First $B_0 = 1;$
$T = 7$ $C,B \leftarrow scC,B;$ IF $(I = n)$ $T \leftarrow 9;$
$T = 8$ IF $(B_0 = 0)$ $C \leftarrow \bar{A} + C + 1; I \leftarrow I + 1; T \leftarrow 7;$
$T = 9$ HALT;

Figure 10.9 Program for Finding the Product $(C = B \times A)$ of 2's Complement Numbers.

‡10.3 FAST MULTIPLICATION

Although much has been accomplished in the area of fast multiplication since 1950, the number of recent journal articles indicate that advancements in theory as well as in practice have continued to make this a rich field for research and development. (The progress in MSI and LSI have certainly had an important influence.)

*Overflow is considered in detail in Section 10.5.

Methods for obtaining multiplication speedup can be classified into the following three main areas:

1. *Combinational logic.* Methods of this type employ combination circuits to directly yield the product of two numbers.
2. *Adder speedup.* These methods decrease multiply time by reducing the time required to add the partial products.
3. *Multiple bit processing.* These are techniques in which multiplication time is decreased by simultaneous processing of more than one multiplier bit.

This section will present representative members of each of the three areas. (For simplicity, most techniques will consider only positive numbers.)

Combinational Logic

It is certainly possible to find the logic function for each bit of a product by constructing a truth table for all combinations of the multiplier and the multiplicand. This brute-force approach requires a truth table with 2^{2n} rows, where n is the number of bits per data word, and it is not a reasonable design technique for most practical values of n. On the other hand, another combinational approach called the *array multiplier* does permit a design description which can easily be specified for any desired value of n.

Employing the requirements segmentation heuristic, examine the manual multiplication process, for binary numbers, shown in Figure 10.10. It is obvious that

					A_3	A_2	A_1	A_0
					B_3	B_2	B_1	B_0
				P_{03}	P_{02}	P_{01}	P_{00}
			P_{13}	P_{12}	P_{11}	P_{10}	
		P_{23}	P_{22}	P_{21}	P_{20}		
	P_{33}	P_{32}	P_{31}	P_{30}			
R_7	R_6	R_5	R_4	R_3	R_2	R_1	R_0

Figure 10.10 Manual Multiplication—Partial Products and Results.

each partial product, P_{ij}, can be formed from the individual multiplicand, A_j, and multiplier, B_i, bits simply by means of an AND gate. These gates and the array of full adders required to sum the partial products into the result bits, R_i, may be systematically arranged as shown in Figure 10.11. The first row of full adders in the figure basically combines the top two rows of partial products. As expected, the sum outputs from one row of full adders provide one input to the full adder one row below, but the carry outputs are shifted one bit to the left. The last row of adders provides a ripple carry and picks up the last, p_{nn}, partial product.

A photomicrograph of the MPY-16HJ 16 × 16-bit array multiplier manufactured by TRW LSI Products is shown in Figure 10.12. The chip forms the 32-bit 2's complement product of two 16-bit numbers in 100 ns. Moreover, this multiplier is designed with expansion capability so that several chips can be employed to produce a 32 × 32-bit multiplication and even greater expansion is possible.

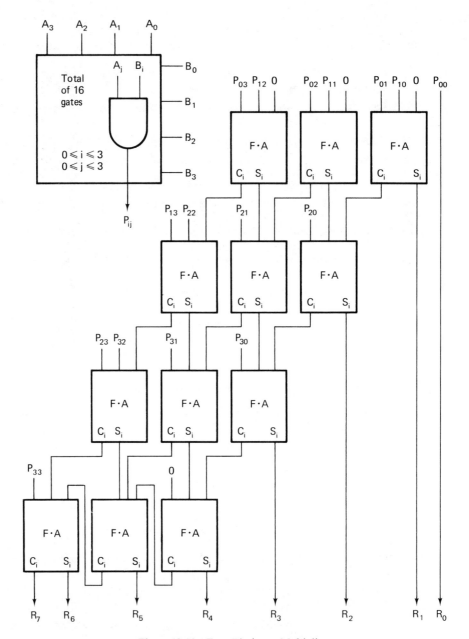

Figure 10.11 Four-Bit Array Multiplier.

Adder Speedup

Since the time required for the carry bits from the addition of each partial product to ripple through the adder is a major function of the total multiply time, anything that can be done to make addition more efficient is beneficial. Obvious improvements are to employ fast circuitry and to implement the techniques mentioned earlier for

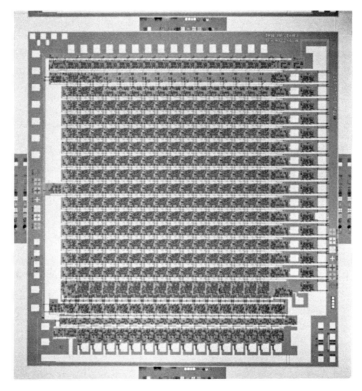

Figure 10.12 A Fast Multiplier Chip: Photomicrograph of the MPX-16J. (Courtesy of TRW LSI Products.)

fast addition, e.g., carry look-ahead and similar methods. On the other hand, there are aspects of the multiply process that, in themselves, provide an opportunity for adder speedup, and one of the best of these is the *carry-save method*. Here, the carries from the addition of two partial products are saved in a set of flip-flops and used as one of the adder inputs when the next partial product is determined. By employing this scheme, it is not necessary to wait for the carries to ripple through the adder.

Figure 10.13 shows the block diagram for a carry-save multiplier. Except for the adder itself and a few new control signals, the diagram is the same as that for the conventional multiplier shown in Figure 10.4. It is in Figure 10.14, however, that the details of the carry-save principle are evident. In this figure, the full adders are standard; but the carry-in terminals, C_{i-1}, receive their signals from register R, which actually stores the carry information from the previous addition.

The multiply cycle consists of two modes: carry-save, *CS*, and carry-propagate, *CP*. The carry-save mode employs steps similar to those in our previous multiply algorithms in that a shift-and-add scheme is used; but in this mode, the carry is temporarily stored in R. After the last partial product is completed, R will usually contain some nonzero bits. These must be combined with the final partial product before the true product is obtained. Thus, it is the purpose of the carry-propagation mode to perform the operation: $E \leftarrow E + R$. This time, however, the carry-out of

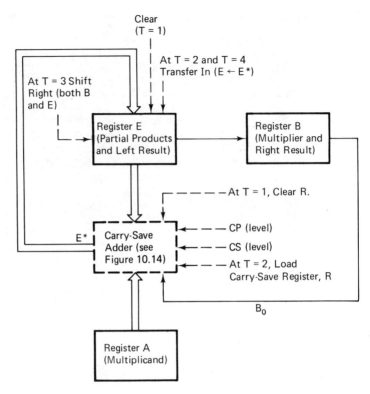

Figure 10.13 Carry-Save Multiplier.

each adder is not transferred to R, but is passed to the next most significant bit in the conventional ripple-add procedure. In Figure 10.14, note how the CS level controlling gate 1 of each bit places the system in the carry-save mode, while the CP level controlling gate 2 places it in the carry-propagate mode. (Recall that since the two modes are mutually exclusive, $CS \neq CP$.) It might at first seem that the carry-out signal for each bit, CO_i, should be transferred to the R_{i+1} carry-save flip flop ($R_{i+1} \leftarrow CO_i$); but it must be remembered that the partial product register, E, is shifted one bit right before the next addition. Therefore, the transfer $R_i \leftarrow CO_i$, with no shift of R, places the carry information in the correct position for the next step.

The detailed operations of Figure 10.14 can be understood more completely after a careful examination of the RTL program, for positive numbers, shown in Figure 10.15.* This figure is similar to other RTL multiplication programs; but it should be understood that the EXOR function employed at $T = 2$ produces modulo-2 addition of each bit. [In $E \leftarrow E \oplus R \oplus A$ the individual bits follow the expression $E_i \leftarrow (E_i \bar{R}_i \bar{A}_i) \lor (\bar{E}_i R_i \bar{A}_i) \lor (\bar{E}_i \bar{R}_i A_i) \lor (E_i R_i A_i)$ and the carry is not allowed to propagate.] For example, if $n = 2$, $A = 10$, and $E = R = 01$, the process arranged

*In Figure 10.15 it is reasonable to assume that B_0 will have been available and the full adders will have completed operation before $T = 2$ occurs, so that the new data can be transferred to registers E and R.

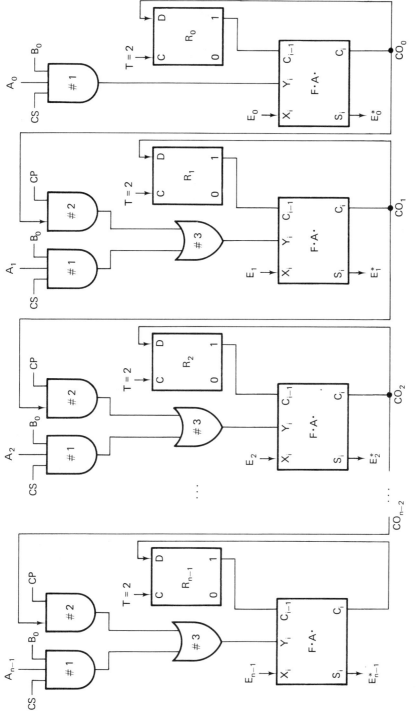

Figure 10.14 Carry-Save Adder. (*Note:* To avoid unnecessary complications, the method for clearing register *R* has been omitted.)

$T = 1$ $E \leftarrow 0$; $CS \leftarrow 1$; $CP \leftarrow 0$; $A \leftarrow$ (Multiplicand);
$B \leftarrow$ (Multiplier); $I \leftarrow 1$; $R \leftarrow 0$;
$T = 2$ IF $(B_0 = 1)$ $E \leftarrow E \oplus R \oplus A$ ELSE $E \leftarrow E \oplus R$;
$R \leftarrow CO$; $I \leftarrow I + 1$;: Carry-save phase;*
$T = 3$ $E,B \leftarrow \text{sr} E,B$; IF $(I \neq n + 1)$ $T \leftarrow 2$ ELSE $CS \leftarrow 0$, $CP \leftarrow 1$;
$T = 4$ $E \leftarrow E + R$;: Carry propagation phase;
$T = 5$ HALT;

Figure 10.15 RTL Program for the Carry-Save Multiplier (positive numbers).

in the form for conventional addition is

$$A = 10$$
$$E = 01$$
$$\underline{R = 01}$$
$$E = 10$$
$$R = 01$$

where the numbers shown below the horizontal line represent the new values of E and R. By contrast, the equation at $T = 4$, $E \leftarrow E + R$, indicates conventional carry-propagation addition, which the algorithm demands and which the circuit in Figure 10.14 realizes when $CP = 1$. It should also be understood that the CS and CP signals are actually levels originating in flip-flops, so that once they are set at state $T = 1$, they remain constant until externally changed at state $T = 3$.

Table 10.2 shows the individual steps in the carry-save multiplication of two positive numbers. It is instructive to follow the program in Figure 10.15 while verifying the table. Note that the table lists the contents of the registers as well as the T, I, CP, and CS signals for all steps of the multiplication process. Examine each line of the table and check the calculations shown in the rightmost column. (The values of T and I are key references to the program in Figure 10.15.) When performing the calculations, particular attention should be given to the fact that, through the quantity $A \wedge B_0$, multiplier bit B_0 controls whether the contents of register A or a zero is added to the partial product. Also, study the way $(A \wedge B_0)$, E, and R are combined to produce the noncarry sum E and the carry information R.

Multibit Processing

There are actually several methods that fall under the classification of multibit processing; three of the most popular will be considered here. Instead of the usual test, add, and shift cycle for each multiplier bit, considerable time can be saved if advantage is taken of certain simplifying multiplier bit patterns.

Shifting over zeros. Probably the most obvious technique is to keep checking the remaining portion of the multiplier for zero as the multiplication process continues. When this condition is detected, no further additions to the partial product will be required, and the final result can be quickly established. (In practice, it often happens that the multiplier is a small positive number so that the left half, or more, of its bits are zeros and a considerable savings is achieved.)

TABLE 10.2 CARRY-SAVE MULTIPLIER EXAMPLE
($1011 \times 1111 = 11 \times 15 = 10100101 = 165_{10}$)

At End of State, T	I	A	E	B	R	CS	CP	Calculations* (for boxed values of E and R)
1	1	1 1 1 1	0 0 0 0	1 0 1 1	0 0 0 0	1	0	
2	2	1 1 1 1	1 1 1 1	1 0 1 1	0 0 0 0	1	0	$A \wedge B_0$ = 1 1 1 1 E = 0 0 0 0 R = 0 0 0 0 E = 1 1 1 1 R = 0 0 0 0
3	2	1 1 1 1	0 1 1 1	1 1 0 1	0 0 0 0	1	0	$A \wedge B_0$ = 1 1 1 1 E = 0 1 1 1 R = 0 0 0 0 E = 1 0 0 0 R = 0 1 1 1
2	3	1 1 1 1	1 0 0 0	1 1 0 1	0 1 1 1	1	0	
3	3	1 1 1 1	0 1 0 0	0 1 1 0	0 1 1 1	1	0	$A \wedge B_0$ = 0 0 0 0 E = 0 1 0 0 R = 0 1 1 1 E = 0 0 1 1 R = 0 1 0 0
2	4	1 1 1 1	0 0 1 1	0 1 1 0	0 1 0 0	1	0	
3	4	1 1 1 1	0 0 0 1	1 0 1 1	0 1 0 0	1	0	$A \wedge B_0$ = 1 1 1 1 E = 0 0 0 1 R = 0 1 0 0 E = 1 0 1 0 R = 0 1 0 1
2	5	1 1 1 1	1 0 1 0	1 0 1 1	0 1 0 1	1	0	
3	5	1 1 1 1	0 1 0 1	0 1 0 1	0 1 0 1	0	1	E = 0 1 0 1 R = 0 1 0 1
4	5	1 1 1 1	1 0 1 0	0 1 0 1	0 1 0 1	0	1	E = 1 0 1 0 (carries propagate)
5 (HALT)	5	1 1 1 1	1 0 1 0	0 1 0 1	0 1 0 1	0	1	

Final result = 165_{10}, as expected

*The numbers above the horizontal line are input values from the previous row and those below the line are results.

For the elementary multiplication process in Figure 10.4, a particularly simple case occurs, and once the multiplier zero is detected, the final result is immediately available in register C. Figure 10.16 shows the minor modifications needed in the RTL program of Figure 10.4(a) to convert it to an elementary multibit processor. (Note that here we are using our standard notation with B_0 as the LSB.)

$T = 1$ $C \leftarrow 0; I \leftarrow 1; A \leftarrow$ (Multiplicand); $B \leftarrow$ (Multiplier);
$T = 2$ IF $(B = 0)$ HALT; IF $(B_0 = 1)$ $C \leftarrow A + C$; IF $(I = 6)$ HALT;
$T = 3$ $A \leftarrow$ slA; $B \leftarrow$ srB; $I \leftarrow I + 1$; $T \leftarrow 2$;

Figure 10.16 Elementary Multibit Processor.

On the other hand, the multiplication process in Figure 10.5 is a somewhat more complicated case because registers B and C must still be shifted as many times as there are bits in the multiplier in order to get the result into the proper position. Also, since register B contains bits from the result as well as from the multiplier, it is somewhat more difficult to test the remaining multiplier bits. (A multibit processing version of Figure 10.5 is left as an exercise.)

Sec. 10.3   Fast Multiplication

Of course, it is not necessary that all the remaining multiplier bits be zero before the above technique can be employed; as long as a group of zeros occurs in a string, a savings can be achieved. The method of simultaneously examining two or more bits and, if they are zero, performing a multiple shift over them is known as *shifting over zeros*. (A similar method for simultaneously processing groups of ones is called *shifting over ones*.)

Consider a multiplication unit similar to one in Figure 10.5(b) but with n-bit registers, where n is of the magnitude employed in large-scale computers—say between 30 and 50. We will evaluate a shifting-over-zeros scheme in which the next four multiplier bits, B_0 through B_3, are examined, and if they are all zero, the registers are shifted by four bits.*

On the other hand, if the four bits are not zero, or if there are less than four multiplier bits remaining, the usual single-bit processing is initiated. (A practical multiplication system would probably employ a more extensive version of this method by including a shifting-over-ones scheme and perhaps also simultaneous examination of other groups of multiplier bits, with the largest group satisfied being the one that determines the actual shift.) Since one cannot afford the circuitry to mechanize shifts of all possible magnitudes, a shift of four is chosen in this example as being sufficiently probable while saving a considerable amount of multiply time. Figure 10.17 illustrates the RTL program for the shift-of-four algorithm. To account for the adder operating time, an extra state, $T = 5$, is introduced to provide a delay when the multiplicand is added to the partial product, but not when shifting over one or more zeros. Note that C^* is the raw output of the adder, which is developed during the delay and is transferred to register C at $T = 5$.

$T = 1$ $C \leftarrow 0$; $I \leftarrow 1$; $A \leftarrow$ (Multiplicand); $B \leftarrow$ (Multiplier);
$T = 2$ IF $(B_{0-3} = 0$ AND $I \leq n - 3)$ $C,B \leftarrow$ sr3C,B, $I \leftarrow I + 3$;
         IF $(B_0 = 1)$ $C^* \leftarrow A + C$, $T \leftarrow 5$;
$T = 3$ $C,B \leftarrow$ srC,B; $I \leftarrow I + 1$; IF $(I \neq n)$ $T \leftarrow 2$;
$T = 4$ HALT;
$T = 5$ $C \leftarrow C^*$; $T \leftarrow 3$;: Delay for adder to reach completion;

Figure 10.17 Shift-of-Four Algorithm for Positive Numbers.

It might at first appear that there is a contradiction at $T = 3$, since the variable I is simultaneously incremented and checked with an IF statement. However, if I is implemented by means of master-slave flip-flops, both actions can take place without mutual interference. Furthermore, the close location of these operations is dictated by the requirement for a fast and efficient algorithm. When a string of four zeros is found by the first IF statement in $T = 2$, only three shifts are produced because the fourth shift is accomplished at $T = 3$, where I is checked for multiplication complete. If the number 2 is the multiplier in a 12-bit machine, the arrows and brackets in the following diagram indicate how the shifting-over-zeros scheme of Figure 10.17 groups the individual bits of the multiplier and what the processing sequence is:

---

*Multibit shift registers can be constructed in a manner similar to the one-bit versions discussed in Chapter 5.

Multiplicand multiples. A slightly more complicated scheme called *multiplicand multiples* employs precalculated values of the multiplicand which are gated into the adder when the proper pattern of multiplier bits occurs. Since it is expensive to store many precalculated values, advantage can be taken of the fact that a multiple that is a power of 2 greater than one already stored can be produced simply by shifting the original number to the left the proper number of places as it enters the adder. This method can be employed when two, three, or more bits are processed, but for purposes of our discussion the three-bit version, shown in Table 10.3, will be considered. For three bits, there are eight cases, and in all but two of them the full three bits are processed simultaneously. Since 101 and 111 would require storage of multiples not used in any other case, only two of the bits are processed, B_0 and B_1, while B_2 is left for the next group. (The fact that only two bits are processed is indicated by a double rather than a triple shift in column 4.) With the above simplification, column 2 shows that only one special multiple need be calculated; this is designated by the symbol T, since it is three times the contents of the register A.

TABLE 10.3 MULTIPLICATION BY TRIPLE BIT PROCESSING*

Multiplier Bit Pattern $(B_2B_1B_0)$	Adder Input	Shift into Adder (Bits Left)	Shift of B and C (Bits Right)
000	0	0	3
001	A	0	3
010	A	1	3
011	T	0	3
100	A	2	3
101	A	0	2
110	T	1	3
111	T	0	2

*Where register A contains the multiplicand, register T contains 3 times A, and 0 represents a zero input.

Consider the bit pattern 100. The table indicates that it can be satisfied by an input from register A, which has been shifted left two bits before entering the adder, in order to multiply by 4, and that once the partial product has reached register C, registers B and C are shifted three places to the right. In the same way, verify the other lines in the table.

Observe that the shift-left operations of column 3 require versions of the adder and register C that are two bits wider than normal.

The Booth algorithm. A very efficient multiplication algorithm for all 2's complement numbers was conceived by A. D. Booth.[2] His scheme simultaneously tests two bits of the multiplier but shifts the partial product only one place at a time; it requires both an add and a subtract capability. The algorithm is based on the fact that a multiplier consisting of n consecutive 1's can be implemented by means of a sequence of n additions and shifts; or since n 1's equals $(2^n - 1)$,* first a subtraction, then n shifts, and finally an addition can be performed. This means that if the multiplier, B, is a string of 1's, the following sequence will be produced:

1. The right end of the string is detected—even a single 1 satisfies the criterion for a string.
2. The multiplicand, A, is subtracted from the partial product, E. Initially assume $E = 0$; then $E \leftarrow -A$.
3. Registers B and E are shifted right and B_0 is examined each time. (Since the string contains n 1's, B and E are shifted n times.)
4. The left end of the string is detected; the contents of register A are added to the shifted E register, generating the term $2^n A$; and the final result is $E \leftarrow 2^n A - A = A(2^n - 1)$. (It is interesting to note that if the multiplier had been all 1's, which is (-1) in 2's complement, register A would *not* have been added, and the final result would be $E = -A$. However, this is exactly what is expected for the product $-1 \times A$.)

Through superposition, the above approach also applies to multipliers consisting of groups of 1's separated by groups of 0's, eg., 11100 would be represented by one subtraction and one addition as follows: $-2^2 + 2^5$.

In order to facilitate implementation of this algorithm, an extra bit, B_{-1}, is added to the right end of register B. Thus, $B_0 B_{-1} = 10$ indicates that the right end of a string of 1's has been detected; $B_0 B_{-1} = 01$ indicates that the left end of a string has been detected; $B_0 B_{-1} = 11$ indicates the continuation of a string; and $B_0 B_{-1} = 00$ indicates the absence of a string, in this region. From the above concepts, a general proof of the Booth algorithm may be derived;[2] a concise summary of the algorithm, however, appears in Table 10.4.

TABLE 10.4 STATEMENT OF THE BOOTH ALGORITHM

Case	Condition	Required Action
1	$B_0 B_{-1} = 00$ or $B_0 B_{-1} = 11$	Shift B and E right
2	$B_0 B_{-1} = 10$	Subtract A from E; then shift B and E right
3	$B_0 B_{-1} = 01$	Add A to E; then shift B and E right

The basic block diagram, Figure 10.18, for this multiplier is quite similar to previous ones except that circuitry is required to implement the logic of Table 10.4 and a dual add/subtract unit is needed to operate between the partial product and multiplicand registers.

Table 10.5 demonstrates the step-by-step operation of the algorithm for the

*For example: 1111 has $n = 4$; therefore, $(2^4 - 1) = 16 - 1 = 15$ as expected.

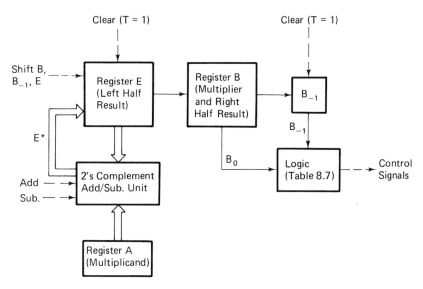

Figure 10.18 Booth Multiplier.

TABLE 10.5 BOOTH MULTIPLICATION EXAMPLE: 9 × 5 = 45*

T	I	Case	Register A Multiplicand	Register E	Register B	B_{-1}	Calculations
1	0	—	0|0 1 0 1	0|0 0 0 0	0|1 0 0 1	0	0|0 0 0 0
2	0	2	0|0 1 0 1	1|1 0 1 1	0 1 0 0 1	0	1|1 0 1 1
5	0	—	0|0 1 0 1	1|1 0 1 1	0 1 0 0 1	0	1|1 0 1 1
3	1	—	0|0 1 0 1	1|1 1 0 1	1 0 1 0 0	1	1|1 1 0 1
2	1	3	0|0 1 0 1	0|0 0 1 0	1 0 1 0 0	1	0|0 1 0 1
5	1	—	0|0 1 0 1	0|0 0 1 0	1 0 1 0 0	1	0|0 0 1 0
3	2	—	0|0 1 0 1	0|0 0 0 1	0 1 0 1 0	0	
2	2	1	0|0 1 0 1	0|0 0 0 1	0 1 0 1 0	0	
3	3	—	0|0 1 0 1	0|0 0 0 0	1 0 1 0 1	0	0|0 0 0 0
2	3	2	0|0 1 0 1	1|1 0 1 1	1 0 1 0 1	0	1|1 0 1 1
5	3	—	0|0 1 0 1	1|1 0 1 1	1 0 1 0 1	0	1|1 0 1 1
3	4	—	0|0 1 0 1	1|1 1 0 1	1 1 0 1 0	1	1|1 1 0 1
2	4	3	0|0 1 0 1	0|0 0 1 0	1 1 0 1 0	1	0|0 1 0 1
5	4	—	0|0 1 0 1	0|0 0 1 0	1 1 0 1 0	1	0|0 0 1 0
4	4	HALT	0|0 1 0 1	0|0 0 1 0	1 1 0 1 0	1	

Result = 55_8 = 45_{10} as expected.

*Notes:
1. $-A = -(0|0101) = 1|1\ 011$
2. Once the algorithm starts, the sign of register B is meaningless. Thus, the ⌋ symbol is dropped; but an underline is employed to mark the location of the original MSB as it is shifted.
3. In accomplishing the right shift operation, at each T = 3, again observe that the scale, instead of the shift, operation is employed to preserve the sign bit.
4. The column labeled "Case" refers to the information in Table 10.4.

product $9 \times 5 = 45$. The reader should examine the table in detail while following the RTL program of Figure 10.19. (The first three columns of the table provide a convenient set of references to the program and the algorithm statement.) Again, it should be noted that the RTL inserts a delay between bit testing and register shifting, *only* in those situations (cases 2 and 3) where an arithmetic operation requires time for the bits to ripple down. Thus, considerable time will be saved for multipliers containing large strings of consecutive 1's or 0's. Each line of the table represents conditions after completion of the action indicated by the RTL state in column 1.

$T = 1$ $E \leftarrow 0$; $B_{-1} \leftarrow 0$; $I \leftarrow 0$; $B \leftarrow$ Multiplier; $A \leftarrow$ Multiplicand;
$T = 2$ IF ((B_0, B_{-1} = 0,0 OR 1,1) AND $I = n$) $T \leftarrow 4$;
         IF (B_0, B_{-1} = 1,0) $E^* \leftarrow E - A$, $T \leftarrow 5$;
         IF (B_0, B_{-1} = 0,1) $E^* \leftarrow E + A$, $T \leftarrow 5$;
$T = 3$ $E,B,B_{-1} \leftarrow scE,B,B_{-1}$; $I \leftarrow I + 1$; $T \leftarrow 2$;
$T = 4$ HALT;
$T = 5$ IF ($I = n$) $T \leftarrow 4$ ELSE $T \leftarrow 3$; $E \leftarrow E^*$;:   Delay for add/subtract circuit to operate;

Figure 10.19 RTL Program for the Booth Algorithm.

The final result contains eight bits, not counting sign, since two 4-bit numbers were multiplied. It should be noted from column 1 that five B_0B_{-1} test and add/subtract operations were performed ($T = 2$), but only four right-shifts ($T = 3$). The fact that the number of shifts equals the number of bits in each number is not surprising since this is the only way for the LSB of the first partial product to reach the LSB of the 2n-bit result. On the other hand, the multiplier contains five bits, including sign, and if only four (B_0B_{-1} test and add/subtract) operations were performed (none after the last shift), the sign bit of the multiplier would not have a role in determining the result.* Thus it is apparent that, in general, two n-bit numbers required $n + 1$ test operations, but only n right-shifts.

‡10.4 DIVISION

Our discussion of addition and multiplication has already provided an adequate introduction to hardware design from algorithms. Moreover, many of the multiplication techniques are also applicable to division. Thus, this section will be of moderate length and will only consider positive numbers; however, after completing the material, the reader should have a good basic understanding of division methods.

The three approaches to division to be considered differ mainly in the way it is decided whether or not the divisor goes into the partial remainder:

1. *Comparison.* In the comparison method, the divisor is compared with the partial remainder one bit at a time, starting with the MSB, much as is done in manual arithmetic.

---

*If only four operations were performed for the table's example, the last (case 3) addition would not appear. This results in $111011101 = -43_8 = -35_{10}$, which, of course, is incorrect.

2. *Restoration.* The divisor is subtracted from the partial remainder to determine go/no-go, and if the result is negative the old remainder is restored before continuing.
3. *Nonrestoration.* When divisor subtraction produces a negative result, the old remainder is not restored; but an alternate quicker process is followed.

Comparison

The manual division below should be studied and note taken that it is basically a subtraction and shifting process; but a comparison is always made between the partial remainder and the divisor to insure that the new remainder will always be positive. For example, in the first step the comparison

$$
\begin{array}{r}
01011. \\
110\overline{\smash{)}1000010.} \\
\underline{110} \\
1001 \\
\underline{110} \\
110 \\
\underline{110} \\
000
\end{array}
$$

100/110 is made; since the first number is smaller than the second, a 0 is added to the quotient and the divisor is shifted. The new comparison is 1000/110; since the divisor goes into the partial remainder, a 1 is added to the right end of the quotient. Subtraction is performed to get a new partial remainder ($N \leftarrow N - D$), where N is the partial remainder and D is the divisor. Then the process repeats.

It is not difficult to implement the comparison method as the block diagram in Figure 10.20 shows. Contrary to the procedure for the manual method, here it is more convenient to shift the dividend to the left while the divisor remains fixed. In many other respects, the block diagram in this figure is similar to the previous block diagrams for multiplication except that subtraction is achieved by adding the 2's complement of D, a comparison between D and N is required, and as already mentioned, the partial result (along with the remainder) is shifted left.

Of course, the comparison circuit is at the heart of this type of divider. Let us demonstrate the design of a comparator by employing the heuristic method for a four-bit circuit. The requirements segmentation and information analysis heuristics can be employed to achieve a module for each bit, and then combine the modules with logic which may be iterated to produce a circuit for as many bits as desired.

Consider the manual comparison process during division; the leftmost bits for the dividend and its nonzero divisor are examined. If $N_{n-1} = 1$ and $D_{n-1} = 0$, we have a match, M, and immediately know that the divisor will go into the dividend, completing the comparison process. On the other hand, if $N_{n-1} = 1$ and $D_{n-1} = 1$, or if $N_{n-1} = 0$ and $D_{n-1} = 0$, the situation is undecided, U, indicating that the next place

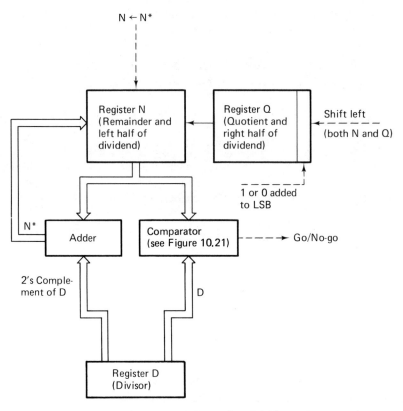

Figure 10.20 Comparison Divider.

TABLE 10.6 LOGIC CONDITIONS FOR A COMPARATOR MODULE

N_i	D_i	M_i	U_i
0	0	0	1
0	1	0	0
1	0	1	0
1	1	0	1

on the right must be compared. These basic comparator module conditions are summarized in Table 10.6 for an arbitrary bit i.

Once the M_i and U outputs are available for each of the n bits, all that remains is to provide the logic between them for developing the final go/no-go signal. The go-condition, as bits further to the right are considered, depends on all previous positions having developed an undecided signal and the current position indicating a match. (Of course, when the above go condition exists, it is immaterial what the values for M_i and U_i are for all positions further to the right.) The logic just described, which is actually for the relation $N \geq D$, is realized by the circuit shown in Figure 10.21.

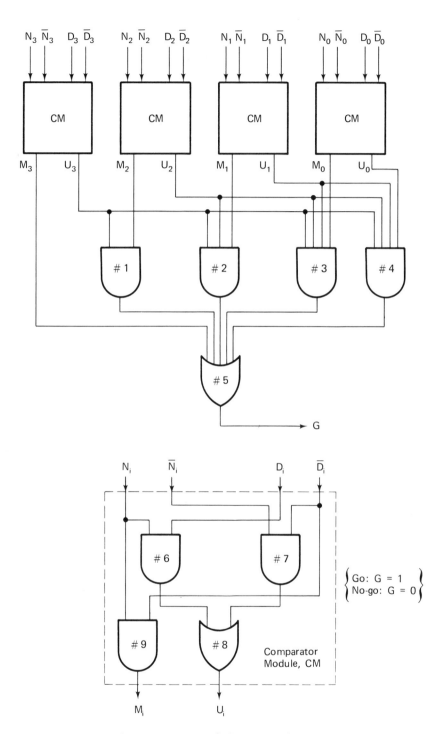

Figure 10.21 Four-Bit Comparator ($N \geq D$).

The following points concerning the figure should be studied:

1. Iteration of the circuit for as many bits as desired is a straightforward process. For each additional bit, another comparator module and another AND gate must be added.
2. In completing the heuristic design process (the step of checking for circuit simplification), it will be observed that the AND gate producing M_2 can be combined with AND gate 1, resulting in the saving of one gate and one input. Moreover, the same savings can be achieved for M_1 and M_0. (The construction of a simplified version of Figure 10.20 is left as an exercise.)
3. The output of gate 4 indicates whether the contents of the two registers are equal.
4. Some typical register contents and the resulting M_i, U_i, and go/no-go signals are:

$$
\begin{array}{lll}
N = 1011 & N = 1000 & N = 0101 \\
\underline{D = 1000} & \underline{D = 1000} & \underline{D = 1000} \\
M = 0011 & M = 0000 & M = 0101 \\
U = 1100 & U = 1111 & U = 0010 \\
G = 1, \text{from} & G = 1, \text{from} & G = 0, N < D \\
\quad \text{gate 2} & \quad \text{gate 4} & \\
& \quad \text{(equality)} &
\end{array}
$$

Restoring Division

The basic system for *restoring division* is presented in Figure 10.22. The process is similar to comparison division except that the go/no-go decision is based on the sign of register N after the divisor is subtracted. Thus, a negative sign indicates that D did not go into N, and the register N must be restored to its original value, through the step $N \leftarrow N + D$, before the zero can be entered in the quotient and the left-shift produced. The figure reveals that addition is accomplished by entering D directly into the adder, and subtraction is accomplished by entering the 2's complement of D. The 2's complement is produced by the same method employed previously: $\bar{D}$ is entered into one set of adder inputs (N is entered into the other set), and a 1 is entered into the carry input. Note how the sign bit of N is used to make the go/no-go decision and how each new bit of the quotient is entered into the LSB of Q. The hardware shown is for integer numbers as is the basic division instruction for such machines as the IBM System 360/370 and the PDP-11. Thus, it is the programmer's responsibility to keep track of the binary point if the numbers contain a fractional part, and to perform any scaling which is desirable. It can be shown that the relationship between the bits in the various registers in the division process is as follows:

$$(Q) = (NQ) - (D) + 1 \tag{10.23}$$

where (Q) is the number of bits in the quotient register, (NQ)* is the bits in the

---

*NQ is employed to represent the dividend because the bits are divided between the N and Q registers.

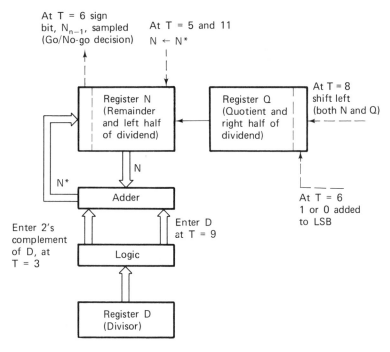

Figure 10.22 Restoring Divider. (The T values shown are discussed later.)

dividend, and (D) is the bits in the divisor. Thus, with a $2m$-bit dividend and an m-bit divisor, the quotient requires

$$(Q) = 2m - m + 1 = m + 1 \text{ bits} \tag{10.24}$$

The details of the restoration algorithm are shown in the program of Figure 10.23. It is assumed that the registers N, D, and Q each contain n bits (bits 0 to $n - 1$); but the fact that N can go negative in the subtraction process requires that the MSB bit be reserved for sign, leaving $n - 1$ bits for the magnitude. When $m = n - 1$ is substituted into Equation (10.24),

$T = 1$ $I \leftarrow 0; E \leftarrow 0; N \leftarrow$ (Left half of dividend); $Q \leftarrow$ (Right half of dividend);

$T = 2$ IF $(N_{n-1}, N_{n-2} \neq 0,0$ OR $D_{n-1} \neq 0)$ $E \leftarrow 1$, $T \leftarrow 12$ ELSE $N,Q \leftarrow \text{sl}N,Q$;: Test for legal dividend and divisor;

$T = 3$ $N^* \leftarrow N + \overline{D} + 1$;: Subtract using 2's complement addition;

$T = 4$: Delay for adder bits to ripple down;

$T = 5$ $N \leftarrow N^*$;

$T = 6$ IF $(N_{n-1} = 1)$ $T \leftarrow 9$, $Q_0 \leftarrow 0$ ELSE $Q_0 \leftarrow 1$; $I \leftarrow I + 1$;

$T = 7$ IF $(I = n)$ $T \leftarrow 12$;

$T = 8$ $N,Q \leftarrow \text{sl}N,Q$; $T \leftarrow 3$;

$T = 9$ $N^* \leftarrow N + D$; IF $(I = n)$ $T \leftarrow 12$;

$T = 10$: Delay for adder bits to ripple down;

$T = 11$ $N \leftarrow N^*$; $T \leftarrow 8$;

$T = 12$ HALT;

Figure 10.23 RTL Program for Restoring Division.

$$(Q) = (n-1) + 1 = n$$

which means that the magnitude bits occupy the entire Q registers. Now from Equation (10.23),

$$(NQ) = (Q) + (D) - 1 = n + (n-1) - 1 = 2n - 2$$

and since we assume that the dividend is right-justified (the number is as far right in the register as possible), the number of dividend bits (B) originally in the N register is

$$(B) = (NQ) - (Q) = (2n - 2) - n = n - 2$$

Therefore, there is an extra bit between the MSB and sign bit of the dividend, which means that the divisor and dividend are misaligned by one bit. For these reasons, at $T = 2$ in the program, N and Q are shifted left, which also conveniently makes room for the first bit of the quotient.

Observe that the IF statement at $T = 2$ checks that the dividend and divisor are both positive—sign bit equals 0. In addition, it checks that the extra bit which will become the sign bit, after shifting, is also 0. When these conditions are violated, the error bit, E, is set to 1. (The error bit is not shown in Figure 10.22, but in most computers there are several types of error flags (bits) which are automatically set by the hardware and can be read with a special instruction.)

In the sequence of instructions, $T = 3, 4$, and 5, the transfer to the adder output, N^*, at $T = 3$ is merely a way of indicating that the 2's complement of D is switched to the adder input and the sum will start to be formed. After the bits ripple down at $T = 4$, the actual transfer into N occurs at $T = 5$. (A similar sequence occurs at $T = 9$, 10, and 11.) Carefully study the individual steps of Figure 10.23—the comments within the program should be helpful in following the algorithm.

As a further guide to understanding the restoration algorithm, Table 10.7 has been prepared. Again, the first two columns provide a convenient reference to the RTL program, and these two should be studied together. Note that:

1. Underscores are used to show the progress of the MSB of the quotient as it is shifted along.
2. An X is used as a temporary filler in bit Q_0 after shifting and before the next bit of the quotient is ready. (The hardware would actually place a zero there after shifting, but the X is a useful device for explanation purposes.)
3. The E bit is not shown explicitly; but it would be zero in this example since the given input data would not generate an error.
4. In order to keep the table as concise as possible, states 3–5 (also 9–11) are shown on a single line with the entries representing results of the action initiated by $T = 5$ ($T = 11$).
5. The remainder always appears in register N, while the quotient always appears in register Q.

Nonrestoring Division

The method of *nonrestoring division* is very similar to restoring division. Except for the states, T, it has the same block diagram as Figure 10.22; but since negative partial remainders are not explicitly restored, the system runs somewhat faster.

TABLE 10.7 RESTORING DIVISION EXAMPLE: 111100/110 = 60/6 = 10_{10}, $n = 4$ [2'S COMPLEMENT OF $D = -(0110) = 1001 + 1 = 1010$]

T	I	Register D (divisor)	Register N (Remainder and left half of dividend)	Register Q (Quotient and right half of dividend)	Calculations
1	0	0|1 1 0	0|0 1 1	1 1 0 0	
2	0	0|1 1 0	0|1 1 1	1 0 0 X	0|1 1 1
3-5	0	0|1 1 0	0|0 0 1	1 0 0 X	1|0 1 0 / 0|0 0 1
6	1	0|1 1 0	0|0 0 1	1 0 0 1	
7	1	0|1 1 0	0|0 0 1	1 0 0 1	
8	1	0|1 1 0	0|0 1 1	0 0 1 X	0|0 1 1
3-5	1	0|1 1 0	1|1 0 1	0 0 1 X	1|0 1 0 / 1|1 0 1
6	2	0|1 1 0	1|1 0 1	0 0 1 0	1|1 0 1
9-11	2	0|1 1 0	0|0 1 1	0 0 1 0	1|1 1 0 / 0|0 1 1
8	2	0|1 1 0	0|1 1 0	0 1 0 X	0|1 1 0
3-5	2	0|1 1 0	0|0 0 0	0 1 0 X	1|0 1 0 / 0|0 0 0
6	3	0|1 1 0	0|0 0 0	0 1 0 1	
7	3	0|1 1 0	0|0 0 0	0 1 0 1	
8	3	0|1 1 0	0|0 0 0	1 0 1 X	0|0 0 0
3-5	3	0|1 1 0	1|0 1 0	1 0 1 X	1|0 1 0 / 1|0 1 0
6	4	0|1 1 0	1|0 1 0	1 0 1 0	1|0 1 0
9	4	0|1 1 0	0|0 0 0	1 0 1 0	0|1 1 0 / 0|0 0 0
12	4	0|1 1 0	0|0 0 0	1 0 1 0	
(HALT)			↑	↑	
			Remainder = 0, as expected.	Result = $12_8 = 10_{10}$, as expected.	

The following is a summary of basic steps in the nonrestoring process:

1. Subtract the divisor from the partial remainder.
2. If the result is positive, make $Q_0 = 1$, shift N and Q left, and return to step 1.
3. If the result is negative, instead of restoring, make $Q_0 = 0$, shift N and Q left, and add D.
4. If the partial remainder continues to be negative, repeat step 3; otherwise, make $Q_0 = 1$, shift N and Q left, and return to step 1.

Consider the mathematical rationale for nonrestoring division. Let NQ represent the $2m$-magnitude bits of the dividend, i.e., the m bits of register N, not including the sign bit, and the m bits of register Q. (Recall that since there are $m + 1$ bits in the quotient, Q_0 is not included in the NQ bits.) When we subtract (add the 2's complement), the divisor from the remainder the mathematical operation actually is $NQ - 2^m D$, where the 2^m factor accounts for the fact that the LSB of the D register is really shifted m bits to the left of the LSB of NQ.

Assume that the subtraction leads to a negative result. Then, according to the above steps for nonrestoring division, the new remainder, R, is formed by shifting the old remainder left and adding D:

$$\begin{aligned} R &= 2(NQ - 2^m D) + 2^m D \\ &= 2NQ - 2^m \cdot 2D + 2^m D \\ &= 2NQ - 2^m(2D - D) = 2NQ - 2^m D \end{aligned} \qquad (10.25)$$

Therefore, shifting and adding D has had the effect of shifting NQ one bit to the left and subtracting D, which is equivalent to restoring NQ, shifting, and subtracting D. Of course, if the remainder of Equation (10.25) is still negative, the process may be repeated by shifting and adding D as many times as necessary. It can easily be seen that if k shifts and additions are employed, the result is:

$$R = 2^k NQ - 2^m D \qquad (10.26)$$

which is exactly the desired result.

Now, when the divisor is subtracted from the remainder and a positive result occurs, NQ is shifted to the left and the effect is the same as starting a new problem with a dividend of $NQ - 2^m D$.

Since nonrestoring division is so similar to restoring division, the RTL program, examples, and other details are included as part of the exercises at the end of this chapter.

‡10.5 OVERFLOW

One potential problem with which the programmer must be concerned is that of register overflow during arithmetic operations. Since the possibilities are so varied, the correction of an overflow condition is usually the responsibility of the software. The detection of overflow conditions can, however, be very efficiently made by the ALU hardware.

The term *overflow* can be somewhat misleading; but in the computer field it is commonly employed to represent the condition where the result of some arithmetic operation is too large for its register. This most frequently occurs when large numbers with like signs are added and when large numbers of opposite sign are subtracted.

Table 10.8 shows examples for each of the eight possible cases that exist for the three sign bits involved in the addition process. Overflow is produced only in cases 2 and 7, which are large-magnitude numbers of the same sign. In these two cases, note that the result is too large to be contained in the three magnitude bits, and this error causes the sign bit to be changed. Overflow, of course, is not correlated with the carry-out of the sign bit alone in that cases 3, 5, 7, and 8 have carry-outs, but all except case 7 yield the correct result. On the other hand, the overflow condition may be detected either from the values of the three sign bits involved or from a function of the

TABLE 10.8 THE EIGHT POSSIBLE COMBINATIONS OF ADDER DATA TYPES (2'S COMPLEMENT NUMBERS)

Case 1	Case 2	Case 3	Case 4
$\begin{bmatrix}\text{Input: Small}\\(+)\text{ numbers.}\end{bmatrix}$	$\begin{bmatrix}\text{Input: large}\\(+)\text{ numbers.}\end{bmatrix}$	$\begin{bmatrix}\text{Input: large}\\(+)\text{ and large}\\(-)\text{ numbers.}\end{bmatrix}$	$\begin{bmatrix}\text{Input: small}\\(+)\text{ and large}\\(-)\text{ numbers.}\end{bmatrix}$
$x_i = 0\|011 = 3$	$0\|110 = 6$	$0\|110 = 6$	$0\|011 = 3$
$y_i = 0\|011 = +3$	$0\|110 = +6$	$1\|010 = +(-6)$	$1\|010 = +(-6)$
$S_i = 0\|110 = 6$	$1\|100 \neq 12$	①$0\|000 = 0$	$1\|101 = -3$
$C_i = 0011$	0110	1110	0010
	Overflow occurs		

Case 5	Case 6	Case 7	Case 8
$\begin{bmatrix}\text{Input: small}\\(-)\text{ and large}\\(+)\text{ numbers.}\end{bmatrix}$	$\begin{bmatrix}\text{Input: large}\\(-)\text{ and small}\\(+)\text{ numbers.}\end{bmatrix}$	$\begin{bmatrix}\text{Input: large}\\(-)\text{ numbers.}\end{bmatrix}$	$\begin{bmatrix}\text{Input: small}\\(-)\text{ numbers.}\end{bmatrix}$
$x_i = 1\|101 = -3$	$1\|010 = -6$	$1\|010 = -6$	$1\|101 = -3$
$y_i = 0\|110 = +6$	$0\|011 = +3$	$1\|010 = +(-6)$	$1\|101 = +(-3)$
$S_i = $ ①$0\|011 = 3$	$1\|101 = -3$	①$0\|100 \neq -12$	①$1\|010 = -6$
$C_i = 1100$	0010	1010	1101
		Overflow occurs	

carry-in and carry-out of the sum sign bit. (The adder carry-out bits, C_i, are listed under the sum bits in Table 10.8.)

Table 10.9 is the truth table which shows all possible combinations of the three sign bits correlated with the eight cases presented in Table 10.8. Thus, the Boolean equation to detect the overflow condition, Ov, is given by:

$$Ov = \bar{x}_3\bar{y}_3 s_3 + x_3 y_3 \bar{s}_3 \tag{10.27a}$$

Detection of overflow from the carry-in and carry-out states of the sum sign bit is represented in the truth table of Table 10.10 and the resulting equation is:

$$Ov = \bar{c}_3 c_2 + c_3 \bar{c}_2 \tag{10.27b}$$

TABLE 10.9 OVERFLOW CONDITIONS BASED ON THE VALUE OF SIGN BITS

Case	x_3	y_3	s_3	Overflow
1	0	0	0	0
2	0	0	1	1
3	0	1	0	0
4	0	1	1	0
5	1	0	0	0
6	1	0	1	0
7	1	1	0	1
8	1	1	1	0

TABLE 10.10 OVERFLOW CONDITIONS BASED ON CARRY-IN AND CARRY-OUT OF THE SUM SIGN BIT

Case	C_3	C_2	Overflow
1, 4, 6	0	0	0
2	0	1	1
7	1	0	1
3, 5, 8	1	1	0

The selection of Equation (10.27a) or (10.27b) as the equation to be mechanized depends on the detailed nature of the other circuits employed in the ALU. In either case, however, the application is straightforward.

10.6 BIBLIOGRAPHY

1. Atkins, D. E., and H. L. Garner (Guest eds.), "Special Section on Computer Arithmetic." *IEEE Transactions on Computers*, Vol. C-22, 1973, pp. 549–610.
2. Booth, A. D., "A Signed Binary Multiplication Technique." *Quarterly Journal of Mechanics and Applied Mathematics*, Vol. 4, Pt. 2, 1951, pp. 236–40.
3. Habibi, A., and P. A. Wintz, "Fast Multipliers." *IEEE Transactions on Computers*, Vol. C-19, 1970, pp. 153–57.
4. Hill, F. J., and G. R. Peterson, *Digital Systems: Hardware Organization and Design* (2nd ed.). New York: John Wiley & Sons, Inc., 1978.
5. Hwang, K., *Computer Arithmetic: Principles, Architecture, and Design.* New York: John Wiley & Sons, Inc., 1979.
6. Lewin, D., *Theory and Design of Digital Computers.* New York: John Wiley & Sons, Inc., 1972.
7. Oberman, R. M. M., *Digital Circuits for Binary Arithmetic.* New York: John Wiley & Sons, Inc., 1979.
8. Rhyne, V. T., *Fundamentals of Digital Systems Design.* Englewood Cliffs, N.J.: Prentice-Hall, Inc., 1973.
9. Shively, R. R. (guest ed.), "Special Issue on Computer Arithmetic." *IEEE Transactions on Computers*, Vol. C-19, 1970, pp. 679–757.

10.7 EXERCISES

10.1 Design a modified full-adder stage similar to that in Figure 10.1 using the fewest possible NAND gates. (Assume that inverted inputs are *not* available.)

10.2 *Derive* the simplified equations for an half-subtractor.

10.3 Employ the usual add and shift process indicated by Equation (10.18) and a *correction term* to develop an algorithm for 2's complement multiplication of $(-b) \times (a)$. Show that this gives the proper results for $(-5) \times (3)$.

10.4 Prepare a table similar to Table 10.2 for the carry-save multiplication of 3×7.

10.5 *Prove* the rule, given above Equation (10.17), for multiplication of a 2's complement number by 2.

10.6 Employing the Booth algorithm, complete a table showing the processing steps for the example $(-3) \times (2)$. (Use three bits, including signs for the E and B registers.)

10.7 Employing the Booth algorithm, complete a table showing the processing steps for the example $(-3) \times (5)$.

10.8 Repeat Exercise 10.7 for the example $(3) \times (5)$.

10.9 Design a comparator circuit that detects the condition $A < B$, for an arbitrary number of bits. Employ the comparator module of Figure 10.21.

10.10 Write an RTL program for comparator division.

10.11 Construct a table showing the processing steps of comparator division for the example $Q = 10000100/0110$. (The table should correlate with the RTL program of Exercise 10.10.)

10.12 Register A contains $\underline{1}|\,1101$.
  (a) What will the result be after the operation scA?
  (b) What will the result be after $sc2A$?
  (c) What will the result be after rlA?
  (d) Are the expected results obtained for division and multiplication? Explain. (Assume 1's complement numbers.)

10.13 Using restoring division, construct a table similar to Table 10.7 for $Q = 110111/101$.

10.14 Write the RTL program for nonrestoring division. Include a check for improper divisor and dividend.

10.15 Construct a table similar to Table 10.7 for nonrestoring division using $Q = 111100/110$.

10.16 Verify (prove) the sequential 2's complement method given above Equation (10.22).

Chapter 11

Computer Interface Design

As digital computers become increasingly popular, there is a growing demand for specialized interfaces which allow them to be *directly connected* to an ever-wider variety of peripherals. These external devices can be standard computer units such as disks, tape drivers, CRT displays, and printers; they can be various electrical, medical, or other instruments that employ the computer for *on-line* data processing; or they can be industrial equipments that employ the computer to supervise operations such as an automatic oil refinery or a computerized steel-rolling mill.

The process of selecting an interface for a particular application can be grouped into the following classes:

1. Purchase of a complete interface system from the computer manufacturer. (Even here there is often a significant engineering problem in selecting the proper system from the many that are available.)
2. Assembly of the interface from commercially available printed circuit cards.
3. "From-scratch" design and construction of the interface with individual integrated circuits.

Typical sources for these three levels of interface design are found in the bibliography.[3,5,6] Although further study and experience will be required for the engineering of complex interfaces,* it is our purpose, in this chapter, to employ the tools previously developed (particularly from Chapters 3, 6, and 8) to describe the basic principles of interface design and to prepare the reader for more advanced work.

*This is particularly true for item 3 in the above list.

11.1 DATA TRANSFER TECHNIQUES

There are several methods for transferring data between a computer and the outside world. This area has been one of intensive development, and there is every indication that this activity will continue into the foreseeable future. Fortunately, the subject consists of a few basic ideas, and when the reader masters these, he should be in a position to understand the complexities of both present and future systems.

Data transfer methods may be divided into the following three major areas:

1. Programmed transfers
2. Interrupts
3. Direct memory access

In the *programmed transfer* method, the peripheral is serviced at the convenience of the computer, i.e., whenever the I/O instructions happen to be reached in the *normal* program sequence. By contrast, the *interrupt* method places the data transfer at the convenience of the external device, i.e., service is initiated through what amounts to an I/O subroutine *as soon as the peripheral requests it* instead of when the computer gets around to it.

Direct memory access, DMA, is a very quick and efficient method for transferring large blocks of information such as when data is passed to or from a disk. It is similar to the interrupt method in that *service is initiated immediately*; but instead of having a moderate-size software subroutine supervise the transfer, it is done entirely by special hardware. Thus, the data are transferred much quicker at the expense of more complicated and costly hardware.

11.2 PROGRAMMED TRANSFERS

Let us now consider programmed data transfer from a peripheral into the LINC* accumulator. For I/O operations we divide the 12-bit instruction word into the three sections shown in Figure 11.1. As usual, the three most significant bits are reserved for the operation code, with 7XXX having been selected to represent the entire class of I/O instructions. The command section of the word indicates the type of I/O operation being performed, e.g., 4_8 signifies the data is to be read by the accumulator from the device selected. Finally, the least significant six bits provide coding capability for selecting one of $2^6 = 64$ peripheral devices. For example, the instruction word 7410

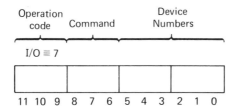

Figure 11.1 I/O Instruction Word Format.

*For tutorial reasons, we return to the elementary LINC Systems, but once the principles are learned, in the current chapter, they should be easy to apply to more complex modern computers.

indicates that information is to be transferred from the *data buffer*, DB, of device number 10 onto the data bus where it will be read by the accumulator. REad data from Device number n, 74 n, will be given the mnemonic RED n. (For those I/O instructions, to be described shortly, which do not require a device number, bits 0–5 will be held at 00_8. Here the machine code will be of the form 7c00, where c is one of eight possible commands represented by a single octal digit. Thus, there are available eight commands employing a device number; eight different commands without a device number; and $64 - 1 = 63$ possible devices.)

The logic for implementing the RED instruction is shown in Figure 11.2. There

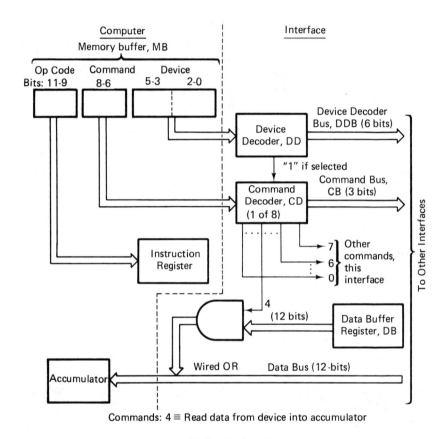

Commands: 4 ≡ Read data from device into accumulator

(a) Required Logic

$$T = 0 \quad DDB \leftarrow MB_{0-5}; CB \leftarrow MB_{6-8};$$
$$T = 1 \quad A \leftarrow DB;$$
$$T = 2 \quad S \leftarrow 0; T \leftarrow 0; DDB \leftarrow 0; CB \leftarrow 0;$$

(b) RTL Program

Figure 11.2 Implementation of the RED Instruction.

are three separate buses connecting the various interfaces to the computer. The six-bit word on the device decoder bus (through the device decoder, DD) controls which one of the 63 possible interfaces is to be connected to the computer through the command bus and the data bus. With the device decoder actuated, the command decoder, CD, is free to take the information from the command bus and, in this case, place the word contained in the data buffer register onto the common data bus so that registers from many other interfaces may be simultaneously attached to the bus. (Only a single register, however, is actually delivering information to the bus at one time.)

What type of RTL program is required by the RED instruction? Assume it is known that the device is ready to deliver the needed data word. Then all that the I/O instruction need do is place the word from the device onto the data bus and then actuate the transfer from the bus into the accumulator. Figure 11.2(b) shows the execution portion of the RTL program for accomplishing this. It should be understood that once the transfers are made from the MB to the decoders (DD and CD), the levels will be held until the end of the $T=2$ state. The RTL for the RED instruction in Figure 11.2(b) should be compared with that for the other instructions shown in Table 7.6.

The minor changes required in Figure 11.2 to implement the companion instruction WED n, WritE the information from the accumulator into Device n, is left as an exercise. (WED n will have the numerical code 73 n.)

Implementation of Flags and Branch Instructions

In describing the RED instruction, it was assumed that the peripheral was always ready to transmit the needed information. Frequently, this is not the case, and the computer must wait until the device is ready; otherwise, erroneous data will be received. A typical case where the data are not immediately available is just after a tape controller has been instructed to read from a remote section of tape. Here, the computer must wait until the tape drive has found the data before the ready signal can be given. Often device readiness is indicated by a flag flip-flop, which is gated onto the common flag but by a level from the command decoder. (Since wired-OR logic is employed where connection is made to the bus, only one flag line is required for all 63 interfaces.) Figure 11.3(a) shows the logic required to implement the flag instruction. The reader should carefully study the diagram noting such things as:

1. The method of setting and resetting the flag
2. The fact that command 5 controls both the flag's presence on the bus and its reset line
3. The way that flag signals enter the computer logic

Usually the flag concept is employed in conjunction with program branching; for this reason, we will define a new instruction called Skip On flag High (from device n), SOH n. Thus, the sequence of instructions required to make the computer wait until the flag of device 15 produces a ready signal is:

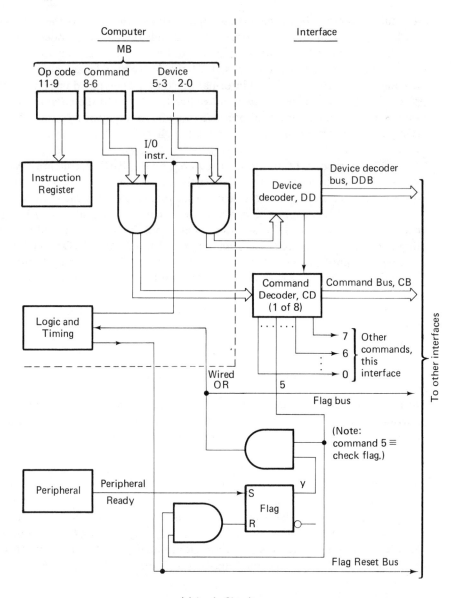

(a) Logic Circuit

T = 0 DDB ← MB_{0-5}; CB ← MB_{6-8};
T = 1 IF (Flag Bus = 1) PC ← PC + 1;: Flag high therefore skip;
T = 2 S ← 0; T ← 0; (Flag of selected device) ← 0;
       DDB ← 0; CB ← 0;
T = 5 M(MA) ← MB; PC ← 1; S ← 0; T ← 0;

(b) Execution Cycle for SOH Instruction

Figure 11.3 Implementation of the Flag Instruction.

SOH 15

JMP $p - 1$: Flag is low; therefore, go back and try again.

$\begin{Bmatrix} \text{Device Service} \\ \text{Routine} \end{Bmatrix}$: Flag is high; therefore, service routine is started.

In order to gain insight concerning the implementation of the SOH n instruction, refer to Figure 11.3(b). In particular, study the similarities and differences between this RTL program and the one for RED shown in Figure 11.2(b). [It is also instructive to see how the RTL productions in part (b) of Figure 11.3 are mechanized in part (a) of the figure.]

11.3 INTERRUPT TRANSFERS

Frequently it is necessary to concurrently service one or more peripherals during the running of a main program, or a peripheral may require immediate service once it becomes ready. Both of these situations are conveniently handled by means of interrupt logic which, at the end of each fetch cycle, checks whether any device has called for service. If one has, the current value of the program counter, PC, is automatically stored in a specified location, and the starting point of the service routine is transferred to the PC. (Somewhat arbitrarily, we choose memory location 0 as storage for the PC and begin the service routine at location 1.)

Figure 11.4(a) shows how the logic for an elementary interrupt facility may be implemented. When the peripheral requires service, it sets the interrupt call flip-flop in the interface. This information is placed on the interrupt call bus; and if the computer's interrupt allowed flip-flop has previously been set, the interrupt request is received by the computer's logic and timing unit. In order to save the calculation in progress at the occurrence of the interrupt, the request is recognized only at the end of each execution cycle—usually producing only a few microseconds delay. At that point the current value of the program counter is stored and a software service routine is initiated. The service routine must perform the following tasks:

1. Disable the interrupt flip-flop (with the Interrupt OFf, IOF $\equiv$ 7100, instruction) so that the computer is able to finish servicing one device before being interrupted again.
2. Save the register information from the original program so that it may be properly started again.
3. Determine which interface caused the interrupt. One way to do this is to poll the devices by means of the Skip if Interrupt call flip-flop of device n is Low $\equiv$ SIL n $\equiv$ 77 n.

[From Figure 11.4(a) it can be seen that the result of the test is wired-OR connected to a bus; but with only one of the gates selected at a time, a unique determination of the calling device is made.]

4. Carry out the required data exchange with the interface.

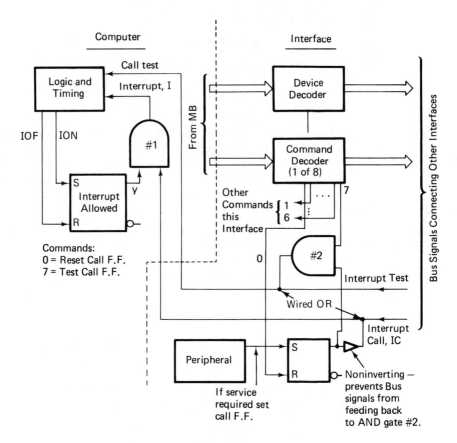

(a) Logic Circuit

```
T = 0   MA ← MB_{0-8};
T = 1   MB ← M(MA);
T = 2   A ← A+MB; M(MA) ← MB;
        IF (Interrupt = 1) MA ← 0 ELSE S ← 0, T ← 0;
T = 3   MB ← M(MA); :Must read core location before writing;
T = 4   MB ← PC;
T = 5   M(MA) ← MB; PC ← 1; S ← 0; T ← 0;
```

(b) ADD Instruction — Modified Execution Cycle

Figure 11.4 Interrupt Implementation.

5. Reset the call flip-flop (with the instruction flip-flop n REset Call ≡ REC n ≡ 70 $\underline{n}$), thereby indicating device service completed.
6. Set the interrupt flip-flop (with the Interrupt ON, ION ≡ 7200, instruction).
7. Restore the program counter, the accumulator, and any other registers so that the original program can continue.

In order to recognize the interrupt request during execution of a program, some statements must be added to the original RTL description of each machine language instruction. Since ADD is a typical instruction, its RTL implementation (originally found in Table 6.6) has been modified as shown in Figure 11.4(b). Attention should be given to the way original information in the PC is saved and how the computer program is forced to branch when the interface calls for service. The production at $T = 3$ is necessary since core memory must be read before writing into it. Moreover, it is not possible to combine $T = 3$ and $T = 4$ because they produce conflicting changes in the MB.

The interrupt service routine is presented in Table 11.1. The reader should study the way the six device-service tasks listed above are accomplished by the program. For brevity, only two interrupting devices are checked in this program; but the extention to additional devices should be obvious. Since the interrupt allowed flip-flop is disabled at the beginning of the program and not set until the end, we see that

TABLE 11.1 INTERRUPT SERVICE ROUTINE

Memory Address	Mnemonic	Memory Contents Machine Code	Remarks
0	—	(X)	Address where the interrupt occurred
1	IOF	7100	Interrupt Off. Disable interrupt allowed flip-flop.
2	STC 600	4600	Store accumulator contents in 600
3	SIL 1	7701	Skip If device 1 flip-floop is Low
4	LMP 701	6701	Jump to device 1 service routine at 701
5	SIL 2	7701	Skip if device 2 flip-flop is low
6	JMP 711	6711	Jump to device 2 service routine at 711
7	CLR	0011	← Return here after calling device is serviced
10	ADD 0	2000	$A \leftarrow X$
11	ADD 601	2601	$A \leftarrow$ JMP X
12	STC 15	4015	$M(15) \leftarrow$ JMP
13	ADD 600	2600	$A \leftarrow$ original $C(A)$
14	ION	7200	Set interrupt allowed flip-flop
15	(JMP X)	(6 X)	Jump back to original program
...			
			← (Interrupt occurs immediately before this instruction. Hardware stores X in location 0 and sets PC for address 1.)
600	—	—	Accumulator contents stored here
601	JMP	6000	
...			Device service routines. At the end of each are the instructions: REC n followed by JMP 7, where n is the device number.
701			
...			
777			

Interrupt Occurs Here — Original Program

Hardware produces this jump (from 1 to entry above)

devices, perhaps of higher priority, requiring attention during the service routine are not recognized until service to the first device is complete. Moreover, our scheme always tests the devices in numerical sequence and does not take into account device priority or length of time a device has been waiting. These requirements are properly handled, at the expense of greater complexity, in many contemporary systems. More will be said on this subject in Section 11.5.

Another problem concerning more than one interrupt call existing at a time is that a new interrupt must *not* be recognized until the program counter again points to location X. If an interrupt were recognized by our program during execution of the instruction in location 14, the hardware would automatically write 15 into location 0, destroying X, which is where the original program should be restarted. Then it would not be possible to return to the original program. One solution, which is employed in some contemporary computers and the one we will adopt, is to prevent ION from taking effect until execution of the next instruction following it. Thus, in the program of Table 11.1, ION is executed at 14 but JMP X will be in the process of completing execution and PC $= X$ before the interrupt is recognized. For this reason, the return to the original program will *not* be lost no matter how many interrupt service cycles are completed.

‡11.4 DIRECT MEMORY ACCESS

When large amounts of data are to be transferred to or from sequential memory locations in concurrence with the running of a main program, it is not feasible to use an interrupt system because it requires too much time for service by the software subroutine. This problem is solved in many computers through the use of *Direct Memory Access, DMA*. Here, each time the peripheral is ready to transfer a data word, it merely signals the computer; and without disturbing the PC or the accumulator, the main program instruction stream is *delayed one memory cycle* while hardware logic completes the data transfer. (For this reason, the method is frequently called *cycle stealing*.) Thus we see that, once DMA is initiated, the entire full data transfer will take place, whether it be 100 or 10,000 words, independently of the main program and without use of software. Although all three types of data transfers—programmed, interrupt, and DMA—can be active concurrently in a modern computer, for simplicity we will draw logic diagrams and write RTL programs here as if only DMA exists in our computer. Through this assumption, the principles of DMA can be quickly grasped; and after that it will be easier to understand systems composed of two or more transfer methods operating simultaneously.

Since DMA does not employ a software subroutine, the following aspects must be implemented entirely in hardware:

1. An address register, AR, pointing to the next memory location which is to receive or transmit data
2. A method for incrementing AR so that a sequential group of memory locations can take part in the data transfer operation
3. A method of determining when the last memory location has been served

4. A data buffer, DB, to serve as a temporary storage register for data as they are transferred between the memory buffer, MB, and the peripheral
5. Control signals to let the computer know when device service is required and to indicate whether data are being transferred in or out

The circuitry required to implement this list is presented in Figure 11.5. Note

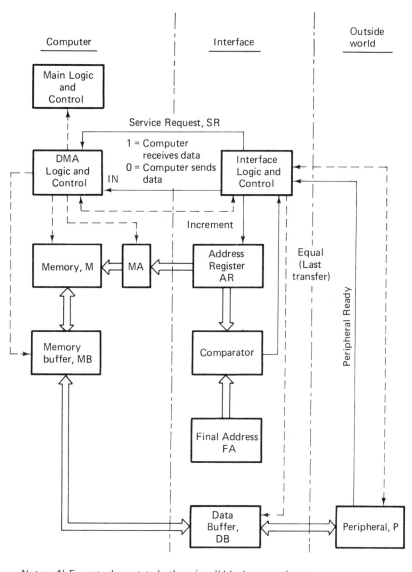

Notes: 1) Except where stated otherwise all blocks are registers
2) The connecting lines employ the following notation:
   – – → General timing and control signals
   —→ Specific control signals (labeled)
   ⇒ Data lines
3) Only DMA Logic shown

Figure 11.5 Direct Memory Access System.

389

how each item on the list is realized by one or more logic blocks and how these interact with the major computer circuits on the left side of the figure. A comparator is included within the interface to determine when the address register is pointing to the final memory address so that the interface logic will know when the DMA operation is complete. (For simplification. the methods of loading the address and FA registers, as well as several other details, have been omitted.)

The RTL program required to implement DMA within the ADD instruction is shown in Table 11.2. (Refer to Table 6.6 for the RTL of the normal ADD instruction.)

TABLE 11.2 ADD INSTRUCTION: EXECUTION CYCLE FOR DMA

$T = 0$ $MA \leftarrow MB_{0-8}$;
$T = 1$ $MB \leftarrow M(MA)$;
$T = 2$ $A \leftarrow A + MB$; $M(MA) \leftarrow MB$;
         IF $(SR = 0)$ $S \leftarrow 0$, $T \leftarrow 0$ ELSE $MA \leftarrow AR$;
$T = 3$ $MB \leftarrow M(MA)$;
$T = 4$ IF $(IN = 0)$ $DB \leftarrow MB$ ELSE $MB \leftarrow DB$;
$T = 5$ $M(MA) \leftarrow MB$; $S \leftarrow 0$; $T \leftarrow 0$;

The reader should study the correspondence between the information given in the table and that in Figure 11.5. Note that, when no service is required ($SR = 0$), the ADD execution cycle ends, as previously implemented, at $T = 2$. On the other hand, when service is required, states $T = 3$ through $T = 5$ are executed and the IF statement, at $T = 4$, controls whether the computer receives or sends data. Thus, $T = 3$ through $T = 5$ essentially constitutes the memory cycle which is stolen from the main program in order to transfer one data word. The significant amount of time saved by DMA compared to the interrupt method should now be obvious. (Observe that in attempting to eliminate a state from the table, it is not possible to move the $MB \leftarrow DB$ production from state $T = 4$ back into state $T = 3$ because there would then be a contradiction on the contents of MB.)

Frequently there is a millisecond or more delay between the time one data word transfer takes place, through the DMA channel, and the time the peripheral is ready for the next word, which means that 500 or more instructions from the main program could be executed. Thus, the increase in efficiency by using DMA to run a program concurrently with data transfer, rather than doing the two operations separately, presents a tremendous saving—one much larger even than the savings between DMA and the interrupt method.

Can the logic required by the DMA interface be described by RTL? Certainly, it is feasible, and as a matter of fact, the required program is shown in Table 11.3. The symbol T^* is used to represent the program states of the interface in order to distinguish them from the T states of the computer (Table 11.2); but it can be seen that synchronization is established between the two state generators at $T^* = 0$ and $T^* = 2$.

It should be obvious that $T^* = 0$ is the state where the interface waits until the peripheral is ready, and that the computer must be starting the fetch state before

TABLE 11.3 INTERFACE CONTROL SIGNALS

$T^* = 0$ IF (Peripheral Ready AND $S,T = 10$) $SR \leftarrow 1$, $AR \leftarrow AR + 1$ ELSE $T^* \leftarrow 0$;
$T^* = 1$ IF (Computer Receives Data) $IN \leftarrow 1$, $DB \leftarrow$ Peripheral Data;
$T^* = 2$ IF $(T \neq 4)$ $T^* \leftarrow 2$;
$T^* = 3$ IF $(IN = 0)$ Peripheral $\leftarrow DB$;
        IF $(AR \neq$ Final Address) $T^* \leftarrow 0$;
$T^* = 4$ HALT;

the interface actually begins the next DMA cycle. The reason the interface pauses at $T^* = 2$ until the computer is at $T = 4$ is to prevent premature transfer of data from DB to the peripheral. Thus, as well as several other principles, will be clarified by an examination (in sequence) of the productions in Table 11.3 within the context of both Table 11.2 and Figure 11.5. Since the content of the address register is incremented before it is sent to MA, the initial value of AR should be one less than the starting address of the data in memory.

‡11.5 INPUT/OUTPUT SYSTEM VARIATIONS

The previous discussion focused on single examples of each main type of data transfer method. We will now present some of the many variations which can be employed. No attempt will be made to be exhaustive in this treatment since there are a vast number of alternate schemes in existence and many more likely to be developed.

Interrupt Polling Revisited

In the discussion of interrupt data transfers, a single interrupt bus connected all interfaces to the CPU. Thus, the computer had to poll the individual interfaces until the first one requiring service was identified; this can take a long time if there are many interfaces. There are several ways to solve this problem; but we will choose a solution that services the interrupts on a priority basis under program control.

The system is based on the following principles:

1. Each interface interrupt line is separately connected to the computer rather than being combined on a bus. Thus, the requesting device's identity is not lost.
2. A decoder receives the interrupt lines and produces an output, in binary, which is the number (address) of the device requiring attention. If two or more devices are requesting service, the one with the lowest device number is given priority.
3. When the computer executes the instruction Load Accumulator from Decoder $\equiv$ LAD $\equiv$ 7300, the original content of the accumulator is replaced by the decoder's output. This number may then be employed as the address part of various instructions to perform the actual data transfers between the computer and the device. For example, if 5 were the interrupting device with its binary number in the accumulator, and if location 200 contained 7400, execution of the instruction ADD 200 would yield 7405 in the accumulator, which is the

instruction RED 5. (In order to execute RED 5, ADD 200 would be followed by STC $p + 1$.)

4. The interrupt call signal is developed from an OR combination of all the service requests. The CPU is then interrupted much as was done in Figure 11.4.
5. The ordering of the device signals entering the decoder gives priority to the lowest-number device requesting service. To provide program control over the priority, however, the bits of a *mask register* are ANDed with the corresponding device lines before they reach the decoder. When a particular mask bit output is zero, that interrupt is inhibited and the next interrupt in sequence will be selected by the decoder's output. The contents of the mask are controlled by the instruction: Transfer data from the Accumulator to the Mask ≡ TAM ≡ 7400. (Since n is never equal to zero, RED $n = 74$ n is easily distinguished from TAM.)

A logic diagram for the interrupt identification and priority system is shown in Figure 11.6. The reader should carefully study how each of the above features is realized by a specific part of the circuit. It should be clear that the mask register has no effect on the interrupt itself since AND gates 0 through 11 occur after the interrupt call signal is developed in the OR gate. Note that the $\bar{y}_i$-output of each mask bit is employed as an input to an AND gate; thus, a cleared mask register, has 1's on the mask output lines and allows *all* interrupt requests to reach the decoder.

How is the decoder implemented in order to select the lowest-numbered calling device and to produce a corresponding binary code? In Figure 11.6 there are 12 devices, which means that the decoder must have four output bits ($2^3 < 12 < 2^4$). For simplification, however, the design of a three-device decoder is presented in Figure 11.7. The first step, part (a) of the figure, requires that all possible combinations of the device inputs be expressed in a truth table. Then, it is only necessary to scan each line of the table from right to left and determine the bit number (in binary) of the first logical 1. This number becomes the decoder output in the second half of the table. Having completed the truth table, the information is transferred to Karnaugh maps, and the output functions IA_i (for all i) are read.

A Multilevel Interrupt System

Although the circuit in Figure 11.6 has removed some of the deficiencies of the previous interrupt system, it does not permit higher-priority devices to interrupt the service routine of lower priority devices. Thus, an urgent interrupt (such as a power failure warning occurring just after the service initiation of a low-priority device) would be forced to wait until the computer was finished with the first device before the new interrupt could be considered. Frequently such a delay would be unacceptably long. Fortunately, this problem can be solved by a modification of the interrupt logic. The major change is a circuit on the output of the decoder which determines whether a new device, of higher priority, is requesting service; and if it is, a new interrupt is immediately generated. One way to accomplish this is to store the number of the device currently being served in a register and continuously compare the register's contents with the decoder's output. When the comparator generates a not-equal signal, a new

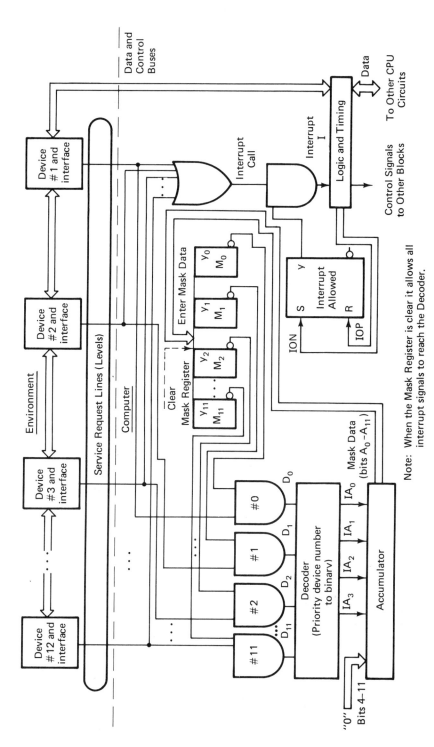

Figure 11.6 Interrupt Identification and Priority System.

D_2	D_1	D_0	IA_1	IA_0
0	0	0	0	0
0	0	1	0	1
0	1	0	1	0
0	1	1	0	1
1	0	0	1	1
1	0	1	0	1
1	1	0	1	0
1	1	1	0	1

(a) Decoder Specification
(Priority given to least significant on-bit)

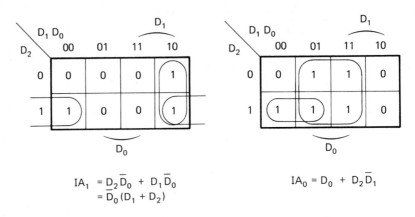

(b) Interrupt Address Functions

Figure 11.7 Decoder Design.

service request has occurred with higher priority than the one currently being handled. (Recall that the decoder, similar to the one shown in Figure 11.7, develops a binary output which is the lowest-number device currently requesting service; hence, it is the device of highest priority.)

A block diagram of the multilevel priority interrupt system is shown in Figure 11.8; note how the interrupt is generated by the comparator from the decoder and current device register signals. The similarity between this circuit and the one in Figure 11.6 will be quickly recognized. Actually, the only major difference is that, in Figure 11.8, a new interrupt is produced whenever a request for service is received from a device of higher priority than the one currently being processed. In contrast, the system in Figure 11.6 disables the interrupt capability, by means of the interrupt allowed flip-flop, whenever the service for a device starts, and it is not enabled again until the first service request is completed.

Figure 11.8 should be given careful consideration. Starting with the device interfaces, we see that the service request signal is a level. [It comes from an interrupt call flip-flop; see Figure 11.4(a).] Thus, each time the CPU has finished servicing a device, its request signal must be removed by the REC n instruction. Since the decoder's output is zero when there are no service requests, it is necessary that the current device register be cleared before processing starts and at the end of each cycle in which *all*

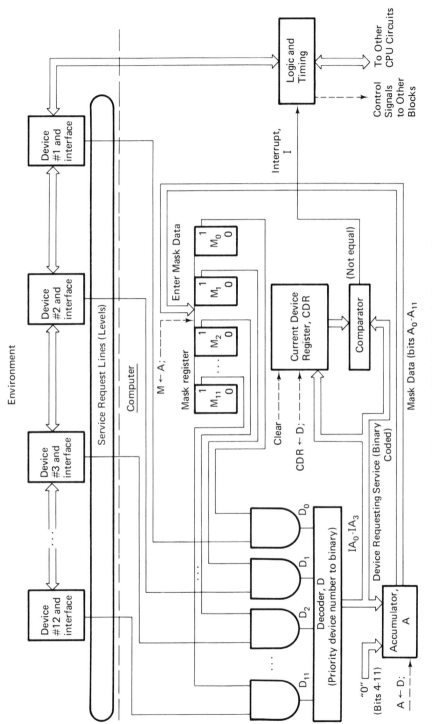

Figure 11.8 Multilevel Priority Interrupt System.

interrupt services are completed. If this is not done, the current device register will not match the decoder's output, resulting in a false interrupt signal being generated. When a service request occurs, the decoder's output becomes a binary representation of the device number, and the comparator's output (i.e., not-equal signal) immediately causes an interrupt signal to be generated. Included with the electronics initiating the interrupt service routine is automatic hardware to transfer the new device number to the current device register. Thus, the interrupt signal, I, is turned off after service begins; but should the decoder's output change during the service routine, indicating a higher-level priority request, a new interrupt would be developed. In contrast to the previous interrupt systems, the multilevel method does not require an interrupt allowed flip-flop; but similar to previous systems, a machine language instruction is available to transfer the device number from the decoder to the accumulator.

One question which should immediately come to mind is: When a service routine is interrupted, how is the value of the program counter stored? With the proper combination of service requests, a chain of interrupted service routines would exist and a whole series of PC, accumulator, and perhaps other register values would need to be stored so that eventually the CPU could correctly accomplish each request within its proper priority. This can produce a rather long and complicated program; but some machines, such as the PDP-11, have a special hardware register called a *pushdown stack* that stores numbers in sequence and then allows them to be removed in last-in first-out order. Such a stack is the key to quickly matching an interrupted service routine with its correct PC value and other information.

Advanced I/O Systems

Significant improvements are currently being made in I/O methods. Two typical advanced systems will now be briefly mentioned in order to complete our discussion of data transfer methods.

Data channels. Large-scale, third- and fourth-generation computers employ what amounts to several small independent processors, sometimes called *data channels*, to handle I/O transfers with the outside world. In this scheme, the main CPU is never burdened with routine I/O operations, but it is free to devote its full time to computational tasks. While the CPU is finishing computational work currently in memory, the data channels load new jobs into the main memory from disk or other sources. When the CPU finishers a job, it signals the I/O controller which then assigns a free data channel to output the results on the specified peripheral. Thus, parallel processing is achieved with several I/O jobs simultaneously in progress and overlapping in time with the actual computation work.

The Control Data CYBER-70 series of computers is representative of this approach. It contains as many as 20 of these independent computers called *peripheral processors*. Each has its own small memory, which can execute programs independently of the main processor and can communicate with the main memory as well as with the CPU.

Certain models of the IBM System 370 (also System 360) provide for high-speed peripherals (such as disk and tape) on *selector channels* and for low-speed peripherals (such as printers and card readers) on *multiplexer channels*. Because of the speed requirement, selector channels are devoted exclusively to servicing a single peripheral at one time. As the name implies, however, multiplexer channels are designed to work with several peripherals at once by rapidly going through an operation cycle in which each is served for a short time in sequence. Of course, special precautions must be taken to insure that the data for the various peripherals are not mixed.

Recent development in microcomputer technology will certainly lead to even more efficient ways of handling data transfers; and they will no doubt allow parallel I/O processing to be achieved on much smaller-scale systems.

Memory-increment and add-to-memory. The two features which are proving to be very popular additions to computer DMA units are *memory-increment* and *add-to-memory*. Memory increment is an operation in which the current contents of a selected location are read, 1 is added, and the result is written back into the original location. This process is very handy for producing histograms from statistical information, e.g., a voltage frequency spectrum from a noise source. The main additions required to our original DMA units are a circuit to do the incrementing and one to set the address register in the interface to the required histogram bin number. (Usually this number will be from an A/D converter, which is sampling at a rate specified by a a clock in the interface.)

Add-to-memory is similar to memory-increment except that other number instead of just 1 may be added to the original memory contents. One of the main uses for this process is to average repetitive signals which are corrupted by noise. Here sequential memory locations are employed to store increasing time samples of the signal. The address register in the interface is always reset to the first sample's memory location before each new signal occurs. When many corresponding (in time) measurements of the noisy signal are added at each sample point, the random noise cancels, leaving the uncorrupted signal.

A detailed discussion of memory-increment and add-to-memory is found in Reference 4. Moreover, References 2, 4, and 7 are broad and rich sources of material on all facets of computer interfacing.

11.6 BIBLIOGRAPHY

1. Davies, D. W., and D. L. A. Barber, *Communication Networks for Computers.* New York: John Wiley & Sons, Inc., 1973.
2. Falk, H., "Linking Microprocessors to the Real World." *IEEE Spectrum*, 1974, pp. 59–67.
3. *Integrated Circuits: Digital-Linear-MOS.* Sunnyvale, Calif.: Signetics Corporation. (Consult most recent edition.)
4. Korn, G. A., *Minicomputers for Engineers and Scientists.* New York: McGraw-Hill Book Company, 1973.
5. *Logic Handbook.* Maynard, Mass.: Digital Equipment Corporation. (Consult most recent edition.)
6. *PDP-11 Peripherals and Interfacing Handbook.* Maynard, Mass.: Digital Equipment Corporation. (Consult most recent edition.)

7. Schoeffler, J. D., and R. H. Temple, *Minicomputers: Hardware, Software, and Applications.* New York: IEEE Press, 1972.
8. Sharp, D., *Handbook of Interactive Computer Terminals.* Reston, Va.: Reston Publishing Company, Inc., 1977.
9. Souček, B., *Microprocessors and Microcomputers.* New York: John Wiley & Sons, Inc., 1976.
10. Wakerly, J. F., "Microprocessor Input/Output Architecture." *Computer*, Vol. 10, No. 2 Febuary 1977, pp. 26–33.
11. Zissos, D., and F. G. Duncan, *Digital Interface Design.* London: Oxford University Press, 1973.

11.7 EXERCISES

11.1 Propose a logic circuit to be contained in the device decoder block of Figure 11.2(a).

11.2 In a manner similar to that in Figure 11.2, show how the WED n instruction may be implemented.

11.3 Instead of employing a skip instruction, it is desired to write a new instruction, FAC, to cause the <u>F</u>lag to make the <u>AC</u>cumulator positive for flag set and negative for flag not set. Write the RTL program to accomplish this.

11.4 If everything is the same except a SOL $n \equiv$ <u>S</u>kip <u>O</u>n flag <u>L</u>ow instruction, instead of SOH n, is the only flag test available, write the sequence of instructions to test flag 15.

11.5 Twelve interrupt interfaces are to have a flag which can be identified quickly without polling. Explain your method and draw the logic diagram for the flag portion of the system.

11.6 Rewrite the RTL program in Figure 11.4(b) assuming that memory location zero is a semiconductor register with none of the limitations of core but which is controlled by the MA and MB registers.

11.7 Describe the logic employed by the ION instruction so that an interrupt is not recognized until the instruction following it is being executed.

11.8 Estimate the approximate percentage of savings in time achieved by using DMA instead of an interrupt to read data into memory.

11.9 With a computer having a 12-bit accumulator, is it feasible to employ the general technique of Figure 11.6 to service 24 or more interrupts? Explain in detail.

11.10 Consider a decoder for an interrupt identification system with four interrupt lines and with highest priority on the most significant bit. Find the logic function which bit *IA* must satisfy.

11.11 If a large number of interfaces exist in a system, the decoder technique of Figure 11.7 can become unwieldy. Present a heuristic approach for decoder design which does not have this limitation. Break your design into a priority chain circuit having:
(a) The same number of outputs, n, as it has inputs.
(b) A logical-1 output *only* for the lowest-number device requesting service.
Follow the priority circuit by logic which accepts n input lines and produces the binary code of the device requesting service. (If no device is requesting service, output the number 0.) Demonstrate your method with a system having four devices.

11.12 For the interrupt system in Figure 11.6, write a pseudo-LINC program for the device identification portion of an interrupt service routine having the following requirements: If the device requiring service has high priority (as defined by the original mask register contents), be sure the device number is in the accumulator and branch to location 300 (service routine 1). If the device has low priority, place the device number in the accumulator and branch to location 350 (service routine 2). Assume that the mask register was originally loaded with a nonzero value. (You may want to employ the instruction Skip if Accumulator is ZEro ≡ AZE ≡ 0450.)

Chapter 12

Digital Electronics

The discrete-component networks of a few years ago have, to a large extent, been replaced by integrated circuits, ICs. In order to use these new devices effectively, the computer designer needs to have an understanding of basic digital electronics, including the fundamentals of the various IC families and not just a *black-box* concept.

Diodes and transistors will be represented by an approximate circuit model in the discussion that follows because this approach is easily understood and because our present motivation stems more from an interest in the use of electronic building blocks for logic design rather than detailed circuit design itself. On the other hand, a reader whose long-range interests are mainly in computer hardware or in electronic circuits will eventually want to become knowledgeable in the more fundamental approach based on the exact physical behavior of the devices themselves.[4, 5, 6, 7, 9] In Chapter 4 we learned how to construct logic diagrams from AND, OR, NOT, and other types of gates. Here the realization of such gates by means of circuit elements will be presented.

It is assumed that the reader has an elementary acquaintance with electrical networks. This chapter will extend this knowledge into the field of digital electronics in order to briefly explore the circuit level of computer design. We will consider a wide range of topics, from the design of logic gates to flip-flops and from the mechanization with discrete components to integrated circuits.

12.1 DIODE GATES

Although gate circuits constructed exclusively from diodes are no longer extensively used, they exhibit most of the same properties as the more complex, currently popular gate types. In an elementary treatment such as this, it is much easier to derive these properties for diode circuits. Moreover, diodes are the key to understanding transistors and ICs, and diode–transistor combinations are still employed in many digital circuits. For these reasons, we now begin with an elementary consideration of diode physical properties and then devote an extensive discussion to their gate circuits.

Silicon, Si, is the most common semiconductor material from which diodes, transistors, and integrated circuits are constructed. By adding a small amount of an impurity with a valence of three (such as aluminum or indium) to pure silicon, a so-called *p-type* semiconductor is produced. *Holes* are the majority carriers in this type of material, and a memory aid here is to think of *p* as standing for *positive* charge. Conversely, a small amount of an impurity with a valence of five (such as arsenic or phosphorus) may be added to pure silicon to produce an n-type semiconductor. *Electrons* are the majority carriers in this type of material, and the corresponding memory aid is to think of *n* as standing for *negative*. As a first-order approximation, one may view a p-n junction diode, like the one shown in Figure 12.1(a), as consisting of only free positive charge carriers on the p-side of the junction and only free negative charge carriers on the n-side of the junction. With this concept, it can be seen from Figure 12.1(b) that a negative voltage applied to the p-terminal and a positive voltage applied to the n-terminal pull both charge carriers away from the junction. This results in no current flow. On the other hand, the opposite voltage applied to the terminals, as shown in Figure 12.1(c), force both charge carriers across the junction, resulting in

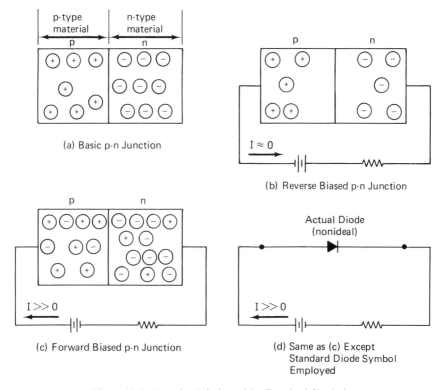

Figure 12.1 Junction Diode and Its Standard Symbol.

significant current flow. This simple picture predicts the basic rectifying action of a diode, but it does not account for the fact that silicon diodes require a *cut-in voltage*, V_γ, of about 0.7 V before appreciable current will flow. In Figure 12.1(d), note the standard diode symbol employed and the fact that the arrow on the symbol points in the direction of easy conventional current flow. (If the battery were reversed, the symbol and current arrows would be in opposition, indicating high diode resistance or nearly zero current flow.)

For analyzing diode networks, the above physical behavior is best represented in terms of the circuit model shown in Figure 12.2(a) and the diode response shown in Figure 12.2(b).

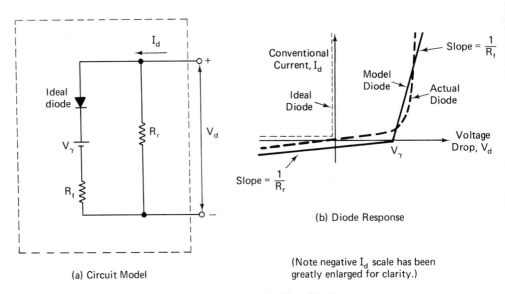

(a) Circuit Model

(b) Diode Response

(Note negative I_d scale has been greatly enlarged for clarity.)

Figure 12.2 Diode Circuit Model and Its Response.

Note the degree to which the model characteristic approaches the actual, and how they both differ from the ideal. The abrupt change in the model diode response at $V_d = V_\gamma$ is due to the fact that the ideal diode, which is an integral part of the model, becomes forward biased. This results in a low input (forward bias) resistance, R_f, instead of the very large (reverse bias) resistance R_r. (For silicon, V_γ typically is 0.7 V, $R_f = 20\ \Omega$, and $R_r > 10^6\ \Omega$, which means R_r will be neglected in the analysis that follows.)

OR Gates

An elementary diode OR gate circuit and the corresponding truth table are shown in Figure 12.3(a). By making the simplifying assumptions that $R_f = 0$ and source resistance $R_s = 0$ we can see that with $x = 5$ V and $y = 0$, D_x is forward biased. Thus, 0.7 V is lost across V_γ and the remainder, $5 - 0.7 = 4.3$, appears at the output, z.

Similar reasoning leads to the other three sets of voltages in the table. In producing logic values from the voltage values in the truth table, note that the most positive voltage is assigned the logic value 1 and the most negative voltage is assigned the logic value 0. Such an assignment is known as *positive logic*—more will be said on this subject shortly. Another fact to observe concerning the truth table is that there is a 0.7-V offset between the maximum input voltage and the maximum output voltage, which of course is due to the diode drop.

AND Gates

A very similar arrangement of diodes can be used to produce an AND gate, with the elementary circuit and truth table (using positive logic) as shown in Figure 12.3(b).

In analyzing the circuit in Figure 12.3(b) with $x = 5$ V, $y = 0$ V, we might be tempted to say that the input $x = 5$ V causes the output to go to 5.7 V. However, further consideration leads us to the realization that the y diode will now have 5.7 V across it, which will cause it to draw a large amount of current until the output drops

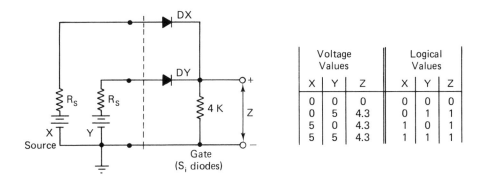

(a) OR Gate and Truth Table

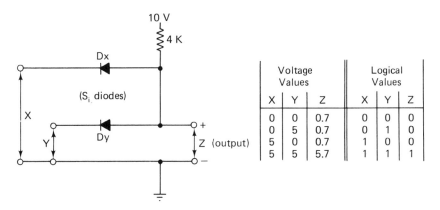

(b) AND Gate and Truth Table

Figure 12.3 Simplified Diode Gate Circuits.

to 0.7 V. At this time, the x diode would be reverse biased. Thus, the output is stable at $z = 0.7$ V. Of course, when both inputs reach 5.0 V, the output will be stable at 5.7 V.

Positive versus Negative Logic

Logic blocks are named according to the function they perform with positive logic. As shown in Figure 12.3 the block name describes the electrical behavior of the device. However, the AND block can be made to produce other logic functions when negative logic is defined on either the input or the output or both. Positive logic has already been defined and *negative logic* is merely the converse, i.e., the most positive level is defined to be a logical 0 and the most negative level is defined to be a logical 1. Table 12.1 shows how the AND block in Figure 12.3(b) can be used to produce the AND, OR, NAND, and NOR functions.

TABLE 12.1 THE FOUR FUNCTIONS PRODUCED BY THE AND BLOCK

Voltage Values			Logic Values											
			(+) Logic			(−) Logic			(+) Logic (−)			(−) Logic (+)		
									Logic (−)			Logic (+)		
x	y	z	x	y	z	x	y	z	x	y	z	x	y	z
0	0	0.7	0	0	0	1	1	1	0	0	1	1	1	0
0	5	0.7	0	1	0	1	0	1	0	1	1	1	0	0
5	0	0.7	1	0	0	0	1	1	1	0	1	0	1	0
5	5	5.7	1	1	1	0	0	0	1	1	0	0	0	1
			AND Function			OR Function			NAND Function			NOR Function		

When we draw a logic diagram, the gate symbols will continue to represent the function performed by that gate and not necessarily the electrical performance. The only time an AND function is produced by an AND block is when positive logic is used on both the input and output.

Gate Analysis

Having analyzed simplified diode gates, we will now consider practical gates. Most of the observed effects also carry over to transistor gates, but the principles can be more easily visualized with the simpler diode circuits.

Let us return to the OR gate shown in Figure 12.3(a). The original circuit analysis included V_γ, but the voltage drop $I_d R_f$ was neglected as was the voltage drop across the internal resistance of the source $I_d R_s$. Typical values are $R_f = 20\ \Omega$, $R_s = 1\ k\Omega$; thus, $R_f \ll R_s$, and R_f can be neglected for the circuits of interest here. A more practical OR circuit, which will now be analyzed, is shown in Figure 12.4. Note that the x input has been tied to zero volts so that diode D_x may be neglected. This has no effect on the conclusions of our analysis, but it does simplify the work. By showing

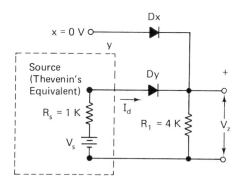

Figure 12.4 OR Circuit to Be Analyzed.

the OR gate input in the form of a Thevenin's equivalent, we can analyze the effect of a general source whether the actual source is a pulse generator, an AND gate, or even another OR gate. Now, when $I_d > 0$, the effect of the diode offset voltage V_γ is to make the source appear to be of voltage $V_s - V_\gamma$. Therefore, the output voltage V_z is easily determined from the voltage-divider rule.

$$V_z = \frac{(V_s - V_\gamma)R_1}{R_1 + R_s} \qquad (12.1)$$

Substituting the above-mentioned values for the parameters in Equation (12.1), we find that

$$V_z = \frac{(5 - 0.7)4 \times 10^3}{(1 + 4) \times 10^3} = 3.44 \text{ V} \qquad (12.2)$$

The result of Equation (12.2) brings up an interesting question: What logic value should be assigned to an output of 3.44 V? It has been previously stated that with positive logic the most positive voltage level is defined to be a logical 1 and the most negative, a logical 0. While this definition is clear to us, it is too general for use in circuit construction. Gates and other digital circuitry are most easily designed on an absolute basis rather than on the above relative scale. If 0 and 5 volts are the minimum and maximum extremes, the question then becomes: How low can a voltage go (positive logic) before it ceases to be called a logical 1? The actual numbers used in defining logic values are somewhat arbitrary, but the method commonly used in practice employs a voltage margin between the low threshold, V_L, and the high threshold, V_H. Figure 12.5 shows this method for defining both negative and positive logic. For positive logic, any voltage up to V_L is interpreted as a logic 0 and any voltage greater than V_H is interpreted as a logic 1. Voltages between V_L and V_H are not assigned a logic value. In the design of gates and other logic circuits, the usual practice is to accept inputs at the threshold but to develop outputs which are removed from the threshold by a safety factor allowing for circuit attenuation, noise, loading, and other sources of degradation. For example, if values of $V_L = 1$ V and $V_H = 3$ V are employed, the design goal might typically be to achieve output voltages of 0.5 V and 4 V.

Assuming $V_L = 1$ V and $V_H = 3$ V, and returning to our original question concerning the logic value to be assigned to the result of Equation (12.2), we immediately see that $V_z = 3.44$ V represents a logical 1. Thus, the OR gate in Figure 12.4 generates the correct logical output. On the other hand, it does introduce a consider-

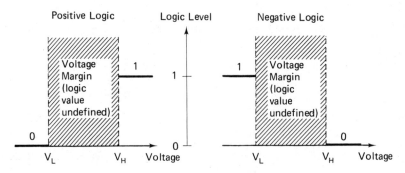

Figure 12.5 Logic/Voltage Relations.

able loss (from $V_s = 5$ V to $V_z = 3.44$ V). Now Equation (12.1) reveals that all OR gates of this form possess an offset of V_γ and an attenuation of $R_1/(R_1 + R_s)$. Later we will discuss methods for eliminating the offset; but the attenuation is a physical limitation of all passive gates. This is the reason that transistors are introduced in the next section as a means of producing gain to overcome the attenuation. If the numerator and denominator of Equation (12.1) are divided by R_1, a way of minimizing attenuation is suggested:

$$V_z = \frac{(V_s - V_\gamma)}{1 + R_s/R_1} \qquad (12.3)$$

If $R_s/R_1 \ll 1$, the attenuation can be neglected. Often the designer does not have control of R_s; thus, from the standpoint of attenuation, it is desirable to make $R_1 \gg R_s$.

Propagation delay. A large value of R_1, as shown in Figure 12.4, can produce an unacceptable *propagation delay* as the following analysis demonstrates. (Propagation delay is defined as the finite time required for a logic circuit's output to change state in response to a change of state at the input.)

Consider the circuit shown in Figure 12.6. It is the same as in Figure 12.4, except the capacitance C_1 shunting R_1 is explicitly shown. The major contributor to C_1 is stray wiring capacitance, which typically has a value of $C_1 = 50$ pF unless special precautions are taken. If initially $x = 0$ and $V_s = 5$ V, and then later the input changes to $V_s = 0$ V, both diodes will be off and the only discharge path for C_1 is through R_1. (It is assumed that any output load, R_L, satisfies the equation: $R_L \gg R_1$.) From elementary network theory, recall that at time t the parallel resistor-capacitor combination, shown in Figure 12.6, produces the response

$$V_z = V_0 e^{-t/R_1 C_1} \qquad (12.4)$$

where V_0 is the initial voltage across the capacitor terminals. If $V_0 = 3.44$ V, which was the value of the circuit output from Equation (12.2), and if $V_z = V_L = 1$ V is substituted into Equation (12.4) along with the values of R_1 and C_1, we find:

$$1 = 3.44 e^{-t/(4 \times 10^3 \times 50 \times 10^{-12})} \qquad (12.5)$$

which results in $t = 247$ ns. Therefore, the propagation delay in turning the gate off becomes $Pd_{\text{off}} = t = 247$ ns, which is much too long for many digital applications.

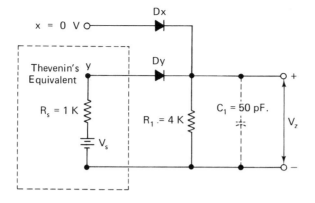

(a) Circuit Model

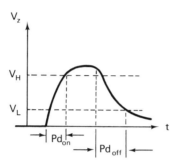

(b) Time Response

Figure 12.6 OR Gate Performance.

To reduce Pd_{off}, it is necessary to reduce R_1; but from Equation (12.3), this increases the attenuation problem. Frequently, a satisfactory compromise value for R_1 cannot be achieved, and this is one of the main reasons that transistor logic (discussed in Section 12.2) has replaced diode logic in digital systems.

In the circuit of Figure 12.6(b), Pd_{on} is the delay in turning the gate on. $Pd_{on} < Pd_{off}$, which is a consequence of the fact that a low-resistance charging path, through R_s, is provided for Pd_{on} as opposed to the higher-resistance discharge path through R_1. Thus, Pd_{off} is the major contributor to the delay produced by the diode OR circuit. (Many authors calculate a single propagation delay parameter for a gate or other logic device by averaging Pd_{on} and Pd_{off}.)

AND-OR Circuits

As indicated in Chapter 4, it is very common to develop a logic function in sum-of-products form; thus, we seek a circuit design to mechanize a set of AND gates followed by an OR gate. This situation is represented in Figure 12.7. The dotted sections indicate other portions of the circuit, which are not important for the following *worst-case design*. (This is a conservative design philosophy in which the most pessimistic values possible are assumed for the inputs, the component tolerances, and other circuit parameters.) Consider Figure 12.7 with the V_{st} inputs all at 5 V.

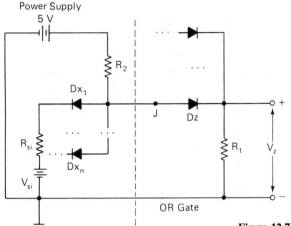

Figure 12.7 AND-OR Logic.

The circuit to the right of point J can be replaced by an equivalent resistor R_e. Then the power supply, in conjunction with R_e and R_2, forms a voltage divider. Thus, the voltage drop, V_J, across R_e is $V_J \leq 5$ V; and with $V_{si} = 5$ V for each i, all of the D_{xi} diodes are reverse biased and may be omitted from further consideration.

Now consider the voltage divider from the positive side of the power supply, through R_2, D_z, and R_1 and back to the supply. The resulting output voltage becomes:

$$V_z = \frac{(5 - 0.7)R_1}{R_1 + R_2} \tag{12.6}$$

If we continue to use $R_1 = 4$ kΩ and require a nominal value of $V_z = 4$ V, the result from Equation (12.6) is $R_2 = 300$ Ω.

Fan-Out

Having designed the AND-OR circuit with $R_1 = 4$ kΩ and $R_2 = 300$ Ω, we must now determine how many OR gates could be connected to the output of the AND gate, point J. *Fan-out* is a term employed in situations such as this where we desire to have a measure of the number of other circuits a particular circuit can drive. The equivalent circuit for this case is shown in Figure 12.8 where, for worst-case conditions, only one input of each OR gate is assumed to have a nonzero value. Now with every diode D_{zi} on, all points K_i will be at the same voltage and all the resistors R_1 will be effectively in parallel. The resulting equivalent resistor is $R_e = R_1/n$, where n is the fanout, the number of OR gates being supplied by the AND gate.

The general output voltage can now be determined from

$$V_{zi} = \frac{(5 - 0.7)R_1/n}{(R_1/n) + R_2} \tag{12.7}$$

If we are willing to accept an output value as low as $V_{zi} = V_H = 3$ V, Equation (12.7) may be solved for n (assuming the previous values of $R_1 = 4$ kΩ and $R_2 = 300$ Ω). The result is $n = 5.77$. We round this to the next *lowest* whole number to obtain a fan-out of 5. This means the AND gate can drive five OR gates and the output will still be $V_{zi} \geq V_H$.

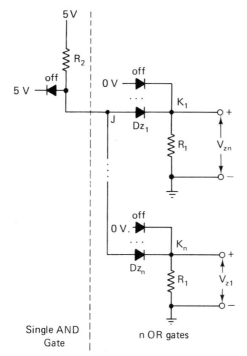

Figure 12.8 Circuit for Fan-Out Analysis.

Diode Current Constraint

The constraints imposed by the desired frequency response, the voltage margin, and the desired fan-out have already been considered. Now the restriction imposed by maximum allowable diode current will be analyzed. Consider the circuit shown in Figure 12.9 with the worst-case conditions of only one gate input at 0 V and negligible source resistance, $R_s = 0$. Under these conditions, diode D_1 will be forward biased and point J will go to 0.7 V. With point K near 0 V and with R_1 of moderate size, I_1 will be negligible; thus, all of I_2 will flow through D_1 ($I_d = I_2$), resulting in the equation

$$I_d R_f + I_d R_2 = 5 - 0.7$$

Solving for R_2, we obtain

$$R_2 = \frac{4.3}{I_d} - R_f \quad (12.8)$$

For a maximum diode current of $I_d = 10$ mA and the usual value of $R_f = 20\ \Omega$, Equation (12.8) yields

$$R_2 = 410\ \Omega$$

as the minimum allowable resistance.

Employing this R_2, $V_{zt} = V_H = 3$ V, and the previous OR gate resistance, i.e., $R_1 = 4$ kΩ, Equation (12.7) now yields

$$3 = \frac{(4.3)4/n}{4/n + 0.41}$$

Therefore, $n = 4.22$, which means that the diode current limitation has reduced the fan-out from 5 to 4.

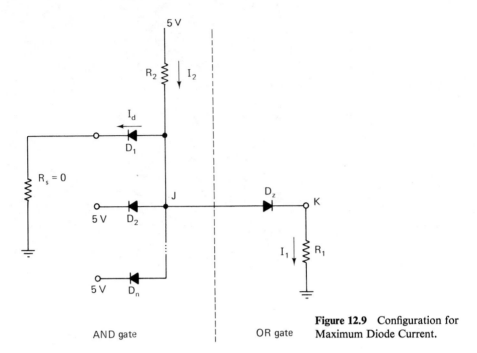

Figure 12.9 Configuration for Maximum Diode Current.

AND gate | OR gate

We will consider the way other constraints influence diode gate design in the exercises.

Elimination of Offset Voltage

Frequently the offset of 0.7 V (between input and output) observed in both the OR and AND gates (Figures 12.2 and 12.3) is objectionable. The circuit shown in Figure 12.10 provides a simple solution to this problem through the use of another diode to produce downshifting. In order to gain a basic understanding of the gate, assume that all voltage drops associated with the input circuit are negligible except the diode cut-in voltage, V_γ. Then, if 5 V sources are placed at the x and y inputs, and if diodes D_x

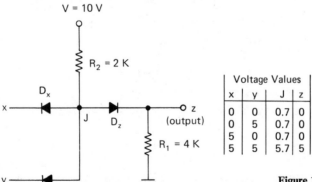

Voltage Values

x	y	J	z
0	0	0.7	0
0	5	0.7	0
5	0	0.7	0
5	5	5.7	5

Figure 12.10 Modified AND Gate and Its Truth Table.

and D_y are forward biased, point J goes to 5.7 V. Under these conditions, D_z will be forward biased with a 0.7 V drop across it. This results in the output being at 5.0 V, which is what the last line of the truth table in Figure 12.10 shows. The other three lines of that truth table follow in a similar way, with the result that the AND gate is achieved without offset or attenuation. The offset has been eliminated by D_z which, through its back-to-back connection with the other diodes, counters the effect produced by D_x and D_y. This shifting technique is very useful and is employed in several commercial gates, which will be discussed later. On the other hand, the absence of attenuation is simply due to the fact that the source resistance was assumed sufficiently low that the input current produced a negligible voltage drop. (As previously mentioned, this frequently is not a good assumption in practice.)

One of the critical points concerning the proper operation of the modified AND gate is that D_x and D_y always remain forward biased. This can be assured if the open-circuit voltage, V_J, at point J, without D_x and D_y connected, is at a higher potential than V_s. The open-circuit voltage can be calculated from the voltage-divider rule:

$$V_J = \frac{(V - 0.7)R_1}{R_1 + R_2} + 0.7$$

where V is the power supply voltage. Therefore, the requirement on V_s is

$$V_s < V_J - 0.7$$

or

$$V_s < \frac{(V - 0.7)R_1}{R_1 + R_2} \tag{12.9}$$

For the parameters of Figure 12.10,

$$V_s < \frac{(10 - 0.7)4}{4 + 2} = 6.2 \text{ V}$$

which means that the circuit should operate satisfactorily since the source employed was $V_s = 5$ V. (A similar OR circuit which eliminates the offset voltage will be considered in the exercises.)

12.2 TRANSISTOR GATES

In this section, a simple transistor model is developed, and then a number of practical gate circuits are analyzed on the basis of the model.

Transistor Functional Models

Junction transistors, such as those shown in Figure 12.11, are now mass-produced at a very low cost—many common types sell for under 50 cents each. The base-emitter junction, shown in Figure 12.11(a), resembles the diode in Figure 12.1(a) in many ways. It conducts current only when the p-terminal is made positive by at least 0.7 V (in silicon) with respect to the emitter. (Note that the corresponding circuit symbol has an arrow marking the emitter terminal, pointing in the direction of easy current flow.) The collector-base junction also resembles a diode, so that it becomes forward biased when the base is more positive than the collector. (These polarities are exactly

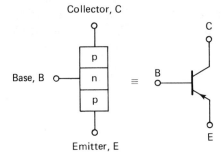

Figure 12.11 Junction Configurations and Standard Symbols for the Two Transistor Types.

reversed for the pnp transistor in Figure 12.11(b).) The major advantage, however, that a transistor has over a diode is that the transistor exhibits *current gain*. A small amount of current flowing in the base-emitter circuit can control a large amount of current in the collector-emitter circuit. This will be explained more fully later.

Because of their wide use in digital circuitry and for simplicity, silicon junction transistors in *common-emitter* configuration, as shown in Figure 12.12(a), will be emphasized in our discussion. Only the *cutoff* and *saturation* regions of the transistor's characteristics are employed in most digital circuits. Cutoff is basically the condition where the transistor is fully off; in silicon, cutoff occurs at $V_{BE} = 0$ V and $I_C = I_B = I_E = 0$ A. Saturation occurs when both the base-collector junction and the base-emitter junction are forward biased. In this situation, to a first approximation a transistor can be represented as the two back-to-back diodes shown in Figure 12.13. Since, by definition, both junctions are forward biased, we would expect that $V_{CE,\text{sat}} \approx 0$. For silicon, a better approximation is $V_{CE,\text{sat}} = 0.2$ V, which is the value we will use here. Later we will also refer to the *active* region, which occurs when the base-emitter junction is forward biased and the base-collector junction is reversed biased.

There are several ways of analyzing digital circuits; but from an elementary approach, a circuit model for transistors is very appealing because it provides a concrete basis on which to formulate designs. One must exercise care, however, to remain within the assumptions of the chosen model. The simplified model we intend

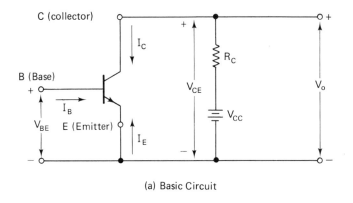

(a) Basic Circuit

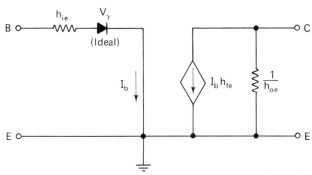

(b) Simplified Common Emitter Model

Figure 12.12 npn Transistor in a Common-Emitter Configuration.

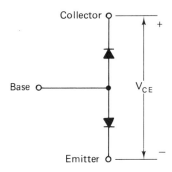

Figure 12.13 Two-Diode Model for an npn Transistor.

to employ is shown in Figure 12.12(b). At first the model may appear somewhat strange and complex, but it is really quite simple. The base-emitter circuit will be recognized as being the same as our diode circuit model with $h_{ie} = R_f$.

The diamond-shaped figure in the collector-emitter circuit is known as a *dependent current source*. This simply means that it is a source of current whose magnitude depends on the current flow in the base circuit, I_b, multiplied by the transistor gain, h_{fe}.* The quantity h_{oe} is an admittance; thus $1/h_{oe}$, shown on the diagram, is the

---

*The symbol β is also frequently employed to represent h_{fe}.

Sec. 12.2   Transistor Gates

corresponding resistance. (The quantities h_{ie}, h_{oe}, and h_{fe}, are formally known as *hybrid parameters*, with the *e* subscript representing the common-emitter configuration, the *i* subscript representing input, and the subscript *o* representing output. These standard electronic symbols are employed so that we may find their values for specific transistors in reference manuals. Do not let the formalism create any mystery about their nature—they are simply two conventional resistors and a gain.)

Shown in Figure 12.14 is a typical inverter circuit including reasonable values for the transistor model parameters. Note that R_c and $1/h_{oe}$ are effectively in parallel and that $1/h_{oe} \gg R_c$, which is almost always true for our cases of interest. Therefore, $1/h_{oe}$ will be omitted from most future diagrams and the output voltage, V_o, can be obtained from the equation

$$V_o = V_{cc} - I_c R_c = V_{cc} - h_{fe} I_b R_c = 5 - (30)(3)I_b \qquad (12.10)$$

where all currents are in milliamperes (mA) and the resistance is in thousands of ohms (kΩ).

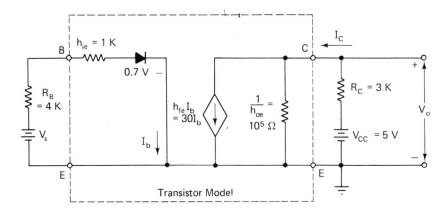

Figure 12.14 Typical Inverter.

If $V_o = 0.2$ V when the transistor is in saturation, Equation (12.10) may be solved for the I_b which brings the transistor to the edge of saturation:

$$I_b = \frac{0.2 - 5}{-90} = 0.053 \text{ mA}$$

This results in a voltage drop at saturation, $V_{i,\text{sat}}$, across h_{ie} of

$$V_{i,\text{sat}} = (1)(0.053) = 0.053 \text{ V}$$

As the transistor is driven further into saturation, I_b increases causing $V_{i,\text{sat}}$ to increase; and a reasonable approximation is $V_{i,\text{sat}} = 0.1$ V. For this reason, it is common practice in the design of digital circuits to neglect h_{ie}, but to assume $V_{BE,\text{cut-in}} = 0.7$ V and $V_{BE,\text{sat}} = 0.7 + 0.1 = 0.8$ V.

Following the above considerations, the very simple transistor models shown in Figure 12.15 can be constructed.

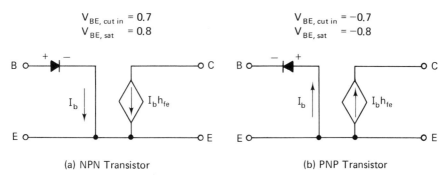

Figure 12.15 Approximate Models for the Common-Emitter Configuration.

Diode-Transistor Logic (DTL)

Since considerable gain can be obtained from an inverter, a straightforward method for improving the performance of the previously discussed diode gates is to follow each of them with an inverter. The AND gates then become NANDs and the OR gates become NORs. This is not a problem because it has already been proven in Chapter 4 that either of these gates can be employed to produce any desired function. However, our main interest is in gates which can be economically mechanized by integrated circuits, where transistors and diodes are very inexpensive to implement but large capacitors and resistors, $C \geq 100$ pF and $R \geq 10$ kΩ, are expensive. (These conditions are the opposite of those observed in discrete-component circuits.) For this reason, a modified DTL NAND gate, such as that shown in Figure 12.16, is the starting point for IC design.

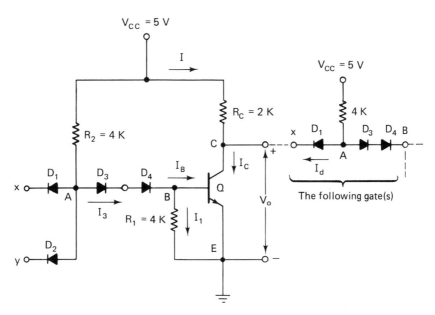

Figure 12.16 DTL NAND Gate.

Sec. 12.2 Transistor Gates

We will first verify that the circuit is indeed a NAND gate. Then we will determine the operating regions for the gate and calculate the fan-out. In analyzing the circuit of Figure 12.16, we will assume the usual diode cut-in potential of 0.7 V and neglect the small voltage drop across R_f. Now, if both inputs are at 0 V, D_1 and D_2 will be forward biased, and point A will reach 0.7 V. This will not be sufficient for D_3 and D_4 to come on. Point B will go to 0 V, Q will be cut off, and V_o will then reach the supply potential of 5 V. These conditions are summarized in the first line of Table 12.2.

TABLE 12.2 NAND GATE PERFORMANCE

				Voltage* and State Values					Positive Logic Values		
x	y	D_1	D_2	A	D_3 & D_4	B	Q	V_o	x	y	V_o
0	0	On	On	0.7	Off	0	Cutoff	5	0	0	1
0	5	On	Off	0.7	Off	0	Cutoff	5	0	1	1
5	0	Off	On	0.7	Off	0	Cutoff	5	1	0	1
5	5	Off	Off	2.2	On	0.8	Saturation	0.2	1	1	0

*Ground reference.

On the other hand, if it were thought that 0 V at the inputs would turn Q on, the following conclusion would be true: point B would require 0.7 V; diodes D_3 and D_4 would need to turn on, requiring point A to be at $B + 2(0.7) = 2.1$ V, With x at 0 V, point A cannot be at 2.1 V, which means that the assumption of Q turning on has led to a contradiction.

In a similar way, the other three lines of Table 12.2 may be verified. In the fourth line of the table, note that the assumption of Q in saturation leads to point A being at $0.8 + 2(0.7) = 2.2$ V, which is consistent with D_1 and D_2 being off. (Here, sufficient current must be provided by the power supply through the resistor R_2 to saturate the transistor. This point will be verified shortly.)

Gate Operating Region

Table 12.2 provides some insight concerning the operation of the DTL gate. But the fourth line indicates $V_o = 0.2$, and with the assumed input values of 0 and 5 V, one might think that some of the offset (and perhaps attenuation) problems might be present here as was found in diode gates. Using a method similar to that employed in analyzing Figure 12.16 and applying the same increasing voltage to both inputs, we see that the voltage at point A follows the input but is biased ahead by a diode drop of 0.7 V. Now, in order for Q to reach cut-in, point A must be at $0.7 + 2(0.7) = 2.1$ V, which means that the input $x = y = 2.1 - 0.7 = 1.4$ V. Within the model, saturation is reached at 0.1 V higher than cut-in, which leads to $x = y = 1.5$ V. Moreover, all input voltages higher than this will merely drive Q further into saturation; thus, a first-order approximation of the gate's performance is given in Figure 12.17.

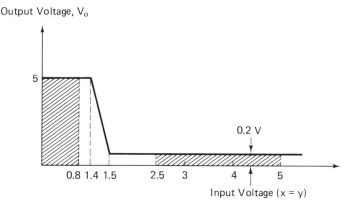

Figure 12.17 DTL Gate Transfer Characteristics (shaded areas show typical operating characteristics).

The operating regions are marked on the diagram to show typical conditions rather than absolute limits. Thus, $V_o = 0.2$ indicated in Table 12.2 will easily drive another stage, like itself, into saturation. This means that DTL gates, in contrast to our previous experience with diode gates, may be employed in multilevel logic without detrimental effects. The DTL gate also provides reasonable resistance to noise. Assume that all of the inputs start at 5 V. Then Figure 12.17 indicates that an input could drop to 1.5 V before Q comes out of saturation. This means that a noise spike as large as $5 - 1.5 = 3.5$ V could exist without affecting gate performance. The term *noise margin** is often used to indicate the magnitude of noise required to cause erroneous operation (3.5 V in this case). On the other hand, if at least one input is driven by the output (0.2 V) of a saturated transistor, Figure 12.17 shows that the input could reach 1.4 V before Q reaches cut-in. This means that a noise spike as large as $1.4 - 0.2 = 1.2$ could exist without affecting gate performance.

Fan-Out

Now consider the fan-out capabilities of the DTL gate in terms of the number, N, of gates (similar to itself) which it can drive. When Q is cut off, $V_o = 5$ V and the input diode D_1 of the next state will be off. Thus, the next gate draws no current and does not load the gate under study. For these reasons, the loading effects on Q need only be considered when the transistor is in saturation. At that time, the collector current, and hence the base current, must be sufficient to place Q in saturation and to sink the current for the N gates it is driving. An equation describing this situation may be obtained from Figure 12.16:

$$h_{fe}I_B = I_C = I + NI_d \qquad (12.11)$$

---

*In a more general sense, the term *noise margin* is usually applied to the worst-case conditions expected for a batch of integrated circuits, e.g., the difference between the minimum expected output from one stage and the minimum input to a similar stage which is expected to drive that stage into saturation. For DTL this is typically 0.7 V.

where I is the current required to saturate Q and I_d is the current from one of the gates being driven. The value for I is easily determined from

$$I = \frac{V_{cc} - V_o}{2} = \frac{5 - 0.2}{2} = 2.4 \text{ mA} \tag{12.12}$$

With $V_o = 0.2$ V driving an identical gate, diodes D_3 and D_4 of the gate being driven will be off; thus,

$$I_d = \frac{V_{cc} - (0.2 + 0.7)}{4} = \frac{5 - 0.9}{4} = 1.025 \text{ mA} \tag{12.13}$$

From the circuit:

$$I_B = I_3 - I_1 \tag{12.14a}$$

but Q is in saturation, yielding

$$I_1 = \frac{V_B}{4} = \frac{0.8}{4} = 0.2$$

Now with the inputs at 5 V, diodes D_1 and D_2 are cut off, and point A is at 2.2 V, which leads to

$$I_3 = \frac{V_{cc} - 2.2}{4} = \frac{5 - 2.2}{4} = 0.7 \text{ mA}$$

When these values are substituted into Equation (6.14a), the result is

$$I_B = 0.7 - 0.2 = 0.5 \text{ mA} \tag{12.14b}$$

After substituting Equations (12.12), (12.13), and (12.14b) into Equation (12.11), we find:

$$(30)(0.5) = 2.4 + N(1.025)$$

Therefore, $N = 12.6/1.025 = 12.3$, which means a fan-out of 12 may be achieved.

Direct-Coupled Transistor Logic (DCTL)

In the cutoff/saturated mode that has been assumed here, the transistor acts much like a conventional switch. Thus, the question arises whether we can directly stack transistors in series or parallel to do logic operations, as can be done with ordinary switches. A NOR circuit following this concept is shown in Figure 12.18. Since it is assumed that the gate inputs come from the outputs of similar gates, values of 0.2 V and 3 V are employed in the truth table in Figure 12.18(b). From the logic portion of the table:

$$V_o = \bar{x}\bar{y} = \overline{x + y}$$

which verifies that a NOR gate is realized. One of the major disadvantages of this type of circuit is that the noise margin is low. Transistor cut-in is at 0.7 V and an input low is 0.2 V; thus, a noise spike of $0.7 - 0.2 = 0.5$ V will cause erroneous operation. In addition, there is danger of *current hogging* with parallel-coupled DCTL. This condition occurs in multilevel logic when one gate on a lower level tries to drive two separate gates on another level. If the two transistors being driven have different values for $V_{BE,\text{sat}}$, the one that has the lowest value will tend to clamp the common input voltage, and the second transistor may remain near its cut-in value. Thus, the second transistor may not get sufficient current to reach saturation, leading to $V_o \gg 0.2$ V.

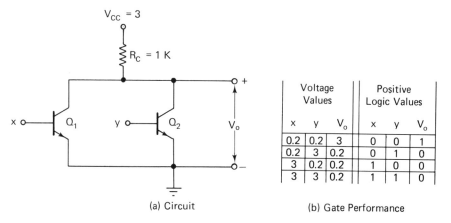

(a) Circuit (b) Gate Performance

Figure 12.18 DCTL NOR Gate.

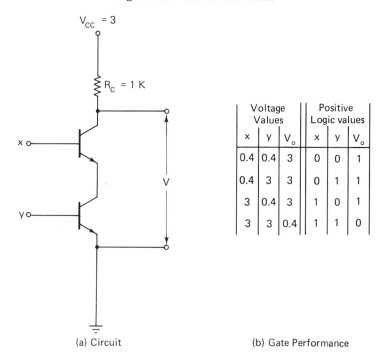

(a) Circuit (b) Gate Performance

Figure 12.19 DCTL NAND Gate.

A DCTL series circuit is shown in Figure 12.19. In the truth table portion of the figure, we see that the $V_{CE,\text{sat}}$ voltages add when both transistors are in saturation, resulting in $V_o = 0.4$ V; thus, input voltages of 0.4 V and 3 V are assumed. From the logic portion of the table, we see that

$$\bar{V}_o = xy \quad \text{or} \quad V_o = \overline{xy}$$

which verifies that a NAND gate is realized.

The noise margin of this gate is even worse than that for the NOR gate in that a noise spike of $0.7 - 0.4 = 0.3$ V will cause erroneous operation. Moreover, if we try to stack more transistors in series to produce a three- or four-input gate, the $V_{CE,\text{sat}}$ buildup becomes intolerable. Another disadvantage of the series gate configuration is the fact that the transistor farthest from ground requires a larger base-to-ground voltage to drive it into saturation. As a result of this, a multilevel logic circuit having a cutoff gate on a lower level supplying two different gates on another level would experience the problem of current hogging if the two transistors being supplied were not the same position above ground. Thus, both the series and the parallel DCTL bipolar configurations have serious limitations which prevent them from being employed in standard integrated circuits. (Later in the chapter we will find that the configuration is much more satisfactory with unipolar transistors.)

‡Resistor-Transistor Logic (RTL)

Having observed in the previous section that DCTL is not usable as a standard bipolar configuration, we will now attempt to eliminate the disadvantages through a simple modification. By placing a resistor in series with each input, we can eliminate the problem of current hogging; in trying to draw a large share of current, the transistor will also produce a voltage drop across its base resistor, causing a larger voltage to appear at the other transistors.

The main advantage of the RTL family of gates is low cost. Of all the standard bipolar configurations employed in integrated-circuit mechanization, it requires the smallest space on the silicon wafer.

A typical medium-power RTL gate employed in practical integrated circuits is shown in Figure 12.20. Note that both inputs at 0 V cut off the transistors and cause the unloaded output to reach the power supply potential of 3 V. On the other hand, one or more of the inputs at 3 V cause that transistor(s) to go into saturation, resulting in $V_o = 0.2$ V. Thus, Figure 12.20(b) shows that $V_o = \bar{x}\bar{y} = \overline{x+y}$, which means that a NOR gate has been realized.

One of the problems with the RTL circuit is that V_o falls quite rapidly as it is loaded with succeeding RTL gates. Figure 12.20(c) shows the equivalent circuit for a gate having a high output and a load of N succeeding gates. From the figure, V_o can be calculated:

$$V_o = 0.8 + \frac{2.2(450/N)}{640 + 450/N} = 0.8 + \frac{(2.2)(450)}{640N + 450} \qquad (12.15)$$

As expected, for no loading ($N = 0$), $V_o = 0.8 + 2.2 = 3.0$; and as N increases, V_o approaches 0.8 V. A graph of this function is shown in Figure 12.21. In practice, the maximum fan-out for this circuit is 5, which means that $V_o \approx 1$ V (actually 1.07 V). Since 0.8 V is the minimum signal which will drive the succeeding gate into saturation, a fairly small noise spike ($1.07 - 0.8 = 0.27$ V) will cause erroneous circuit operation.

Using the calculated input potential of 1.07 V and the usual value of $V_{BE,\text{sat}} = 0.8$, we obtain the required base current:

$$I_B = \frac{1.07 - 0.8}{0.45} = 0.6 \text{ mA}$$

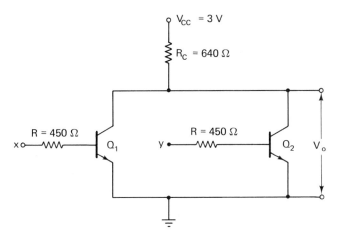

(a) Circuit

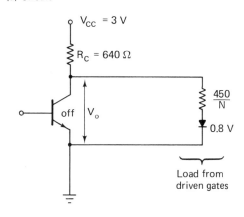

Voltage Values			Positive Logic values		
x	y	V_o	x	y	V_o
0	0	3	0	0	1
0	3	0.2	0	1	0
3	0	0.2	1	0	0
3	3	0.2	1	1	0

(b) Gate Performance

(c) Equivalent Circuit for Fan-Out Calculations

Figure 12.20 RTL NOR Gate.

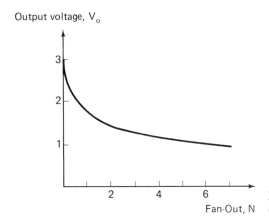

Figure 12.21 Fan-Out Characteristics of a Typical RTL Gate.

But

$$I_C = \frac{3.0 - 0.2}{0.64} = 4.375 \text{ mA}$$

Therefore, the required transistor gain is

$$h_{fe,\min} \geq \frac{I_C}{I_B} = \frac{4.375}{0.6} = 7.29$$

Transistor-Transistor Logic (TTL or T²L)

With integrated circuit technology, it is easy to produce multiple emitters on a single transistor. Configurations of this type are called TTL and have the fastest response of the saturating logic group. A basic circuit for the TTL family is shown in Figure 12.22. Considering the two-diode model of the transistor, we see that Figure 12.22 has the same labeling (points A, F, X, etc.) and the same topology as Figure 12.16,

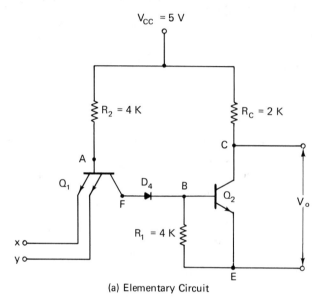

(a) Elementary Circuit

		Voltage and State Values						Positive Logic Values		
x	y	Q_1, B-E Junction	Q_1, B-C Junction	A	B	Q_2	V_o	x	y	V_o
0.2	0.2	on	off	0.9	0	off	5	0	0	1
0.2	5	on	off	0.9	0	off	5	0	1	1
5	0.2	on	off	0.9	0	off	5	1	0	1
5	5	off	on	2.2	0.8	sat.	0.2	1	1	0

(b) Gate Performance

Figure 12.22 TTL Gate Principles. (The voltages shown for V_o are unloaded values.)

where Q_1 replaces D_1, D_2, and D_3. The truth tables for the two circuits are also similar, as a comparison between Figure 12.22(b) and Table 12.2 shows. In Figure 12.22(b), note that with both inputs low the two emitter junctions behave like a pair of forward-biased diodes; hence point A is a diode drop above the input (i.e., $0.2 + 0.7 = 0.9$ V). In order for Q_2 to reach the cut-in voltage, however, it requires that the B-C junction of Q_1, D_4, and the B-E junction of Q_2 be on. This, in turn requires $3(0.7) = 2.1$ V at point A, which is considerably more than the 0.9 V available. Thus, Q_2 remains off and the unloaded output reaches the power supply potential of 5 V. It is easy to see from the figure that the above conclusions continue to hold as long as at least one input is at a low value.

When all inputs are at 5 V, Figure 12.22(b) shows that the B-E junction of Q_1 will be off (reverse biased). Under these conditions, point A will rise toward the supply voltage, turning Q_2, D_4, and the B-C junction of Q_1 on. Thus, point A will reach $0.7 + 0.7 + 0.8 = 2.2$ V, and with Q_2 in saturation, $V_o = 0.2$ V. It is easily seen from the table that the function realized is a NAND ($\bar{V}_o = xy$ or $V_o = \overline{xy}$).

‡Passive versus Active Pull-Up

A circuit such as the one in Figure 12.22 is said to possess *passive pull-up* because the transistor does not carry current when V_o goes high. (V_o is determined only by the passive part of the circuit.) If a circuit containing an "on" transistor is employed to achieve V_o high, *active pull-up* is said to exist. With the same limitations on transistor current and other parameters, active pullup is able to drive heavier loads (from both transistor states); it leads to faster circuit response; and it requires less power.

In order to achieve basic understanding of these two types of output circuits, let us consider a switch model for the transistors, i.e., an on transistor will be considered to have zero resistance (between collector and emitter) and an off transistor will be considered to have infinite resistance. The circuits for passive and active pull-up are shown in Figure 12.23(a) and (b), respectively. Assume that the transistors have a current limit of 10 mA and it is desired to obtain a nominal high output state of $V_o = 4$ V. Let us determine how heavy a load (or how large a fan-out) the circuit can supply, i.e., find the smallest possible value for R_L.

From the circuit in Figure 12.23(a), the $Q_1 = 1$ (transistor on or switch closed) condition for maximum allowable transistor current leads to

$$\frac{5}{R_C} = I_{c1} \leq 10 \text{ mA}$$

(It is assumed that the transistor is not required to sink any current from the load.) Therefore,

$$R_C \geq 0.5 \text{ k}\Omega \tag{12.16}$$

For $Q_1 = 0$:

$$V_o = \frac{5R_L}{R_L + R_C} = \frac{5}{1 + R_C/R_L} = 4 \text{ V}$$

Therefore,

$$R_L = 4R_C \tag{12.17}$$

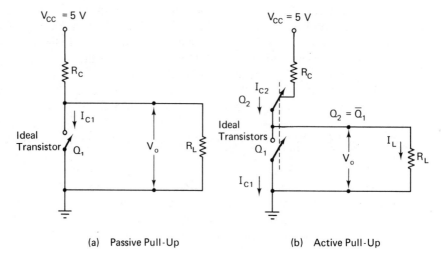

Figure 12.23 Switch Model for Two Types of Output Stages. (Q_1 and Q_2 represent logic states of the corresponding transistors.)

Substituting Equation (12.16) into (12.17):

$$R_L \geq 4(0.5) = 2 \text{ k}\Omega \qquad (12.18)$$

In contrast to the above results, consider the simplified model for active pull-up shown in Figure 12.23(b). With $Q_1 = 0$ (and $Q_2 = \bar{Q}_1 = 1$) and recalling the transistor current limit of 10 mA:

$$\frac{5}{R_C + R_L} = I_{C2} \leq 10 \text{ mA}$$

Therefore,

$$R_C + R_L \geq 0.5 \text{ k}\Omega \qquad (12.19)$$

From the requirement of a nominal upper value for V_o of 4 V,

$$V_o = \frac{5R_L}{R_L + R_C} = \frac{5}{1 + R_C/R_L} = 4 \text{ V}$$

Therefore,

$$R_L = 4R_C \qquad (12.20)$$

Substituting Equation (12.20) into (12.19):

$$R_C + 4R_C \geq 0.5 \text{ k}\Omega$$

Thus,

$$R_C \geq 0.1 \text{ k}\Omega \qquad (12.21)$$

and from Equation (12.20):

$$R_L \geq 0.4 \text{ k}\Omega \qquad (12.22)$$

Comparing Equations (12.18) and (12.22), we see that the active circuit can handle a load which is five times as heavy as passive pull-up. In addition, the active circuit can sink 10 mA when delivering the lower value for V_o (here $Q_1 = 1$ and $Q_2 = 0$). If the passive circuit is required to sink any current from the load under similar conditions, R_C must be increased and the load must be even lighter than specified by Equation (12.17), i.e., $R_L > 2$ kΩ.

‡A Practical TTL Circuit

A practical TTL circuit employing active pull-up is shown in Figure 12.24. Here Q_1 has the same role as it did in the elementary circuit of Figure 12.22(a), but the diode D_4 is replaced by the transistor Q_3, which also assumes the task of providing an inverted input to Q_4. (The requirements of supplying an input to Q_2 and the inverse to Q_4 gives the circuit associated with Q_3, the name *phase splitter*.) Active pull-up is

Voltage and State Values

x	y	A	Q_3	D	Q_4	B	Q_2	V_o
0.2	0.2	0.9	off	≈ 5	on	0	off	≈ 3.6
0.2	⩾ 1.6	0.9	off	≈ 5	on	0	off	≈ 3.6
⩾ 1.6	0.2	0.9	off	≈ 5	on	0	off	≈ 3.6
⩾ 1.6	⩽ 1.6	2.3	sat.	1	off	0.8	sat.	0.2

(a) Circuit Performance
(The voltages shown for V_o are unloaded values.)

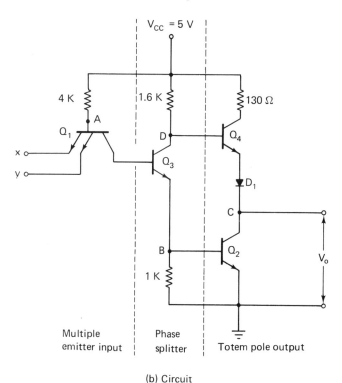

(b) Circuit

Figure 12.24 Practical TTL Gate.

employed in the output circuit with one transistor stacked upon a second, resulting in the term *totem-pole output*.

When comparing the performance tables for the elementary and practical TTL circuits, we see that the voltage values for points A and B are the same,* as is the performance of Q_1. Moreover, the differences observed between the tables are a matter of degree rather than fundamental nature. For example, the totem-pole output circuit leads to the unloaded high output being $V_o \approx 3.6$ V instead of 5 V; but of course, the gate still realizes the NAND function. When transistor Q_3 is off, point D rises toward the supply voltage, turning Q_4 and D_1 on. With the output lightly loaded, Q_4 and D_1 are near their cut-in values, and the very small base current produces negligible drop across the 1.6 kΩ resistor. This results in $V_o \approx 5 - 0.7 - 0.7 \approx 3.6$ V. When transistor Q_3 is on, there is sufficient base current to drive it into saturation. Taking into account that Q_2 is also in saturation, we obtain:

$$D = 0.8 + 0.2 = 1.0 \text{ V}$$

Since point A must be at 2.3 V in order to saturate Q_2 and Q_3, the inputs must be no more than one diode drop less; thus, the table shows values of ≥ 1.6 V. Of course, to allow for noise and other practical considerations, the inputs should actually be considerably larger.

Emitter-Coupled Logic (ECL)

Emitter-coupled logic is the highest-speed circuitry in common use for digital data processing. The main reason for the improved performance is that the delay encountered by driving transistors into saturation, as is done with TTL and other common types of logic, is avoided. Gate delay in the neighborhood of 1 ns can be obtained.

An elementary ECL gate circuit is shown in Figure 12.25. The basic principles employed in ECL are:

1. Transistors are never allowed to reach saturation.
2. The current through a common-emitter resistor, R_E, is supplied either entirely by the reference transistor, Q_3, or by one (or both) of the input transistors, Q_1 or Q_2. (The family name, emitter-coupled logic, comes from this fact.)

If the inputs are low, say -1.58 V, Q_1 and Q_2 will be cut off, resulting in Q_3 being in the *active region*. The active region is defined as the condition where the base-emitter junction is conducting and the collector-emitter junction is cut off. We will assume that the resulting base-emitter voltage drop for an active transistor is 0.75 V; thus,

$$V_E = V_{\text{ref}} - 0.75 = -1.18 - 0.75 = -1.93 \text{ V}$$

From this current, I_E, is found to be

$$I_E = \frac{V_E - V_{EE}}{R_E} = \frac{-1.93 - (-5.2)}{1.18 \times 10^3} = 2.76 \text{ mA}$$

*The voltage at point A shown in the table for the practical gate is 2.3 V instead of 2.2 V because the saturated transistor Q_3 yields approximately 0.1 V greater drop than diode D_4.

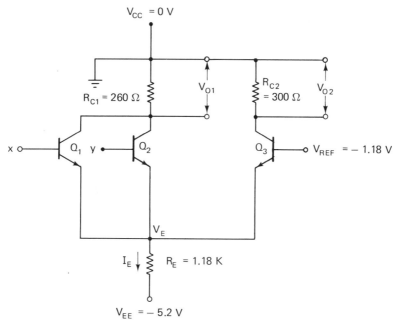

(a) Circuit

								Positive Logic values			
\multicolumn{8}{c}{Voltage and State Values}				
x	y	V_E	Q_1	Q_2	Q_3	V_{o1}	V_{o2}	x	y	V_{o1}	V_{o2}
−1.58	−1.58	−1.93	off	off	active	0	−0.83	0	0	1	0
−1.58	−0.75	−1.50	off	active	off	−0.82	0	0	1	0	1
−0.75	−1.58	−1.50	active	off	off	−0.82	0	1	0	0	1
−0.75	−0.75	−1.50	active	active	off	−0.82	0	1	1	0	0

(b) Gate Performance

Figure 12.25 Elementary ECL Gate.

Since the other two transistors are cut off, all of I_E must pass through R_{C2}, resulting in

$$V_{o2} = -(0.3)(2.76) = -0.83 \text{ V}$$

Also, the cutoff transistors prevent current from flowing in R_{C1}, resulting in $V_{o1} = 0$ V.

Now, if at least one of the inputs goes high, say −0.75 V, its transistor will go into the active region, resulting in

$$V_E = \text{(input voltage)} - V_{BE} = -0.75 - 0.75 = -1.5 \text{ V}$$

Sec. 12.2 Transistor Gates

Then:
$$I_E = \frac{V_E - V_{EE}}{R_E} = \frac{-1.5 - (-5.2)}{1.18 \times 10^3} = 3.14 \text{ mA}$$

Now, all of I_E flows through R_{C1}, resulting in
$$V_{01} = -(0.26)(3.14) = -0.82 \text{ V}$$

Since Q_3 is cut off, no current flows through R_{C2}, resulting in $V_{02} = 0$ V.

The above facts are summarized in the table shown in Figure 12.25(b). Now the circuit actually realizes two functions: $V_{01} = \tilde{x}\tilde{y} = \overline{x+y}$, which is a NOR; and $\bar{V}_{02} = \overline{\tilde{x}\tilde{y}}$ or $V_{02} = \overline{\tilde{x}\tilde{y}} = x+y$, which is an OR. Thus, the gate contains two outputs that are complements of each other, and the gate symbol is represented by:

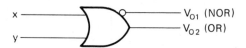

A practical version of Figure 12.25 is shown in Figure 12.26. Although the new figure seems much more complicated, it is exactly the same as the previous one except for a stage (Q_5, D_1, and D_2) which implements the voltage V_{ref} and the emitter-follower output stages (Q_4 for the V_{01} output and Q_6 for the V_{02} output). While V_{ref} could be generated simply by a pair of resistors forming a voltage divider, the diodes provide a measure of compensation for temperature variations. The emitter followers provide a low output impedance; and since they operate in the active region, the previous outputs [from Figure 12.25(b)] are merely shifted down by -0.75 V. Thus, the outputs from the practical circuit are $0 - 0.75 = -0.75$ V for the higher value and $-0.83 - 0.75 = -1.58$ V for the lower value. With these shifts, the gate inputs

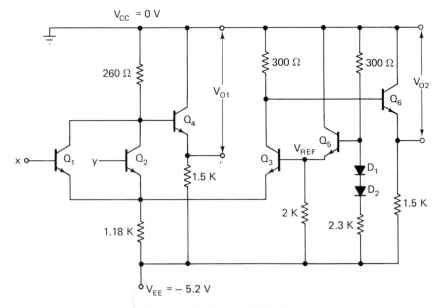

Figure 12.26 Practical ECL Gate.

and outputs are compatible, which means that multilevel logic, with the output of one ECL gate feeding another ECL gate, can be implemented. Moreover, since the emitter followers only shift the output levels, the same functions are mechanized as shown in Figure 12.25(b).

12.3 MOSFET CIRCUITS

A special type of semiconductor device called the *metal-oxide semiconductor field-effect transistor*, or MOSFET for short, is becoming increasingly popular for large shift registers, memories, and other digital applications because of its efficient use of silicon chip area and its ease of manufacture. Thus far, we have considered only bipolar transistors, which employ both holes and electrons in their operation. The MOSFET, however, employs only one current carrier, either holes alone or electrons alone. For this reason, it is known as a *unipolar* device. Although it has a formidable name, the principles of operation of this type of transistor are fairly simple. As shown in Figure 12.27, two separate p-regions are produced on an n-type substrate. A silicon-oxide insulator layer is formed above them, and an isolated metal layer called a *gate*, is formed between the p-regions but above the oxide. (The name MOS comes from the three layers just mentioned.) Since there is n-material between the two p-regions, back-to-back diodes are effectively produced; and with zero potential on the gate, a very high resistance exists from source to drain. It should be observed, however, that the gate, the oxide, and the substrate form a sandwich similar to a parallel-plate capacitor. Thus, the application of a negative voltage to the gate (with respect to the substrate) causes a positive charge to be induced on the upper surface of the substrate as shown in Figure 12.27. This layer of positive charge effectively forms a conducting *p-channel* between the p-regions. The resistance between source and drain, which originally was many megohms, has been reduced to a value in the neighborhood of a few hundred ohms by application of the negative gate voltage. Since the change in resistance covers so many orders of magnitude, the above action is often compared to that of a mechanical switch. Another similarity to a switch is the fact that the device will conduct current equally well in either direction between source and drain.

The MOSFET in Figure 12.27 is classified as an *enhancement* type because the

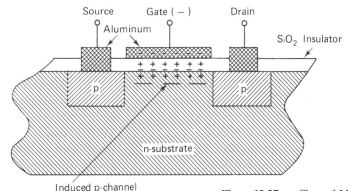

Figure 12.27 p-Channel MOSFET.

current flow increases, is enhanced, by an increasing negative voltage applied to the gate. (A somewhat different structure leads to a MOSFET in which an increase in gate voltage leads to a reduced current flow; hence it is called a *depletion* type. Since the depletion MOSFET is not as important as the enhancement type, it will not be considered here.) By analogy with Figure 12.27, it is easy to see that interchanging the n- and p-type materials leads to an n-channel enhancement-type MOSFET. Of course, the gate voltage must be positive to turn on an n-channel device. Since the p-channel device is actuated by a negative signal between the gate and substrate, its action is similar to a pnp transistor; furthermore, an n-channel device is similar to an npn transistor.

Because a very good insulator is present between the gate and the other elements, except for a small capacitance, the input impedance of the gate is very high—actually as much as 10^{14} Ω in some cases. Incidentally, as we view the action of the gate, it is evident where the name *field-effect transistor* originated.

Circuit symbols for MOSFETs are not as well-standardized as they are for ordinary transistors; but a common set of symbols is shown in Figure 12.28. By analogy with the structure of pnp transistors, the p-channel MOSFET, P-MOS, has the arrow pointing into the transistor at the source terminal. The n-channel device, by analogy with an npn transistor, has the arrow pointing out at the source terminal. Since an external substrate terminal is not shown in parts (a) and (b) of Figure 12.28, *it is assumed that the substrate is internally connected to the source*. When it is necessary to make a different connection to the substrate, however, the symbol shown in part (c) of the figure will be employed.

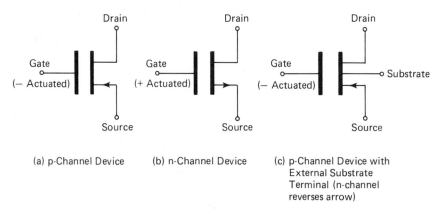

Figure 12.28 Circuit Symbols for MOSFETs.

MOSFET Gates

Direct-coupled transistor logic is used in the construction of MOSFET gates; but there is no possibility of current hogging since the gate terminal has such a high input impedance. Another interesting feature of MOSFET circuits is that considerable room on the silicon chip is saved by using a properly biased transistor to simulate the load resistor. (A MOS transistor requires approximately 1 mil² on the chip, while a corresponding resistor requires more than 10 times that amount.)

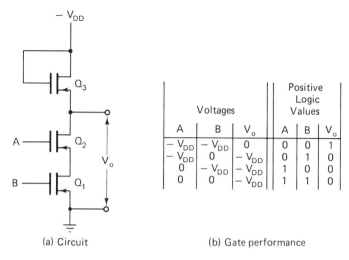

Figure 12.29 p-Channel NOR Gate.

A typical MOSFET circuit is shown in Figure 12.29. Since the gate of Q_3 is tied to the negative supply, the transistor is always on; thus, it serves as a load resistance. The other two transistors serve in their usual capacity as switching transistors. Part (b) of the figure is easily verified if we consider the p-channel gate as a switch, which is closed when a large negative voltage is applied and open when 0 V is applied. Using positive logic, we see from the table that the function realized is

$$V_o = \bar{A}\bar{B} = \overline{A + B}$$

which is a NOR. In the exercises at the end of the chapter, a parallel rather than a series structure will be shown to yield a NAND gate, and other slightly more complicated structures will be shown to yield multilevel logic.

Complementary MOS

By employing both p-channel and n-channel MOS transistors in the same circuit, a family of gates called *complementary MOS* (CMOS), having very low power requirements, may be constructed. The inverter in Figure 12.30 demonstrates the basic principles of the CMOS family. Note that the substrate of Q_2 is shown explicitly connected to the drain terminal. This allows zero volts on input A to place a positive voltage between the gate and substrate, turning the FET on. The remaining aspects of circuit performance are summarized in the table, where it is verified that inverter action is achieved. It is evident that essentially zero static current flows through the unloaded gate when it is in either state. Moreover, connection of the output to the high impedance gate terminals of other circuits provides little additional power dissipation. Thus, the CMOS circuit produces very low power drain.

CMOS devices may, of course, be assembled into NAND and NOR gates as well as into other useful logic elements. A typical CMOS gate is shown in Figure 12.31. The reader should verify that the circuit shown in part (a) of the figure does indeed produce the results shown in part (b). Since positive logic represents the most positive

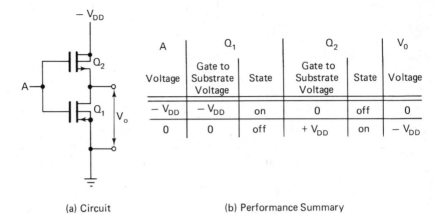

(a) Circuit (b) Performance Summary

Figure 12.30 CMOS Inverter.

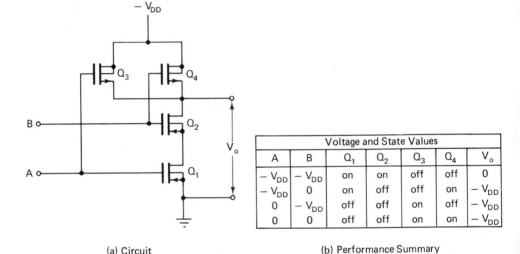

(a) Circuit (b) Performance Summary

Figure 12.31 CMOS NOR Gate.

level as a logical 1, then 0 V = logic 1 and $-V_{DD}$ = logic 0;

$$V_o = \bar{A}\bar{B} = \overline{A + B}$$

which verifies that the circuit is indeed a NOR gate. Other types of CMOS gates will be considered in the exercises.

MOSFET Flip-Flops

There is, of course, no difficulty in constructing conventional gate type flip-flops from MOSFET devices, and this is sometimes done; but other more interesting possibilities exist. Since the input impedance to the gate of a MOSFET is so high, 10^{14} Ω, it is practical to employ the very small gate-substrate capacitance, C_{gs}, to temporarily store the desired state of the flip-flop. Thus, master-slave units can be constructed with much smaller area on the IC chip than with bipolar methods.

The above principle is demonstrated in Figure 12.32 for a so-called *dynamic* flip-flop, which is one that must receive a new input at about a 1-kHz rate to refresh the signal stored on C_{gs}. Note that clock pulse C_{p1} controls the master flip-flop, Q_1 through Q_3; and clock pulse C_{p2} controls the slave, Q_4 through Q_5. The timing diagram for the circuit is shown in Figure 12.32(b), and the basic principles of operation are as follows:

1. $C_{p1} = 1$ turns Q_1 on and allows the input to charge the gate-substrate capacitance, C_{gs2}, of Q_2 to the potential of the input signal. The notation $C_{p1} = 1$

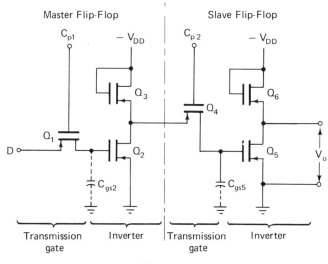

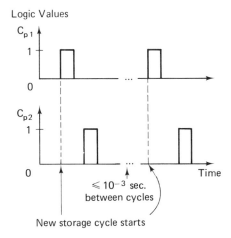

(b) Timing Diagram

Figure 12.32 Dynamic Master-Slave D Flip-Flop.

means that a *logical-1* signal exists at the C_{p1} input, which in turn closes the circuit between the source and drain of the *transmission gate* Q_1. (A transmission gate is merely a MOSFET employed as a switch between two separate sections of a circuit.)

2. The signal is inverted by the Q_2–Q_3 circuit but is blocked by Q_4 because $C_{p2} = 0$. At this point the master has been set to the new state, but the slave still holds the old state.

3. C_{p1} returns to zero. Later $C_{p2} = 1$ and the signal is allowed to pass from the master to the slave and to be stored in the gate substrate capacitance, C_{gs5} of Q_5.

4. The input signal, doubly inverted, now appears at the output V_o, and the master slave action is complete.

Of course, the signal stored on C_{gs5} will eventually decay due to the finite resistance in the circuit; this is why it must be refreshed periodically. Many different techniques could be employed to restore the signal, but one natural situation occurs when the flip-flop is part of a dynamic shift resister. There, many flip-flops, like that shown in Figure 12.32, are connected in a series loop so that data can be circulated and usually picked off, or changed, one bit at a time at a common point.

Another type of MOSFET flip-flop is shown in Figure 12.33. This is a static-type device since refreshing is not necessary; but to keep the signal from degrading, a

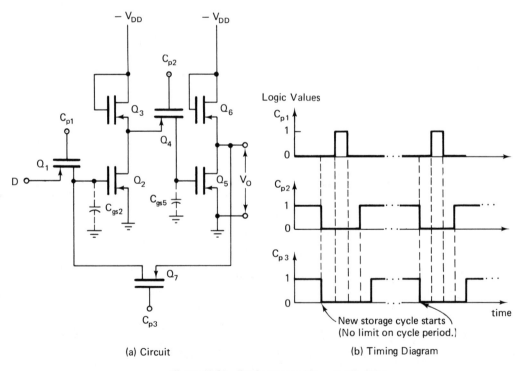

Figure 12.33 Static Master-Slave D Flip-Flop.

feedback path is provided from the output to the input. Note that the transmission gate, Q_7, is the only feature added to the previous figure in order to construct the new flip-flop. Figure 12.33(b) shows the timing sequence required for proper circuit operation. Observe how the following storage cycle is executed:

1. A new cycle begins when transmission gates Q_4 and Q_7 open, i.e., the slave is disconnected from the master and the feedback path is interrupted. The current flip-flop state is still held by C_{gs5}; thus, the output is not disturbed.
2. A short time later $C_{p1} = 1$, and the next state is transferred to C_{gs2}. (Note that C_{p1} returns to zero before any of the other inputs change.)
3. Then $C_{p2} = 1$ and the master-slave transfer occurs.
4. Finally, a short time later, the feedback path is closed by $C_{p3} = 1$. With both Q_4 and Q_7 actuated, the flip-flop is in what can be called steady state. As long as power is supplied through $-V_{DD}$, the circuit will properly hold the given state without any changes in the clock signals. The broken sections in Figure 12.33(b) are employed to represent a steady-state condition of arbitrary length. The MOSFET flip-flops are employed in shift registers, in random-access memories, and in many other types of sequential circuits. Both the static and dynamic flip-flops have similar applications; but the static circuit has the advantage that a refresh operation is not required. On the other hand, the dynamic circuit requires somewhat less chip area and less average power.

Integrated Circuit Families

In previous sections we have considered several digital logic families. It is appropriate at this point to make a detailed comparison of them so that the reader can see why a certain family is chosen for a particular application.

Table 12.3 serves as the main medium in the comparison. Note that the rows in the table indicate seven different parameters and the columns list the six logic families. Of course, the manufacturer's catalogs should be consulted for the actual parameters of a specific circuit. Moreover, subfamilies frequently exist within a given IC type, and these may have different parameter values than the typical ones listed in the table. (This is particularly true for propagation delays, fan-out, and power dissipation.) Also, the catalogs will often carry other important information not considered here. On the other hand, the table does give a fairly good general view of the most popular IC families as they currently exist.

Most of the parameters are self-explanatory, but a few should be discussed:

Row 1. While the positive logic function listed is the basic one implemented, catalogs should be consulted for other functions which may be available.

Row 3. The propagation delay listed is the expected (not worst-case) value for a gate not heavily loaded. Also, when a range of delays exists for different family members, the value given tends toward the short end of the scale. Moreover, wiring delays, which average 1 ns per 15 cm must be added to the table values.

Row 4. The values of fan-out given account for worst-case conditions of output driving capacity and input loading.

TABLE 12.3 CHARACTERISTICS OF IC DIGITAL LOGIC FAMILIES

Parameter	DTL	ECL*	CMOS	P-MOS	RTL	TTL
1. Positive logic function of gate	NAND	NOR/OR	NOR or NAND	NAND	NOR	NAND
2. Wired positive logic function	AND	OR	None	None	AND (some functions)	None†
3. Propagation delay per gate (ns)	30	1	70	300	12	6
4. Fan-out	8	10	$\geq$50	20	5	10
5. Immunity to external noise	Good	Good	Very good	Fair	Fair	Very good
6. Cost per function	Low	High	Medium	Medium	Low	Low
7. Typical power dissipation per gate	10 mW	50 mW plus load	0.01 mW static 1 mW at 1 MHz	0.2–10 mW	12 mW	15 mW

*Slower gates having greater fan-out and lower cost are available.
†Open collector wired-OR is available.
Source: Garrett, L. S., "Integrated-Circuit Digital Logic Families, Part III—ECL and MOS Devices." *IEEE Spectrum*, Vol. 7, December, 1970, pp. 30–42.

12.4 BIBLIOGRAPHY

1. Carr, W. N., and J. P. Mize, *MOS/LSI Design and Applications*. New York: McGraw-Hill Book Company, 1972.
2. Gray, P. E., and C. L. Searle, *Electronic Principles: Physics, Models, and Circuits*. New York: John Wiley & Sons, Inc., 1969.
3. Greenfield, J. D., *Practical Digital Design Using Integrated Circuits*. New York: John Wiley & Sons, Inc., 1977.
4. Grinich, V. H., and H. G. Jackson, *Introduction to Integrated Circuits*. New York: McGraw-Hill Book Company, 1975.
5. Keyes, R. W., "Physical Limits in Digital Electronics." *Proc. IEEE*, Vol. 63, 1975, pp. 740–67.
6. Millman, J., and C. C. Halkias, *Integrated Electronics: Analog and Digital Circuits and Systems*. New York: McGraw-Hill Book Company, 1972.
7. Mitra, S. K., *An Introduction to Digital and Analog Integrated Circuits and Applications*. New York: Harper & Row, Publishers, Inc., 1980.
8. Morris, R. L., and J. R. Miller (eds.), *Designing with TTL Integrated Circuits*. New York: McGraw-Hill Book Company, 1971.
9. Taub, H., and D. Schilling, *Digital Integrated Electronics*. New York: McGraw-Hill Book Company, 1977.

12.5 EXERCISES

12.1 What function is mechanized by the circuit shown in Figure E12.1 (positive logic)? Assume: Inputs = 0 V or −3 V.

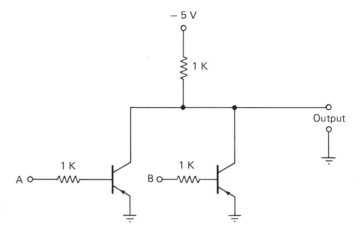

Figure E12.1

12.2 Repeat Exercise 12.1 for Figure E12.2. Also determine what voltages point P can assume (with respect to ground). Use: V_{EB} (forward bias) = 0.5 V, Diode (forward bias) = 0.7 V, V_{CE} (sat.) = 0 V.

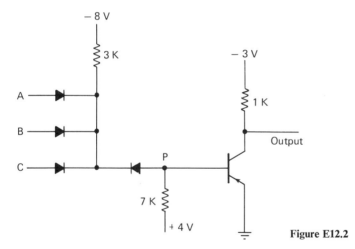

Figure E12.2

12.3 Repeat Exercise 12.1 for Figure E12.3. Also, the inputs are either +5 V or −1 V.

12.4 Refer to the circuit in Figure E12.4. Assume: Positive logic (A & B = +5 V, $\bar{A}$ & $\bar{B}$ = −5 V), V_{diode} (forward bias) = 1 V, V_{BE} (forward bias) = 1 V. Write the voltage at each node.

12.5 If two RTL IC gates have their outputs connected together, what function is produced by the composite gate? Compare wired logic for RTL with that for DTL.

12.6 (a) For the gate circuit shown in Figure E12.6, determine the purpose of the input diodes and the function mechanized.

(b) Propose an expander for use with this gate. Where would it be connected?

12.7 What type of logic block is required, with negative logic on the input and positive on the output, to produce the NOR function?

Sec. 12.5   Exercises

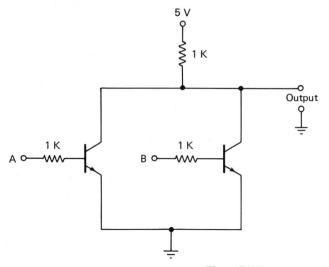

Figure E12.3

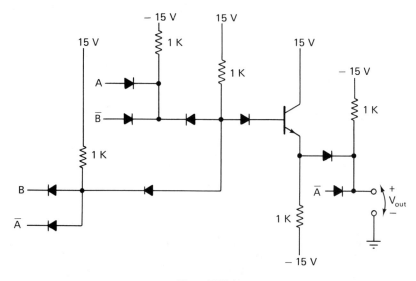

Figure E12.4

12.8 Design a diode OR gate (input voltage levels 0 and 3 volts) which does not show an offset in either the voltage equivalent to a logical 1 or the voltage equivalent to a logical 0. What is the disadvantage of this gate?

12.9 Consider the AND-OR logic of Figure 12.7 and determine the fan-out for a circuit which has the following worst-case limitations:

$$T_f = 200 \text{ ns}, \quad I_{d\,max} = 8 \text{ mA}$$
$$V_H = 3 \text{ V}, \quad V_L = 1 \text{ V}$$

Stray wiring capacitance $= 50$ pF

12.10 In a DTL gate, the noise margin from the nominal input values of 0.2 V and 5 V is

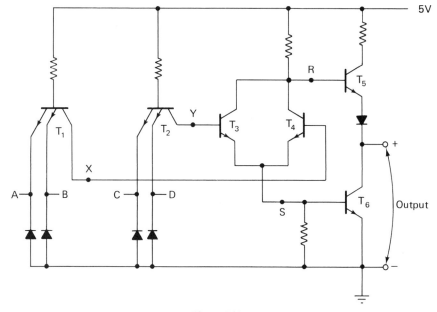

Figure E12.6

smaller for the lower input voltage. What can be suggested to more nearly equalize this, and what is the new performance?

12.11 Using p-channel MOSFETs, draw the circuit, similar to Figure 12.29, for a three-input NAND gate.

12.12 Repeat Exercise 12.11 for an n-channel NAND gate.

12.13 Find the logic block realized by the structure shown in Figure E12.13.

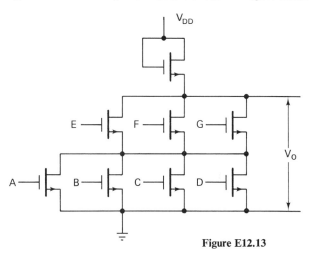

Figure E12.13

Sec. 12.5   Exercises   439

Appendix A

Summary of Pseudo-Linc I/O Instructions

Mnemonics	Machine Code	Meaning
REC n	70 n*	R Eset Call flip-flop n.
IOF	7100†	Turn Interrupt flip-flop O Ff.
ION	7200	Turn Interrupt flip-flop O N.
LAD	7300	Load Accumulator from Decoder.
WED n	73 n	WritE accumulator into Device n.
TAM	7400	Transfer data from the Accumulator to the Mask.
RED n	74 n	R Ead data from Device number n into accumulator.
SOH n	75 n	Skip On flag High from device n.
SIL n	77 n	Skip if Interrupt call flip-flop of device n is Low.

*Device numbers are allowed to have the following range: $01 \leq n \leq 77$.
†The code 7/00 ($0 \leq I \leq 7$) indicates an instruction which does not require a device number.

Appendix B

Answers to Selected Exercises

CHAPTER 2

2.4 (a) 47; (b) 27.375; (c) 1100011001; (d) 0.651; (e) 110011110; (f) 135; (g) 1507.323; (h) 249.543; (i) 112322; (j) 0.3121

2.6 (a) 936; (b) 0.94775; (c) 217.4922

2.7 (a) 2A; (b) 5D; (c) 27.BA; (d) B6C7

2.14 (a) 101.10; (b) 1110.01; (c) 1001110.011; (d) 1010100.011; (e) 10.10011; (f) 1.0001110; (g) 13.043

2.16 (a) 40635.7; (b) 3,030,422; (c) 1,120,033,321

2.18 (a) $\underline{1|}$ 11011; (c) $\underline{1|}$ 11100; (d) 1's complement: $\underline{1|}$ 1111 $= -0$, 2's complement $\underline{0\,|}0000 = +0$, 1's complement allows both ± 0.

2.20 12; 4096; with ≤ 3 holes punched: 299.

2.22 No.

2.24 No. 0 and 5 are adjacent on the map.

2.26 Yes. Weights $= 2, 3, 8, -4; 0$

2.30 Yes. Weights $= 8, 4, 2, 1; -3$

2.34 No; $w_3 w_2 w_1 w_0$; $B = 3\ 2\ 1\ 2$; -2 fails on decimal digit 8.

CHAPTER 3

3.5 (a) 27; (b) 11; (c) $C(24) = 4027$, $C(27) = C(30) = 0$

3.6 (a) 33; (b) 7777

3.12 (a) Yes, 23; (b) 5

441

3.18 (a) Once; yes, 10B; 9, 85; (b) 0, 3; yes, 103; 9, FF

3.23 (a) Halts at 110; (b) once; (c) CCR = 9, C(A) = 08, C(SP) = 106, C(X) = 0A

CHAPTER 4

4.1 AD

4.2 B

4.4 $AB + A\bar{C} + CD$

4.6 $(A + B)(\bar{A} + C)$

4.8 (a) $A\bar{C} + A\bar{B} + \bar{C}D$; (b) $\bar{A}D + B\bar{D} + \bar{C}D$; (c) $AC\bar{D} + \bar{A}B\bar{C}D + \bar{B}\bar{C}\bar{D}\bar{E}$;
(d) $\bar{A}B\bar{D} + \bar{A}BD + \bar{B}\bar{C}D + B\bar{C}\bar{D}$; (e) $\bar{A}\bar{C} + AB + \bar{B}C$; (f) $(A + B)(\bar{A} + C)$

4.11 $[\bar{s} + w(\bar{I} + \bar{T}C)]\bar{H}$

4.17 (a) $\bar{A}C + AB + E + \bar{C}D$; (b) $x\bar{y} + y\bar{z} + \bar{x}z = x\bar{z} + \bar{x}y + \bar{y}z$

4.19 $AB\bar{C} + ABD\bar{E}$

4.21 (a) $\bar{A} \oplus \bar{B} \oplus \bar{C} = A \oplus B + \bar{C}$

4.23 A

CHAPTER 5

5.2 The output, y_2, follows the sequence: 010010, where the left most 0 occurs after the first clock pulse, the 1 after the next clock pulse, etc.

5.5 Redundant states: $0 \equiv 4$ and $3 \equiv 5$.

5.9 $J_1 = K_1 = 1, J_2 = K_2 = \bar{y}_1\bar{y}_3, J_3 = y_1y_2, K_3 = y_1$ where flip-flop 1 is the least significant bit.

5.12 Not consistent: $y_3y_2y_1 = 001, 011, 100$, and 101. The contents of cell 001 is 1, but y_1 is already 1; etc.

5.19 Modules like that shown in Figure B.1 may be connected in series to form as large a ring counter as required. The y_i output of the last module must, however, be connected around to the y_{i-1} input of the first module.

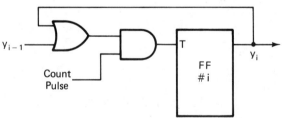

Figure B.1

5.23 Yes. The term $x\bar{y}$ must be added to the original function to avoid the hazard.

5.30 Place an inverter in the clock, C, line.

5.33 The original circuit does not lock in the case of $S = R = 0$. Gates 7 and 8 with the circuit shown in Figure B.2 are added to our original Figure 5.30.

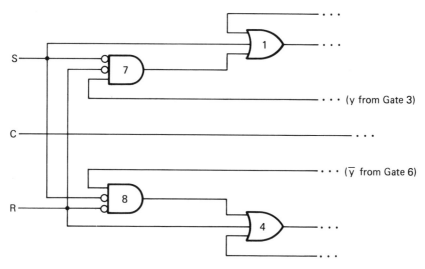

Figure B.2

5.35 $T = \bar{A}B + AB + A\bar{B}\bar{y} = B + A\bar{y}$
5.36 $T_1 = (\text{Input})y_2;\ T_2 = y_1;\ T_3 = \bar{y}_1$
5.38 (See Figure B.3.)

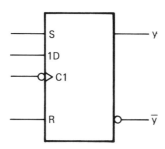

Figure B.3

CHAPTER 6

6.1 $A = 11,\ B = 00,\ C = 01$

6.4 (a) $A = 0011;\ B = 0;\ C = 1011;$ (b) 5; (c) $T = 5$

6.6 $T = 0 \quad MA \leftarrow MB_{0-8};$
$T = 1 \quad MB \leftarrow M(MA);$
$T = 2 \quad \text{IF } (MB_{11} = 0)\ PC \leftarrow PC + 1;\ M(MA) \leftarrow MB;\ T \leftarrow 0;\ S \leftarrow 0;$

6.13 $A = \begin{Bmatrix} \text{First word} = 4437 \\ \text{Second word} = 3744 \end{Bmatrix},\ C = \begin{Bmatrix} 4136 \\ 2241 \end{Bmatrix}$

$E = \begin{Bmatrix} 4577 \\ 4145 \end{Bmatrix},\ 3 = \begin{Bmatrix} 4122 \\ 2651 \end{Bmatrix},\ 4 = \begin{Bmatrix} 2414 \\ 7744 \end{Bmatrix}$

6.19 (a) NAND; (b) $(\bar{A} + \bar{B})(\bar{C} + \bar{D})$; (c) $\overline{A_1 + A_2 + A_3 + A_4} \equiv \text{NOR}$

6.25 Process: $x_1 x_0 + y_1 y_0 \over C_0 S_1 S_0$ From a truth table with inputs x_0, y_0, x_1, y_1 and outputs S_0, S_1, and C_0, three Karnaugh maps are constructed. Reading the maps yields

$$S_0 = \bar{x}_0 y_0 + x_0 \bar{y}_0$$
$$S_1 = \bar{x}_1 \bar{x}_0 y_1 + \bar{x}_1 y_1 \bar{y}_0 + x_1 \bar{y}_1 \bar{y}_0 + x_1 \bar{x}_0 \bar{y}_1 + x_1 x_0 y_1 y_0 + \bar{x}_1 x_0 \bar{y}_1 y_0$$
$$C_0 = x_1 y_1 + x_0 y_1 y_0 + x_1 x_0 y_0$$

6.28 10

6.34 (a) 320; (b) 10

CHAPTER 7

7.2 By inspection of a table like Table 7.1: $a = 1$ and $b = B$, which require no actual multiplexers; but $c = \bar{B}A$, $d = B\bar{A}$, and $e = BA$ require multiplexers wired in a similar way to that shown in Figure 7.4.

7.6 Reduced function: $f = \bar{A}\bar{C} + AC + B\bar{C}$ or $f = \bar{A}\bar{C} + AC + AB$. (See Figure B.4.)

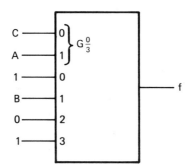

Figure B.4

7.7 Reduced function: $f = B\bar{C} + \bar{A}B + A\bar{B}$. (See Figure B.5.)

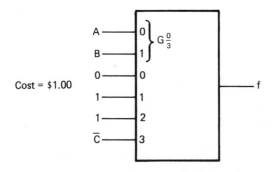

Cost = $1.00

Figure B.5

7.20 Set (a) reduces to $f_0 = A\bar{C}D + AC\bar{D} + CE$, $f_1 = \bar{C}D\bar{E} + A\bar{C}D$. Therefore, the functions are reduced to four input variables and four different product terms, which means that the set is realizable. Set (b) contains five different product terms, which may not be reduced. Thus, it is not realizable with the given PLA.

7.27 From the Karnaugh maps the function required by the flip-flops are $J_0 = K_0 = 1$, $J_1 = K_1 = (U/D)y_0 + (\overline{U/D})\bar{y}_0$, $J_2 = K_2 = (U/D)y_0y_1 + (\overline{U/D})\bar{y}_0\bar{y}_1$, $J_3 = K_3 = (U/D)y_0y_1y_2 + (\overline{U/D})\bar{y}_0\bar{y}_1\bar{y}_2$. Since the output is in binary, $C_0 = y_0$, $C_1 = y_1$, $C_2 = y_2$, $C_3 = y_3$. Where C_i is the counter output for bit i; y_i is the output of flip-flop i; J_i and K_i are the J-K inputs of flip-flop i.

CHAPTER 8

8.1 $T = 1$: $IR \leftarrow M(MA)$; $MA \leftarrow MA + 1$; $PC \leftarrow PC + 1$;: fetch cycle;
  $T = 2$: $A \leftarrow 0$; $N, V, C \leftarrow 0$; $Z \leftarrow 1$; $T \leftarrow 1$;: execution cycle;

8.5 Lowest: 4800 – 49FF; highest: 5E00 – 5FFF

8.7 33

8.12 Words = 4608_{10}

8.14 Addresses correctly reading the switch: 7C12, 1849, 5E31, 7AAA

8.16 Lowest = 0400_{16}; highest = $BDFF_{16}$

8.21 Lowest: 9000–907F; highest: 19B80–19BFF

CHAPTER 9

9.2

Branch Mnemonic	Next Address Function			Next Memory Address	
	11 10 9 8	7 6 5 4	3 2 1 0	Row	Column
FR ≡ Flag Row	0 0 1 1	$d_7d_6d_5-$	$d_3d_2d_1d_0$	$d_7d_6d_5f$	$d_3d_2d_1d_0$
	Not unique but bit 11 must not be a 1.				

9.4 If $CI = 0$, the IR and IM instructions will perform no operation.

9.10 (a) 7, A; (b) 0; (c) 13_{10}; (d) 0

9.13 (a) 1, 8; (b) 4001; (c) 0; (d) 4001

CHAPTER 10

10.2 Process: $\begin{array}{r} x_i \\ -y_i \\ \hline B_iD_i \end{array}$ D_i = difference = $\bar{x}_iy_i + x_i\bar{y}_i$
  B_i = borrow = $\bar{x}_iy_i$

10.3 Correction = $2^n(2^n - a)$, where n and a are defined below Equation (10.12a).

10.9 Implement the function, f: $f = \bar{A}_{n-1}B_{n-1} + U_{n-1}\bar{A}_{n-2}B_{n-2} + U_{n-1}U_{n-2}\bar{A}_{n-3}B_{n-3} + \cdots + (U_{n-1}U_{n-2}\ldots U_1)\bar{A}_0B_0$, where $n \equiv$ the number of bits in each number, A_i is the ith bit of A, and B_i is the ith bit of B.

10.12 (a) $\underline{1|}\ 1110 = -1$, which is correct.
  (b) $\underline{1|}\ 1111 = -0$, which is incorrect since $-2 \div 4 = -\frac{1}{2}$.
  (c) $\underline{1|}\ 1011 = -4$, which is correct.
  (d) Yes, the division in part (a) and the multiplication in part (c) are correct, but the division in part (b) cannot be contained in a five-bit register. Underflow occurs.

CHAPTER 11

11.1 Use an AND gate decoder, e.g., the circuit shown in Figure B.6 selects device 12_8.

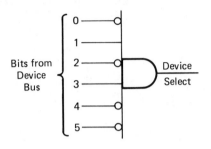

Figure B.6

11.3 $T = 0$: $\quad DDB \leftarrow MB_{0-5}$; $CB \leftarrow MB_{6-8}$;
$T = 1$: $\quad$ IF (Flag Bus $= 1$) $A_{11} \leftarrow 0$ ELSE $A_{11} \leftarrow 1$
$\quad\quad\quad T \leftarrow 0$; $S \leftarrow 0$;

11.7 Simply omit the IF(Interrupt $= 1$) and related productions in this instruction.

11.9 Yes, provide a 24 (or more)-bit mask loaded by a new TAM instruction. The instruction itself contains one or more bits to control which 12 bits of the mask are to be loaded from the accumulator. (Of course, the other circuitry, e.g., the decoder, would also be expanded.)

CHAPTER 12

12.1 NAND
12.3 NOR
12.4 $V_{out} = -6$ V
12.7 AND
12.9 $m = 1.68$; therefore, fan-out $= 1$.
12.13 $V_0 = \bar{A}\,\bar{B}\,\bar{C}\,\bar{D} + \bar{E}\,\bar{F}\,\bar{G}$

Index

Accumulator, 50, 202
A/D, 309
Adder:
   fast, 343-48
   full, 218
   half, 235
   parallel, 219
   serial, 219-20
   speedup, 357-62
Addition, 27-28
Address bus, 284-85
Address decoding (*See* Decoding, address)
Addressing, memory, 288-93
Addressing modes:
   conditional, 61, 66
   extended, 61, 64
   immediate, 61
   implied, 61, 65
   indexed, 61, 64
   relative, 61
Add-to-memory, 397
Aiken, Howard, 9
Algebra:
   Boolean, 81-92
   laws, 86
Algorithm, 46, 55
Alphameric codes, 39-40

ALU (*See* Arithmetic and Logic Unit)
Analysis, sequential, 121-22
AND gate, 82
Arabic notation, 23
Architecture, 183, 348-92
Architecture, microprocessor, 283-88
Arithmetic and Logic Unit, 217-22
Arithmetic operations, RTL, 186
ASCII, 40
Assembler, 57-59
Associative memory, 246-49
Asynchronous, 117, 31
Atanasoff, John, 9

Babbage, Charles, 8-9
Base, 23-24
BCD, 32
Binary adder, 217-18
Binary addition, 27-28
Binary codes (*See* Codes)
Binary multiplication, 30-31
Binary numbers, 24-27
Bit, 24
Boole, George, 81
Boolean algebra, 81-92
Booth algorithm, 365-68
Borrow, 221

Branch and control instructions, 50, 52
Bus, 208-11, 215-16

C-bit, 66
Calculator, 320
Canonical expansion, 89-90
Carry conditions, 344
Carry look-ahead, 344
Cathode ray tube, 196
CCR, 62
Character generator, 196-99
CLA, 345
Clocked flip-flops, 119
Clock pulse, 117
CM (*See* Control memory)
Codes:
   alphameric, 39-40
   ASCII, 40
   BCD, 32
   error-detecting, 38-39
   gray, 36-37, 137-38
   self-complementing, 35
   special, 39-41
   unit-distance, 36-38
   weighted, 33-35
Combinational circuits, 117
Comparator, 195
Comparison division, 369-72

Compiler, 59-60
Complement (for subtraction), 28-30
Complementary approach, 96-98
Computer organization, 201-8
Computers:
  analog, 14-15
  classification of, 14-17
  general-purpose, 16
  special-purpose, 16-17
Conditional addressing, 61, 66
Conditional operations, RTL, 189
Condition code register, 62
Control memory, 268-73, 323
Control memory map, 332-33
Conversion:
  fractions, 26-27
  integers, 24-26
Converter, D/A, 198
Core Memory, 223
Cost, design criteria, 173-74
Counter:
  BCD, 137-39
  gray-code, 144-47
  modulo-5, 143-44
  ring, 143
  ripple, 161-64
  synthesis of, 133-41
  three-bit up, 134-37
CPU, 48
CROM (*See* Control memory)
Cross assemblers, 282
CRT (*See* Cathode ray tube)
Cutoff, transistor, 412
Cycle stealing, 388

D/A converter, 198, 309
Data bus, 284-85
Data channels, 396-97
Data transfer methods:
  DMA, 381, 388-91
  interrupts, 381, 385-88
  programmed, 381-85
Davis, Ruth M., 19
Decoder, 240
  circuits, 240
Decoding, address:
  full external, 288-90
  multiple enables, 291-93
  one-of-N, 290-91
Delay, propagation, 406-7
De Morgan's theorem, 87
Demultiplexer, 240, 251
Destructive readout (*See* Non-destructive readout)
D flip-flop, 120
Dependency notation, 164-69
Design, 109-14, 129-49, 173-74, 239-313, 336-38
Design, structured (*See* Structured design)
Digital computer characteristics, 17-19
Digital computers, evolution, 7-10
Digital logic families, 436
Digital principles, 3-235
Diode:
  characteristics, 401-2
  gates, 400-404
Diode-Transistor Logic (DTL), 415-18
DIP, 170
Direct-coupled transistor logic (DCTL), 418-20
Direct memory access (DMA), 381, 388-91
Disk, 223
Division, 31-32, 369-76
  comparison, 369-72
  DMA, 381, 388-91
  nonrestoration, 369, 374-76
  restoration, 369, 372-74
Don't-care, 96, 130
Dual, 87-88

EBCDIC, 40
ECAP, 17, 57-58
Eckert, J. Presper, 9
ECL, 426-29
Edge triggering, 119, 158
EDSAC, 9
Emitter-coupled logic (ECL), 426-29
Emulation, 330
EN (enable), 240
End-around carry, 29
ENIAC, 9
Error-detecting codes, 38-39
Evolution (heuristic), 142
Excess-3 code, 32-33
Excitation maps, memory, 129
Exclusive-OR, 90-91, 106-9, 251
Execution cycle, 203-5
EXOR, 91
Extended addressing, 61, 64

Fan-out, 408-9
Fetch cycle, 203
Flag, 383-85
Flip-flop, 118-19
  D-type, 120
  dynamic, 432-35
  edge triggering, 119, 158
  elementary, 154
  J-K, 137-38
  master-slave, 157
  MOSFET, 432-35
  realization, 153-61
  R-S, 119-20, 131
  T, 134-36, 144-45
Formats, instruction, 349
FORTRAN, 59
Full adder, 218
Full subtractor, 221
Function of *n* variables, 90-92

Gates, 82-84
  AND, 82, 403
  diode, 400-404
  MOSFET, 430-32
  NAND, 91, 100-106
  NOR, 90, 100-106
  NOT, 83-84
  open collector, 211-13
  OR, 82-83
  transistor, 411-29
  transmission, 434
General-purpose computer, 16
Gray code, 36-37

Half adder, 235
Handshaking, 297
Hardware, 20
Hazard, 150-52
Heuristic design, 141-49
Hexadecimal numbers, 24
h_{fe}, 413
Hold time, 161
Hollerith, Herman, 8
Horizontal microprogramming, 322

IC package, 172
Immediate addressing, 61
Implied addressing, 61, 65
Indexed addressing, 61
Index register, 61
Induction, perfect, 85-86
I/O, 396
Input-output instructions, 396
Instruction, single-address, 51
Instruction cycle, 203
Instruction register, 202
Instructions, LINC, 52
Instructions, Microprocessor, 63-66
Integrated circuit, 169-73
  families, 435-36
Interface designs, 380
Interfacing principles, 293-303
Interrupt:
  circuits, 67, 381, 385-88
  multilevel, 392-96
  service routine, 385

Inverter, 84
Iteration, 141

J-K flip-flop, 137-38

Karnaugh map, 92-99

Languages, programming, 57-60
Large-scale integration, 170
Latch, 154
LIFO, 63
LINC (*See* Pseudo-LINC)
Literal, 100
Logic:
    blocks, 404
    diagrams, 99-100
    exclusive-OR, 106-9
    NAND-NOR, 100-106
    negative, 404
    positive, 403-4
    transistor (*See* specific circuit name)
    tristate, 213-15, 295
    wired, 211-13
Logical design, 81-114, 239-76
Logic design with microprocessors, 280-313
Logic operations, RTL, 187
LSI, 170, 239, 255

MA (*See* Memory address register)
MAC –4, 3
Machine language, 50, 57
Machine organization, 48-50
Macroprogram, 319-20
Manchly, John W., 9
Map, 37-38
Map simplification, 92-99
Mark I computer, 9
Mask register, 392
Mass memory, 222
Maxterm, 92
MB (*See* Memory buffer register)
Mealy model, 126-29
Medium-scale integration, 170
Memory, 201
    address decoding, 288-93
    bipolar, 223-29
    coincidence selection, 223
    control, 268-73, 323
    core, 223
    cycle time, 222
    disk, 223
    excitation maps, 129-31
    linear selection, 223
    map, 290-91
    MOS, 328-30
    nondestructive, 222

PROM, 223
RAM, 223
    random-access, 223
    read-only, 223, 256-63
    ROM (*See* Memory, read-only)
    ROM applications, 258-63
    semiconductor, 223-31
    TU1, 229-31
    volitility, 222
Memory address register, 48, 285-88
Memory buffer register, 202
Memory excitation maps, 129
Memory operations, RTL, 189
Memory-increment, 397
Memory map, 290-91
Microcomputer, 280
Microinstructions, 319, 322-26
Microprocessor, 4, 60, 280-313
    architecture, 283-88
    design examples, 303-13
    instructions, 63-66
    interfacing, 293-303
    memory addressing, 288-93
    microprogrammable, 319-36
    programming, 60-74
    structured design, 303-7, 336-38
Microprogram, 57, 319-20
Microprogrammable microprocessor, 319-36
Microprogramming:
    horizontal, 322
    vertical, 322
Minicomputer, 7
Minterm, 89
Mnemonics, 51
Modeling, 19
Module synthesis, 142
Modulo, 28
Moore model, 126-29
MOS (*See* MOSFET)
MOSFET:
    circuits, 429-35
    complementing, 431-32
    flip-flops, 432-35
    gates, 430-32
    n-channel, 430
    p-channel, 429
μP, 280
MSI, 170-71, 239, 255
Multiple bit processing, 362
Multiplexer, 240-43, 250
Multiplication, 30-31, 349-68
    Booth algorithms, 365-68
    combinational logic, 357
    fast, 356-68
    positive numbers, 349-50

    1's and 2's complement, 351-56
    sign and magnitude, 350
MUX, 266-67

NAND, 91, 100-106
NAND-NOR logic, 100-106
N-bit, 66
Negative logic, 404
Nondestructive readout, 222
Nonrestoration division, 369, 374-76
Nonvolatile memory, 222
NOR, 90, 100-106
NOT operation, 83-84
Number base conversion, 24-28
Numbers:
    binary, 24-27
    hexadecimal, 24
    octal, 24-25

Octal numbers, 24-25
1's complement, 29
Open collector gate, 211-13
Organization, machine, 48-50
OR gate, 82-83
Overflow, 356, 376-78

Parity, 38-39
PC (*See* Program counter)
Perfect induction, 85-86
Peripheral interface unit, 283, 295-303
Peripheral processor, 396
PIU, 283, 295-303
PLA, 263-65, 273-74
Polling, 385, 391
Positive logic, 403-4
Postulates, Boolean, 85
Problem-oriented languages, 17
Processor, peripheral, 396
Production, 184
Product-of sums, 90
Program, 46
Program counter, 48, 202
Programmable components, 239
Programmable logic array, 263-65
Programmed transfer, 381-85
Programming, 57-70
Programming, microprocessor, 60-64
PROM, 223
Propagation delay, 406-7
Pseudo-LINC, 50-57
    I/O instructions, summary, 440
    instructions brief list, 52
Pull-up, active, 423

Index   449

Quotient, 369, 372

Radix, 23
RAM (*See* Memory, random access)
Random logic design, 307
Read-only memories, 246
Redundant states, 125-26
Reflected code, 36-38
Register transfer logic (RTL), 184-90, 206-8
    statements, limitations on, 207-8
    tables of operations, 186-87, 189
Relative addressing, 58-59, 61
Resistor-transistor logic (TRL), 420-22
Restoration division, 369, 372-74
Ripple method, 161-64
ROM (*See* Memory, read-only)
Rotate operation, RTL, 187
R-S flip-flop, 119-220, 131
RTL (*See* Register transfer logic, and Resistor-transistor logic)

Saturation, transistor, 412
Scale operations, RTL, 187
Selector channels, 396
Self-complementing codes, 35-36
Semiconductor memory (*See* Memory)
Sequential analysis, 121-22
Sequential circuits, 117-75
Serial addition, 219-20
Setup time, 161
Shifting over zeros, 362-64
Shift operations, RTL, 187
Shift register, 148, 250
Sign-and-magnitude multiplication, 350-51
Sign bit, 30

Simplification:
    map method, 92-99
    other methods, 99
    theorems, 84-89
Simulation, 19
Software, 46-74
SSI, 170
State, memory, 118
State assignment, 124-25
State diagram, 122-24
State generator, 199-200
State table, 124-26
Stealing, cycle, 388
Stilbitz, George, 9
Structured design, 239-338
    applications, 265-68, 303-13
    component characteristics, 250-51, 336-38
    control memory method, 268-76
    examples, 253-56, 265-68, 303-13
    fixed components method, 246-55
    goals, 247
    implementation comparison, 336-38
    main properties list, 247-48
    microprocessor method, 303-13
    programmable components, 255-65
    programming, 46-47
Subroutine, 69
Subtraction, 24-30
Sum-of-products, 90
Synchronous circuits, 117
Synthesis, circuit, 129-49

Table:
    state, 124-26
    truth, 82
Table-lookup procedure, 305
Teletypewriter, 50

T-flip-flop, 136
Theorems, 84-89
    De Morgan's, 87
    elementary Boolean, 85
    main Boolean, 87
Timing diagram, 119, 192-93
Toggle, 134-36
Top-down, 3, 248, 251-53 (*See* Structured design)
Totem-pole output, 426
Transfer instructions, 50
Transistor:
    characteristics, 411-15
    cutoff region, 412
    gates, 411-29
    hybrid parameters, 414
    models, 412-14
    saturation region, 412
Transistor-transistor logic (TTL), 422-26
Transmission gate, 434
Tristate logic, 213-15, 295
Truth table, 82
TTL (*See* Transistor-transistor logic)
2's complement, 29

Unclocked flip-flops, 144
Unit-distance codes, 36-38

V-bit, 66
Veitch, 92
Vertical microprogramming, 322
Very-large-scale integration, 170, 172
VLSI, 170-73
Volatile memory, 222
Voltage margin, 405-6

Wired logic, 211-13
Worst-case design, 407

Z-bit, 66

Department of Electronic and
Electrical Engineering
University of Strathclyde
Royal College Building
204 George Street
Glasgow G1 1XW